AF804280

Polymers and Nanomaterials for Gene Therapy

Woodhead Publishing Series in Biomedicine: Number 83

Polymers and Nanomaterials for Gene Therapy

Edited by

Ravin Narain

ELSEVIER

AMSTERDAM • BOSTON • CAMBRIDGE • HEIDELBERG
LONDON • NEW YORK • OXFORD • PARIS • SAN DIEGO
SAN FRANCISCO • SINGAPORE • SYDNEY • TOKYO
Woodhead Publishing is an imprint of Elsevier

WP
WOODHEAD PUBLISHING

Woodhead Publishing is an imprint of Elsevier
80 High Street, Sawston, Cambridge, CB22 3HJ, UK
50 Hampshire Street, 5th Floor, Cambridge, MA 02139, USA
Langford Lane, Kidlington, OX5 1GB, UK

© 2016 Elsevier Ltd. All rights reserved.

No part of this publication may be reproduced or transmitted in any form or by any means, electronic or mechanical, including photocopying, recording, or any information storage and retrieval system, without permission in writing from the publisher. Details on how to seek permission, further information about the Publisher's permissions policies and our arrangements with organizations such as the Copyright Clearance Center and the Copyright Licensing Agency, can be found at our website: www.elsevier.com/permissions.

This book and the individual contributions contained in it are protected under copyright by the Publisher (other than as may be noted herein).

Notices
Knowledge and best practice in this field are constantly changing. As new research and experience broaden our understanding, changes in research methods, professional practices, or medical treatment may become necessary.

Practitioners and researchers must always rely on their own experience and knowledge in evaluating and using any information, methods, compounds, or experiments described herein. In using such information or methods they should be mindful of their own safety and the safety of others, including parties for whom they have a professional responsibility.

To the fullest extent of the law, neither the Publisher nor the authors, contributors, or editors, assume any liability for any injury and/or damage to persons or property as a matter of products liability, negligence or otherwise, or from any use or operation of any methods, products, instructions, or ideas contained in the material herein.

ISBN: 978-0-08-100520-0 (print)
ISBN: 978-0-08-100521-7 (online)

British Library Cataloguing-in-Publication Data
A catalogue record for this book is available from the British Library

Library of Congress Control Number: 2015941475

For information on all Woodhead Publishing publications visit our website at http://store.elsevier.com/

ELSEVIER Book Aid International Working together to grow libraries in developing countries

www.elsevier.com • www.bookaid.org

Contents

List of Contributors

M. Ahmed California Institute of Technology, Pasadena, CA, USA

H. Montazeri Aliabadi Chapman University School of Pharmacy, Irvine, CA, USA

C.-S. Cho Department of Agricultural Biotechnology and Research Institute for Agriculture and Life Sciences, Seoul National University, Seoul Republic of Korea

C. Duvall Vanderbilt University, Nashville, TN, USA

M. Ebara National Institute for Materials Science (NIMS), Tsukuba, Japan

P. Kumar University of Alberta, Edmonton, AB, Canada

M.V. Lozano University College of London, London, UK

M. Muthiah Department of Biomedical Science and BK21 PLUS Center for Creative Biomedical Scientists, Chonnam National University Medical School, Republic of Korea

R. Narain University of Alberta, Edmonton, AB, Canada

I.-K. Park Department of Biomedical Science and BK21 PLUS Center for Creative Biomedical Scientists, Chonnam National University Medical School, Republic of Korea

K.I. Papadopoulos THAI StemLife, Bangkok, Thailand

W. Prasongchean Faculty of Pharmaceutical Sciences, Chulalongkorn University, Bangkok, Thailand

S. Quan University of Alberta, Edmonton, AB, Canada

K.C. Remant Bahadur University of Alberta, Edmonton, AB, Canada

M.J. Santander-Ortega University College of London, London, UK

A.G. Schätzlein University College of London, London, UK

B. Thapa University of Alberta, Edmonton, AB, Canada

I.F. Uchegbu University College of London, London, UK

H. Uludağ University of Alberta, Edmonton, AB, Canada

K. Uto National Institute for Materials Science (NIMS), Tsukuba, Japan

S. Uthaman Department of Biomedical Science and BK21 PLUS Center for Creative Biomedical Scientists, Chonnam National University Medical School, Republic of Korea

P. Wattanaarsakit Faculty of Pharmaceutical Sciences, Chulalongkorn University, Bangkok, Thailand

T. Werfel Vanderbilt University, Nashville, TN, USA

Editor and Author

Ravin NARAIN, PhD/PEng, is Professor in the Department of Chemical and Materials Engineering at the University of Alberta. He has published more than 120 articles in peer-reviewed journals and is the editor of three books namely Engineered Carbohydrate-Based Materials for Biomedical Applications (Wiley & Sons), Chemistry of Bioconjugates (Wiley & Sons) and Glycopolymers: Synthesis and Applications (Smithers & Rapra). He has also contributed more than 20 chapters in several books. Dr. Narain's research examines nanomedicine and regenerative medicine, with an emphasis on developing new carbohydrate based biomaterials for drug and gene delivery for cancer therapy. In addition, he has an interest in interdisciplinary research and has extensive experience in developing innovative polymers, nanoparticles, and nanofibers for advanced engineering applications. Before joining the University of Alberta in 2009, he held academic appointment (Assistant and Associate Professor) at Laurentian University for the period of 2004–2009. Dr. Narain completed his B.Sc and PhD from the University of Mauritius and he was then appointed as a postdoctoral fellow at University of Sussex, UK (2001–2003) and as a senior research fellow at the University of Washington, U.S.A (2003–2004).

Contributing Authors

Bindu Thapa

Bindu Thapa is a PhD student at faculty of Pharmacy and Pharmaceutical Sciences, University of Alberta, Canada. She acquired her Masters degree in Pharmaceutical Sciences from The School of Pharmaceutical and Biomedical Sciences, Pokhara University, Nepal. She worked at Department of Drug discovery and Biomedical Sciences, University of South Carolina, U.S., where she explored interdisciplinary research at the interface of engineering and biology particularly focusing on polymeric drug and gene delivery. Her research interest is exploring non-viral gene delivery systems for transgene expression and cancer therapy based on siRNA delivery.

Hamidreza Montazeri Aliabadi, PhD

Dr. Montazeri graduated with a Pharm.D. degree from Tehran University in 1988. He worked in industrial pharmacy for more than 10 years in different capacities (including research and development, formulation, and quality control), before continuing his studies. He received his PhD in pharmaceutics (working on a project on targeted delivery of small molecule drugs to solid tumors) from the University of Alberta (U of A), in 2007. Shortly after graduation, he was awarded a postdoctoral fellowship by Alberta Heritage Foundation for Medical Research (now known as Alberta Innovates Health Solutions) on a project to design a polymeric carrier for siRNA delivery. After completion of the fellowship in the Department of Chemical and Material Engineering, U of A, Dr. Montazeri wrote a research proposal (in collaboration) for the study of signaling pathways to overcome resistance to chemotherapy in breast cancer, which was awarded a 3-year grant from the Canadian Breast Cancer Foundation. He has authored more than 30 peer-reviewed publications and two previous book chapters as well as 27 meeting abstracts. He is currently an assistant professor of Pharmaceutics in Chapman University in California.

Hasan Uludağ, PhD

Dr. Hasan Uludağ has been with the University of Alberta (Edmonton, Canada) since 1997, currently based at the Department of Chemical & Materials Engineering. He holds joint appoints with the Faculty of Medicine & Dentistry and Faculty of Pharmacy & Pharmaceutical Sciences. Dr. Uludağ is directing interdisciplinary research programs on experimental therapeutics, specifically focusing of designing functional biomaterials to realize the potential of new, unconventional therapeutic agents. His research activity is conducted in the context of bone regeneration and anti-cancer therapies. Dr. Uludağ is actively involved in various biomaterials societies around the world and is an elected Fellow of the International Union of Society of Biomaterials Science and Engineering (IUS-BSE). Besides acting as the lead editor for *Frontiers in Biomaterials*, he is serving on the editorial board of six international journals. Dr. Uludağ has published >150 peer-reviewed journal articles. He obtained dual B.Sc. degrees in Biomedical Engineering and Biology from Brown University (Providence, RI) in 1989, specializing in biomedical engineering with a strong emphasis in biological sciences. He then completed his PhD degree in 1993 at the Department of Chemical Engineering & Applied Chemistry at the University of Toronto, where he developed a strong expertise in polymeric biomaterials. He spent 4 years in an industrial setting (Genetics Institute Inc., Boston, MA), where he contributed to development of a tissue engineered bone-inducing device for clinical use.

Remant Bahadur, KC

Remant Bahadur, KC is working as a research associate at the Department of Chemical & Materials Engineering, University of Alberta. His research interest includes the design of novel biomaterials to develop drug/gene carriers for cancer therapy. He has been working in interdisciplinary research at the interface of science, engineering and biology for 15 years as a principle and co-investigator. He received a master's degree in organic chemistry (M.Sc.) from Tribhuvan University, Nepal and a PhD in Bionanosystem Engineering from Chonbuk National University, South Korea (2004–2007). He has developed a strong foundation in designing polymeric material of variable architectures during his PhD. Currently, his research is focused on the synthesis of cationic lipo-polymers for cancer gene and cell therapy.

Mitsuhiro Ebara, PhD

Mitsuhiro Ebara is a senior scientist at World Premier International (WPI) Research Centre for Materials Nanoarchitectonics (MANA), the National Institute for Materials Science (NIMS), Japan. He received his PhD degree from Waseda University. His research interests focus on developing "smart" biotechnologies using stimuli-responsive polymers.

Koichiro Uto, PhD

Koichiro Uto is a senior fellow at the Department of Bioengineering, University of Washington in the United States. He received his PhD (Engineering) from Kagoshima University in 2010. He became a postdoctoral fellow at the National Institute for Materials Science (NIMS). His research interests include preparation of "smart" polymers for biomedical applications.

Manuel J. Santander-Ortega, PhD

Manuel is currently Assistant Professor in Pharmaceutical Technology and Biotechnology at the University of Castilla-La Mancha (Spain). His PhD, at the Applied Physics Department of the University of Granada (Spain), characterized the biophysics of the colloidal behavior of nanoparticles used as drug delivery systems under physiological conditions. Since then his work has focused on linking the fundamental biophysical properties of colloidal systems used in drug delivery to their biological behavior. His work has contributed to the design and optimization of different colloidal carriers used for gene therapy, cancer therapy, or oral delivery of peptides and proteins at the University of Saarland (Germany), University College London (UK), and the University of Santiago de Compostela (Spain). As result, he has published close to 20 peer review articles, several book chapters, and is co-author of a patent licensed by BASF Company based on the design of starch nanoparticles for drug delivery. His current research is focused on the design of lipid nanostructures to improve the bioavailability of neuroprotective antioxidants after oral intake.

M. Victoria Lozano, PhD

María Victoria is Assistant Professor in Pharmaceutical Technology and Biotechnology at the University of Castilla-La Mancha (Spain). She joined the research group *Cellular Neurobiology and Molecular Chemistry of the Central Nervous System* in 2013, which is focused on the development of antioxidant-loaded nanostructures for improving neurodegenerative processes. She obtained her PhD at the University of Santiago de Compostela (Spain) in the research group of Prof. M.J. Alonso, a world leader in the field of drug delivery, working on the design of nanocapsules for anticancer therapy. Her postdoctoral research was at the University College London (UK) in the group of Dr. A.G. Schatzlein and Prof. I.F. Uchegbu aimed at optimizing self-assembling nanostructures for improved blood-brain-barrier transport. Her research has yielded more than 10 peer review articles, several book chapters, and a patent and has helped to implement new technologies in the institutions she has worked in. She has expertise in identifying promising *in vitro* nanoplatforms for adequate *in vivo* results, which has been focused on hurdles such as cancer or central nervous diseases.

Ijeoma F. Uchegbu

Ijeoma is Professor of Pharmaceutical Nanoscience at the University College London. With over 20 years of experience in nanoparticle research, she is the lead inventor on 12 granted patents and the lead inventor on 8 patent families. Her work is focused on the design, engineering, and application of pharmaceutical nanoparticles for drug delivery.

She obtained her PhD from the School of Pharmacy, University of London in 1994, and was appointed to a lectureship within the Department of Pharmaceutical Sciences, Strathclyde University in 1997, and a Chair in Drug Delivery at Strathclyde University in 2002. In 2006, Ijeoma was appointed to the Chair in Pharmaceutical Nanoscience at the School of Pharmacy, London. Ijeoma is the former Scientific Secretary of the Controlled Release Society (CRS), a learned society with over 2,000 members, with interests in the delivery of pharmaceuticals, former Chair of the Academy of Pharmaceutical Sciences of Great Britain, and the former Academia Expert on the Department for Business Innovation and Skills' Science Engineering and Technology Strategy for Women Expert Group. She has been awarded various prizes for her work, notably the UK Department for Business Innovation Skills' Women of Outstanding Achievement in Science Engineering and Technology award. Ijeoma is the founding editor of the journal *Pharmaceutical Nanotechnology*, and of two books. She is also the author of over 100 patents, patent applications, peer-reviewed journal articles, and book chapters.

She is co-founder and Chief Scientific Officer of Nanomerics, a company spun-out from UCL to commercialize pharmaceutical nanotechnology.

Andreas G. Schätzlein

Andreas is Professor of Translational Therapeutics at the UCL School of Pharmacy, University College London. His research interests focus on the discovery and translational preclinical/clinical development of nanomedicines.

He has a track record in medicine development and translational research in industry and academia. As a senior scientist of the biotech start-up IDEA in Munich, he was intimately involved in the scientific, clinical, and commercial development of its transdermal delivery nanotechnology products. In 1996, he joined the Cancer Research UK Beatson Laboratories at the University of Glasgow where he became leader of the Cancer Research UK's Experimental Therapeutics and Gene Medicines Group with a Research Program focused on the development of advanced cancer therapeutics. As head of the center's Analytical Services Unit he introduced good clinical laboratory practice (GCLP) standards into one of Europe's largest centers for Oncology trials, facilitating pharmacokinetics and pharmacodynamics readouts in a range of Translational Clinical Trials, including anti-cancer nanomedicines such as PK1 and DaunoXome.

In 2006, Andreas moved to the School of Pharmacy, London, working on the development of novel cancer drugs and nanomedicines. AGS's research has been featured repeatedly in the media (TV, radio, print) and is supported by the BBSRC, EPSRC, Wellcome Trust, Royal Society, Scottish Enterprise, Tenovus Scotland, CRUK, the Technology Strategy Board, and FP7, with total grant income of over £10 Mio. Andreas is also co-founder and CEO of Nanomerics Ltd., a UCL spin-out company developing pharmaceutical nanotechnology medicines.

Marya Ahmed

Marya Ahmed is a postdoctoral fellow at Faculty of Pharmacy, University of Toronto. She acquired her PhD from the University of Alberta, Canada in Chemical Engineering, with expertise in synthesis of polymeric vectors for gene delivery. She worked as a post-doctoral fellow in the department of Chemical Engineering at California Institute of Technology, and gained valuable experience working on antibody-targeted gene siRNA delivery vectors for *in vivo* applications. Currently, she is working as a post-doctoral fellow at University of Toronto, where her projects are focused on the synthesis of peptide polymer conjugates and the study of their mitochondrial targeting efficacies.

Konstantinos I. Papadopoulos, MD, PhD

Dr. Konstantinos I. Papadopoulos is a Specialist in Endocrinology and Diabetes and holds a PhD in Endocrinology and Immunology from the University of Lund, Sweden. He has held Specialist, Teaching, and Research Positions at the Department of Endocrinology and Diabetes at Malmö University Hospital, University of Lund, Sweden, at the Department of Internal Medicine, Liverpool and Fairfield Hospitals, NSW, Australia where he also has served as a Clinical Tutor in Medicine for the overseas doctors training programs. He has held a Lectureship in Endocrinology at the University of Alexandroupolis, Greece. Currently, he is the Chief of Medical Research and Development at THAI StemLife, an international biotechnology company specializing in umbilical cord blood and peripheral blood stem cell storage and R&D as well as predictive genetic testing located in Bangkok, Thailand. His research interests are in the fields of endocrinology and diabetes, infertility and reproductive endocrinology, HRT, depression, stem cell biology, genetics, obesity, nutrition, and sports medicine. Moreover, Dr. Papadopoulos is a clinical expert in personalized genetic preventive medical programs, longevity and lifestyle planning, and medical fitness.

Phanphen Wattanaarsakit, PhD

Phanphen is a lecturer at Department of Pharmaceutics and Industrial Pharmacy, Faculty of Pharmaceutical Sciences of Chulalongkorn University in Thailand. Her professional interests include pharmaceutical development and pharmaceutical processing technology.

Saji Uthaman, MSc, MTech

Saji Uthaman is currently enrolled in doctoral studies under the supervision of Dr. In-Kyu Park at the Department of Biomedical Science, Chonnam National University Medical School. In 2011 he has obtained his MTech in Nanomedical Science from the Amrita Centre for Nanoscience and Molecular Medicine at Amrita Institute of Medical Science and Research Centre. Currently he is engaged in designing and synthesizing various non-viral polymeric nanomaterials for biomedical applications. He has published eight research articles and two review articles.

Muthunarayanan Muthiah, PhD

Muthunarayanan Muthiah is currently a post-doctoral fellow in interdisciplinary nano-science centre (iNANO) at the Aarhus University, Denmark. He completed his PhD in Biomedical Sciences at Chonnam National University, South Korea under the supervision of Dr. In-Kyu Park. His main area of interest includes gene delivery with modified polymers and imaging with nano-contrast agents. He has 30 research papers, three reviews, one book chapter, and two patents to his credit.

In-Kyu Park, PhD

In-Kyu Park studied natural fiber science at Seoul National University and obtained his PhD in 2002 under the supervision of Prof. Young Hwan Park and Prof. Chong-Su Cho. He pursued his career at the Department of Biomolecular Engineering, Tokyo Institute of Technology with Prof. T. Akaike, and at the Department of Bioengineering, University of Washington with Prof. Suzie H. Pun. He is currently an associate professor in the Department of Biomedical Science, Chonnam National University Medical School. He has over 100 refereed journal articles, five book chapters, and 16 patents to his credit.

Chong-Su Cho, PhD

Chong-Su Cho is a Research Professor, Research Institute for Agriculture and Life Sciences, Seoul National University. He received his PhD from the Department of Polymer Engineering, Tokyo Institute of Technology in March 1979. He worked with Prof. Allan Hoffman at the University of Washington as a post-doctor from 1982 to 1983 and worked with Prof. Sung-Wan Kim at the University of Utah as a post-doctor from 1983 to 1984. His research interests are drug delivery system, gene delivery, agricultural biotechnology, and tissue engineering. Professor Cho has received numerous awards such as the Fellow of Korean Academy of Science and Technology, Fellow of International Union of Societies for Biomaterials Science and Engineering, Korean Science and Technology Minister Research Award, Korean Polymer Society Sangam Award, Korean Health and Welfare Outstanding Research Award, International Controlled Release Society Outstanding Pharmaceutical Paper Award, and Korean National Academy Award. He has published 499 scientific papers and received 40 patents.

Weerapong Prasongchean

Weerapong Prasongchean is a lecturer of pharmaceutical sciences at the Chulalongkorn University, Thailand. He obtained his PhD in stem cells and regenerative medicine at the Institute of Child Health, University College London, UK, where he investigated the potential of somatic stem cells from different sources for regenerative medicine. Previously, he graduated with a BSc in pharmacy from Chulalongkorn University and received a MS degree in integrative neuroscience from Imperial College London School of Medicine, UK. His research interests include developmental biology and regeneration of the choroid plexus and nervous system, development of stem cells-based medicinal products, and pharmaceutical applications of stem cell *in vitro* models. In addition, his current educational initiatives involve development of stem cell cross-curricular learning in high schools and regenerative pharmacy for pharmaceutical education in Thailand.

Piyush Kumar, PhD

Piyush Kumar, PhD is an established basic, translational, and clinical researcher (associate professor) at the Department of Oncology, University of Alberta. He has extensive experience in the development of positron and single photon-emitting radiopharmaceuticals, which is supported by numerous publications in high impact journal, book chapters, and novel inventions in the field of hypoxia management. He has successfully provided two "bench to bedside" radiopharmaceuticals to the nuclear medicine community for clinical imaging of hypoxic tumors and 18F-FAZA, one of the molecule invented by his group, is now a preferred PET imaging agent that is used worldwide for the clinical diagnosis of hypoxic tumors in cancer patients. His current focus is on developing multimodal (nano)theranostic drugs for cancer management.

Stephen Quan, M.Sc

Stephen Quan is a master's student in the chemical and materials engineering program at the University of Alberta. He obtained an undergraduate degree majoring in Biochemistry at the University of Guelph and currently is working toward a master's degree specializing in radiopharmaceuticals, polymer chemistry, and their applications in drug/gene delivery. Currently, his academic research and achievements have been published in four peer-reviewed journals.

Preface

Gene therapy has received significant interest due to its potential for the treatment of genetic and acquired diseases. The primary goals of gene therapy are to correct shortcomings that arise from a disease and to provide an alternative mode of treatment through genetic engineering. The developments of non-viral vectors have drawn significant attention in both academia and industry as viral vectors, and although very effective, they still pose significant challenges and risks in clinical translation. Polymeric- and nanomaterial-based gene delivery systems are the most studied since they provide several benefits. With remarkable recent advances in polymer science and nanotechnology, the fabrication of polymeric and nano-delivery systems has become more versatile. These materials can be designed and synthesized with unique physical, chemical, and biological properties that are essential for gene delivery uses. Improved transfection, specificity, and biocompatibility are some of the parameters monitored in the evaluation of non-viral delivery systems, and these parameters rely significantly on how well the polymers and nanomaterials are tailored to address those challenges. The chapters discuss in detail the different strategies adopted by researchers to address those issues either by engineering the materials or by synthesizing novel materials using advanced techniques.

The main purpose of this book is to provide an update on the non-viral gene delivery systems developed so far with an emphasis on essentially polymeric- and nanomaterial-based delivery systems, since the applications of these materials in biomedicine have become a distinct and particularly dynamic area of scientific and industrial advancement. In addition to all the examples provided in this book, the aim of this text is also to offer readers sufficient details about the mechanisms as well as the challenges and opportunities of gene therapy.

Finally, I would like to thank all the contributors for the efforts in making this book a quality reference for researchers working in this field. In addition, I am grateful to the people at Woodhead Publishing including Glyn Jones and Harriet Clayton for the wonderful support and patience for transforming the concept of this book into a reality.

Ravin Narain
Edmonton, Canada
September 2015

Mechanism, current challenges and new approaches for non viral gene delivery

1

B. Thapa, R. Narain
University of Alberta, Edmonton, AB, Canada

1.1 Introduction

The concept of gene delivery was developed in 1963, and since then it has emerged as an attractive approach to treating acquired and genetic disorders such as cancer, cystic fibrosis, hemophilia, vascular, infectious diseases, and many others [1]. The completion of the Human Genome Project enabled the possibility of treatment of genetic diseases by introducing non-mutated gene or promoting immune response against infectious diseases by administering genes encoding specific antigens [2]. Two major vectors, viral and non-viral, are practiced in both clinical as well as research studies. First clinically practiced in 1989, human gene therapy has since then resulted in several viral and non-viral vectors that have been produced and tested [1]. In spite of efficient delivery capacity, severe limitations of viral vectors such as low DNA/RNA loading capacity, immunogenicity, and toxicity have encouraged the development of non-viral vectors. Non-viral vectors offer non-toxic and highly efficient gene delivery systems due to their facile chemistry, flexible manufacturing, and safe toxicity profiles [1,3,4]. Several cationic polymers, dendrimers, and cationic lipids have been studied. These carriers condense DNA/RNA molecules into nano- to micrometer-sized complexes via electrostatic interaction, and protect the payload from enzymatic/non-enzymatic degradation and enhance cellular communication via electrostatic interaction [5,6].

While significant achievement of non-viral gene vectors has been reported elsewhere, clinical outcomes are still insignificant. It is believed that non-viral vectors are still facing difficulties from administration to localization in the nucleus. Common barriers reported in the literature include instability of the complexes in extracellular spaces, insufficient cellular uptake, low endosomal escape, and their translocation into the nucleus for transcription. In some studies, biological barriers such as immunological responses and DNA sequence have also been reported [7–9]. However, chemical and physical barriers are the most detrimental factors for efficient gene delivery compared to biological barriers, which can be minimized, e.g., by DNA sequence optimization [10,11]. In addition, gene delivery to some extent can affect the growth condition of cells. A detailed understanding of the barriers and mechanisms is required for the development of a stable and effective non-viral gene delivery system.

Polymers and Nanomaterials for Gene Therapy. http://dx.doi.org/10.1016/B978-0-08-100520-0.00001-1
© 2016 Elsevier Ltd. All rights reserved.

Delivery of an exogenous therapeutic gene into target cells is still challenging and complicated for both material and medical scientists. These challenges and the mechanism of non-viral gene delivery system will be the focus of this chapter.

1.2 Current non-viral gene delivery vectors

Gene expression results when therapeutic gene is transported to the nucleus of a cell which requires safe and potential carrier. As of 2012, over 1800 clinical trials have been completed or are in progress with the majority using viral vectors [12]. However, the death of an 18-year-old and reasonably fit gene trial patient has drawn attention to the safety of viral vectors [13]. Non-viral gene delivery systems comprise the use of naked DNA [14], DNA complexes with polymers (polyplexes) [15], cationic lipids (lipoplexes) [16], combination of both cationic lipids and cationic polymers (lipoplexes) [17,18], polymeric vesicles [19], or formulation into liposomes [20].

1.2.1 Naked DNA

Direct injection of free DNA into certain tissues, mainly in muscle to produce gene expression, is the simplest non-viral gene delivery. This approach is used in cancer gene therapy by direct injection into tumor and genetic disease of tissues that are easily accessible such as skin, muscle, etc. The feasibility of this approach was introduced early 1990 when Wolff et al., [14] demonstrated gene expression by direct intramuscular injections of pDNA into mouse skeletal muscle. Later, several studies showed efficient transfection on direct application of naked DNA to liver of rats and cats [21], solid tumors in mouse [22], mouse epidermis and hair follicles [23], mouse corneal stroma and epithelium [24], and human skeletal [25] and cardiac tissues [26]. The transfection ability of naked DNA had directed its use for vaccination purposes where DNA encoding an antigen was administered to develop protective immune response [27,28]. Some studies showed that injection of transgene with carriers such as cationic liposomes hinders gene expression [23,29]. Despite the simplicity of direct injection of naked DNA, poor transfection efficiency as compared to viral or liposomal vectors and incompatibility for systemic administration are its major limitations.

1.2.2 Cationic lipids

Cationic lipids are amphiphilic molecules possessing hydrophilic head and hydrophobic tail. Hydrophobic group may be either one or two fatty acid or alkyl moieties of variable carbon chain, e.g., 12-18 or cholesteryl moiety. These hydrophobic moieties are attached to cationic amine, ensuring exposure of amine head group and shielding hydrophobic tail while forming bilayer vesicles in aqueous media. These two groups are linked together by linker group (Figure 1.1). The cationic amine head group electrostatically binds with DNA/RNA molecules and condenses into small transportable

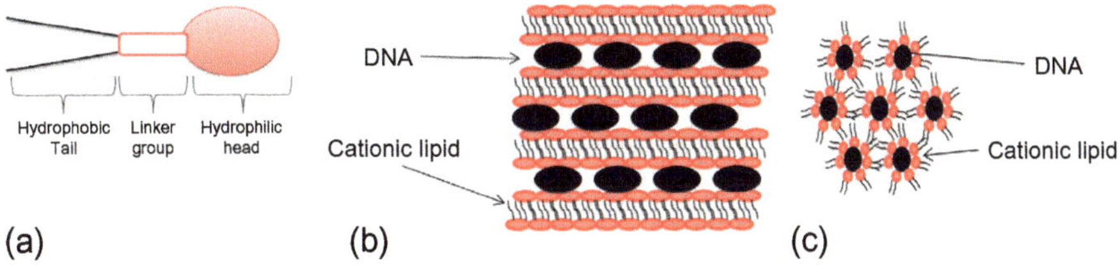

Figure 1.1 (a) General structure of cationic lipids used in gene delivery, with hydrophobic tail and hydrophilic head linked together via linker group. Schematic representation of lamellar (b) or inverted phase (c) structure in the formation of lipoplexes.

units called lipoplexes. Felgner et al. introduced cationic lipids in gene delivery for the first time in 1987 [30], Since then, a large number of cationic lipids has been synthesized and applied in gene delivery [30–33]. Many cationic lipids such as lipofectin (N-[1- (2,3-diolyloxy) propyl] -N,N,N-trimethyl ammonium chloride and 1,2 dioleoyl phosphatidylethanolamine-DOPE, 1:1) are commercially available as a gene delivery kit. It is believed that lipoplexes enter into cells via endocytosis followed by destabilization of endosomal membrane and flip-flop reorganization of phospholipids. Reorganized phospholipids interact with cationic lipids of complexes resulting in the release of DNA into the cytoplasm [34].

Based on structure, cationic lipids used as a gene delivery are categorized in different groups as follows:

1. Monovalent aliphatic lipids with single amine functionality in their head group, e.g., N[1-(2,3-dioleyloxy)propyl]-N,N,N-trimethylammonium chloride (DOTMA),), N-(2-hydroxyethyl)- N,N-dimethyl-2,3-bis(tetradecyloxy-1-propanaminiumbromide) (DMRIE)
2. Multivalent aliphatic lipids with several amine functionalities in head group, e.g., spermine groups, e.g., DOGS (dioctadecylamidoglycylspermine)
3. Cationic cholesterol derivatives, e.g., 3b-[N-(N0, N0-dimethylaminoethane) carbamoyl] cholesterol (DC-Chol), bis-guanidium-tren-cholesterol (BGTC)
4. Neutral helper lipids such as DOPE (1,2-dioleyl-sn-glycerol-3-phosphoethanolamine) or cholesterol, which were added to complex of DNA and cationic lipids to improve transfection efficiency.

In the last few decades, lipids with variable head group functionalities have been introduced. Guanidine, imidazole, pyridinium, piperizine, and amino acid (e.g., lysine, arginine, ornithine, and tryptophan) are the common head groups used in lipid modification [35–37]. Transfection efficiency and toxicity of these cationic lipids generally depends on their structure. Lipid with multiple cationic head group has a strong ability to condense and protect DNA and display higher transfection efficiency compared to single-charged lipids [37]. Some studies have claimed that strong binding results in inhibition of the intracellular release of DNA molecules and increases cellular toxicity [38]. Transfection efficiency of lipid molecules also depends on the distance between cationic head and hydrophobic tail groups. Generally, transfection efficiency increases with the number of cationic head groups and the distance between head and tail groups [39,40].

Cationic lipids are readily modified from other functional groups such as polyethylene glycol (PEG) thereby providing so-called stealth properties to prevent their rapid elimination from blood circulation by macrophages upon systemic injection [41]. Attachment of sugar residues and sugar-linked bio-surfactants offers specific targeting to particular cell receptors allowing cell type-dependent vector specificity [42]. Although hundreds of cationic lipids have been synthesized and used for gene delivery both *in vitro* and *in vivo* [43], their application in gene delivery is limited due to low efficiency. To address this problem new lipids are being synthesized and new approaches for complex formation are being developed continuously.

1.2.2.1 Liposomes

Cationic and helper lipids involve formation of cationic liposomes, which have been used as transfection agent [42]. Cationic liposomes interact with DNA to form lipoplexes via a multistep mechanism. The first step involves the electrostatic interaction between negatively charged phosphate groups of DNA molecules and positively charged amine head groups of cationic lipids. DNA-lipid one-sided interaction results in packing constraint in bilayer, which triggers interaction between adjacent bilayers and breakage of lipid bilayer [44]. This results in the exposure of hydrophobic edges that function as nucleation sites for membrane merging, which leads to lipid mixing and growth. Finally, cationic lipids are wrapped around DNA molecules and further mixing of lipid molecules is ceased, surface becomes smooth with proper DNA packing [42,44]. Then these complexes result in either lamellar or inverted hexagonal phase structure. The lamellar form consists of DNA monolayers sandwiched between cationic lipid bilayers, while inverted hexagonal consists of DNA coated with cationic lipid monolayers arranged on two-dimensional hexagonal lattice (Figure 1.1).

An increased level of protein expression has been observed in many studies upon intravenous injection of cationic liposome/DNA complex, probably due to protection of DNA molecules against plasma degradation [45,46]. However, they were rapidly cleared from plasma to lung tissue and protein expression was observed primarily in lung endothelium [40]. To overcome this problem, many strategies have been applied such as incorporating polyethylene glycol, targeting ligands, and directly applying to the site of pathology [40]. Cationic liposomes with 3β-[N-(N,N-dimethylaminoethane)-carbamoyl] cholesterol (DC-Chol) and DOPE significantly enhanced inhibition efficiency of antisense oligonucleotide in tumor cells [47].

1.2.2.2 Lipid-coated DNA complexes

In some studies, cationic lipids are used as coating material for cationic polymer/DNA complexes. In this strategy, lipid coating prevents the possible interaction between extracellular poly anions and cationic polyplexes [48]. This strategy significantly improves the transfection efficiency of DNA/PEI complexes in liver cells, even in high serum concentration (60% FBS).

1.2.2.3 Niosomes

Niosomes are self-assembly of nonionic amphiphiles and widely studied in gene delivery as an alternative to liposomes. As compared with liposomes, niosomes have higher physico-chemical stability. PEG-modified niosomes have shown improved stability and cellular delivery of oligonucleotides *in vivo* [48].

1.2.3 Cationic polymers

Cationic polymers are exploited as the most attractive material for the construction of non-viral vectors due to their facile chemistry, cost effectiveness, and safe toxicity profiles [5]. Many cationic polymers are commercialized and used to design non-viral vectors, and the most common polymers are polyethylenimine (PEI, branched/linear), poly(L-lysine) (PLL), cationic dendrimers, glycopolymers, poly(amidoamine) (PAMAM), and chitosan (CS). Electrostatic interaction between positively charged amino groups of these polymers and negatively charged phosphate groups of nucleic acid results from the formation of condensed polyionic complexes (polyplexes). These complexes protect nucleic acid from enzymatic/non-enzymatic degradation, enhance cellular uptake via interaction with anionic cell surface proteoglycans, and increase half-life in cytoplasm [49]. However, some reports show that cellular uptake varies significantly depending on the types of cationic vectors, cell types, and amount of cell surface proteoglycans, etc. [50].

1.2.3.1 Polyethylenimine (PEI)

PEI used for gene delivery is a versatile, highly cationic polymer of different geometry, e.g., branched or linear. It possesses three different types of amines, primary, secondary, and tertiary that display a wide range of buffering capacity. This enables PEI to escape from endosome via a so-called "proton-sponge" effect [51]. This polymer can efficiently interact with DNA molecules, resulting in the formation of polyionic complex of homogeneous spherical particles that are endocytosed in many cell types. Condensation of DNA molecules provides incredible protection against degradation by DNase I and DNase II both *in vitro* and *in vivo* [52]. The strong cationic charge density creates unnecessary cellular toxicity. Non-biodegradability of PEI has also been noted as a hurdle in gene delivery [53]. In order to reduce toxicity, charge-neutralizing agents are routinely conjugated to PEI. For example, conjugation PEG to transferrin-PEI/DNA complexes resulted in reduced surface charge and decreased complexes aggregation, interaction with plasma proteins, and toxicity [54]. PEI has also been conjugated with sugar molecules to generate target specific non-viral gene delivery vector. Galactosylated PEI-DNA complexes were prepared for the hepatocyte targeting where transfection can be increased by manipulating the chain of galactose [55].

PEI of a wide range of molecular weight has been studied for its gene delivery efficacy. Transfection efficiency of PEI increased with molecular weight, 0.6 to 70kDa [56]. However, high molecular weight PEI causes higher toxicity due to aggregation and adherence on cell surface, which results in necrosis [5]. Molecular geometry of PEI shows major effect in complex formation and stability. Linear PEI is generally found less effective at condensing DNA as compared to branched PEI for same molecular weights [5]. However, several *in vivo* studies have found linear PEI to be an effective gene delivery agent [5,57].

Branched PEIs (bPEI) were used in the delivery of oligonucleotides, plasmid DNA, Epstein-Barr virus-based plasmid vectors, RNA, and intact ribozymes [58]. Their efficacy and cytotoxicity are determined by material characteristics such as molecular weight, degree of branching, cationic charge density, and buffer capacity as well as their interaction with DNA to form polyplex such as particle size and zeta potential [58]. Although high molecular weight bPEI (up to 800kDa) was found superior to form stable complex with DNA and high transfection efficacy, toxicity associated with them urges to use low molecular weight bPEI (between 5–48kDa) [58,59]. PEI-mediated transfection was observed in 25 different cell types including 18 human cell lines as well as *in vivo* experiments [60]. Some linear derivatives of PEI such as ExGen500 and jetPEI are commercially available transfection agents [5].

1.2.3.2 Poly-L-lysine (PLL)

Poly-L-lysine (PLL) is an effective polypeptide used in gene delivery. PLL is synthesized via solid-phase and automated synthetic procedures allowing polymer preparation of specific molecular weight under controlled dispersity. The primary ε-amine of lysine is protonated in a physiological environment that electrostatically binds to DNA [60]. The efficiency of PLL is dependent on its molecular weight. It linearly increases with molecular weight but its intolerable toxicity at higher molecular weight polymer limits its frequent application. Zhang and Smith showed the relationship between length of PLL and polyplex size, delivery, and duration of transgene expression *in vivo* and *in vitro*, which revealed that higher molecular weight PLL form polyplexes of smaller size, effectively protecting pDNA for nuclease degradation, higher transgene expression in HuH7 cells [61]. The efficacy of PLL alone in gene delivery is relatively poor, therefore, chemical modification by conjugating ligands such as asialoorosomucoid, transferrin, folate, monoclonal antibodies, and basic fibroblast growth factors is frequently used to enhanced its efficacy [40]. For the first time in 1987, PLL conjugated with asialoorosomucoid was utilized in gene delivery to hepatocytes [40]. PLL display a strong DNA binding capacity. These complexes still possess strong cationic surface charge that enable opsonization by plasma protein and are rapidly cleared from plasma [40]. In the last few decades, hydrophilic polymers have been grafted onto PLL to increase stability in serum protein, and polyethylene glycol (PEG) is the most potent candidate for this purpose [62]. Wolfert et al. observed 125-fold higher gene expression in HEK293 cells when using PEGlated PLL as compared to unmodified PLL [63].

1.2.3.3 Polymethacrylate

Polymethacrylate is a vinyl-based polymer that binds with DNA due to its inherent cationic charge. These polymers have been used in gene delivery *in vitro* as well as *in vivo* models. Poly[2-(dimethylamino) ethyl methacrylate] (PDMAEMA) of molecular weight more than 300 kDa offered the highest transfection efficiency along with tolerable toxicity [5]. The ability of PDMAEMA/DNA polyplexes to destabilize endosomes and dissociate from DNA, in cytosol contributed to transfection efficiency [64]. Intravenous injection of PDMAEMA/DNA polyplexes into mice was found to accumulate in lung due to formation of aggregate that facilitate uptake in lung tissue [5]. Methacrylate polyplexes were shown to uptake via both clathrin and caveolae-dependent pathways [65]. Recently, biodegradable methacrylamide has been synthesized for gene delivery purposes. PDMAEMA has been modified to improve gene delivery efficiency and reduce cytotoxicity. Incorporation of guanidinium side group improved transfection efficiency of PDMAEMA polymer [66]. Copolymer of PDMAEMA with ethoxytriethyl glycol methacrylate (triEGMA) or *N*-vinyl-pyrrolidone (NVP) showed reduced toxicity and PDMAEMA-NVP exhibited enhanced transfection [5]. Incorporation of tumor-targeting Fab′ fragment of mAB 323/A3 onto lipid-coated PDMAEMA polyplex as well as folate-conjugated PDMAEMA polyplex showed significantly higher transfection efficiency in human carcinoma (OVCAR-3) [5]. Methacrylate-based polymer with carbonate functionality (pHPMA-DMAE) was synthesized that release DNA at pH7.0 while complexing at pH 5.0 and transfection efficiency was improved in presence of INF-7 [67]. In addition, conjugation of poly(hydroxylethyl methacrylate-*co*-hydroxylethyl methacrylate propargyl alcohol) onto PDMAEMA create brush copolymer, resulting in reduced toxicity and enhanced transfection in the presence of INF-7 [5].

1.2.3.4 Chitosan

Chitosan is a natural biodegradable and biocompatible polysaccharide derived by deacetylation of chitin (polysaccharide found in the exoskeleton of crustaceans and insects [51]). It binds with DNA forming stable particles of size 20–500 nm depending on the molecular weight and degree of deacetylation [68]. Frequent application of chitosan is somewhat limited due to its poor solubility, reactivity, and subtle physical properties such as rigidity and brittleness. Both molecular weight and degree of deacetylation (DD) has been equally attributed to the efficiency of chitosan. Transfection efficiency of chitosan in vitro decreased with decrease in degree of deacetylation [69]. For gene delivery purposes, chitosan is routinely modified with agents such as galactosylated chitosan, galactosylated chitosan-graft-poly(vinylpyrrolidione) (PVP), trimethylated chitosan oligomers, N-docelyated chitosan, and deoxycholic acid modified chitosan [51]. In spite of the solubility issue, higher molecular weight chitosan with higher DD forms small and stable complexes with DNA/RNA molecules and results in better delivery efficiency [70]. Low molecular weight chitosan (Mw – 10 kDa) demonstrated almost negligible endogenous EGFP silencing in H1299 human lung carcinoma cells and murine peritoneal macrophages compared to the higher molecular weight (Mw – 170 kDa, DD 80%) indicating the proportional impact of molecular weight [71].

1.2.3.5 *Dendrimers*

A range of polyamidoamines, poly(propyleneimine), poly-L-lysine, phosphorous-containing dendrimers have been studied in gene delivery. The efficiency of these materials is relatively higher due to crowded cationic charge densities. However, the toxicity corresponding to concentrated cationic charge always limits the application of these materials.

1.2.3.5.1 Polyamidoamine dendrimers (PAMAM)

The first gene delivery using PAMAM/DNA complex called dendriplex was performed in 1993. Since then, several studies have been conducted to explore the transfection efficiency of PAMAM dendrimers as well as their derivatives. In the last few decades, it has been developed as a leading dendrimer utilized in gene delivery [72]. Smith et al. reported that PAMAM dendrimers of high generation (G5-G7) were capable of inducing lipid mixing and leakage from anionic vesicles, which was attributed to cellular penetration [61]. Verkman et al. revealed that the buffering capacity of PAMAM results in a lysosomal lysis via proton sponge effect [73]. The transfection efficiency of PAMAM with G5-G10 was significantly higher, but it was cell-line dependent. PAMAM has also been chemically modified to reduce toxicity and enhance transfection efficiency. A membrane-destabilizing peptide, GALA, was conjugated onto the surface of PAMAM which enhanced gene transfer efficiency of PAMAM upto 2 to 3 fold. Similarly, conjugation of L-arginine onto the surface of PAMAM or PAMAM-PEG-PAMAM triblock copolymer improves transfection efficiency [5].

1.2.3.5.2 Poly(propylenimine) dendrimers (PPI)

PPI has basic amine at peripheral groups (pKa 9-11) and more acidic internal ammonium functionalities (pKa 5-8), which gives it potential for gene delivery. It was proposed that all generation of PPI binds with DNA via electrostatic interactions and higher generations were capable of condensing DNA to form cationic water soluble complex [5]. Low-generation G2,G3 PPI showed the highest transfection with lower toxicity [74]. *N*-quaternization of low-generation (G2) showed improved DNA binding, reduced toxicity, and gene delivery into liver [75].

1.2.3.5.3 Poly(L-lysine) dendrimers

In order to prevent high toxicity of PLL-based polymer, poly(L-lysine)-based dendrons (DPL) have been synthesized and used for gene delivery. Lower-generation DPL (G2,G3) has been shown to be effective for delivering gene to COS7 and BHK cells [5]. Later, high transfection efficiency was found when using higher generation DPL structures (G5,G6) and complex stability of G6 generation was also observed even at high serum (50%) [76]. Later, DPL was used to mediate antisense oligonucleotide [77] and siRNA gene knock-down [78].

1.3 Trajectory and mechanism of non-viral vectors

1.3.1 Basics of non-viral gene delivery system formulation

Delivery of naked DNA into cells via electroporation [79], gene gun [80], or direct injection into target tissue [14] has been reported in several studies, however, the clinical use is insignificant. Cellular uptake of free DNA is limited due to degradation. Alternatively, non-viral vectors are used as an effective tool in gene delivery. Non-viral gene delivery systems are formulated by adding a DNA/RNA solution to the solution of delivery vehicles including cationic polymer, lipid, or liposomes followed by incubation at room temperature [40]. These carriers complex with DNA molecules that not only neutralizes the charge of DNA molecules but also compacts it into colloidal complexes called polyplexes or lipoplexes (Figure 1.2). Non-viral DNA delivery vehicles mainly have cationic charge that electrostatically interacts with the negative charge of DNA. H-bonding between polycations and nucleic acids enhances interaction and contributes to stabilization of polyplex [81].

1.3.1.1 Complexes between cationic lipid and DNA (lipoplexes)

The positive charge of cationic lipids is contributed by the amine group with different degrees of substitution or by other functional groups such as amidine, guanidium, or pyridinium. Complexes of cationic lipids and nucleic acids, called lipoplexes, are prepared by combing preformed cationic liposomes and DNA in an aqueous solution, which results in electrostatic interaction between positive charges of cationic lipid head groups and the negatively charged phosphate backbone of DNA. Concentration, temperature, environment, and kinetics of mixing could affect the characteristics and transfection potential of lipoplexes. In addition, transfection efficiency is higher for slight cationic lipoplexes as their interaction with negatively charged cell surface is

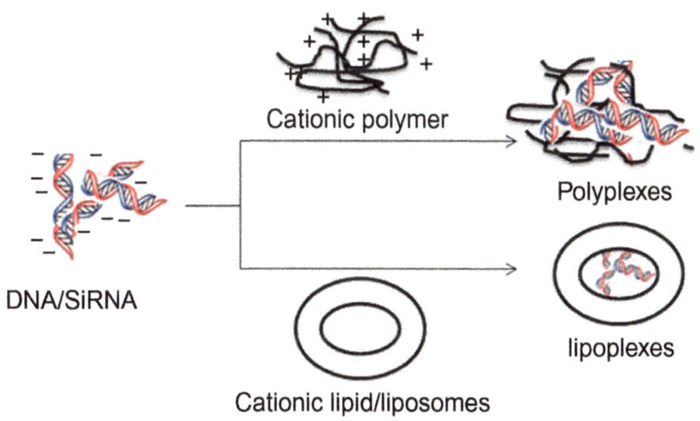

Figure 1.2 Schematic illustration of formation of polyplexes and lipoplexes. Polyplexes and lipoplexes are formed spontaneously via electrostatic interactions when aqueous solution cationic polymer/lipids and DNA/SiRNA are mixed together. Typically, an excess of cationic polymer or lipid is used to generate complexes of positive charge for gene delivery.

strong [82]. Protection of DNA from degradation by cellular nuclease is higher for positively charged lipoplexes [83].

The study of the structure of lipoplexes by freeze-fracture and transmission electron microscopy (TEM) has shown spherical structure in which DNA is surrounded by lipid bilayer [84]. Further, lipoplex formed by unilamellar cationic vesicles and DNA was found in multilamellar structure [84]. These multilamellar structures consist of DNA filaments that are oriented in parallel and sandwiched between lipid bilayers [85]. Formation of lipoplex is thought to be completed in two steps, where the initial step is electrostatic interaction of cationic lipids of liposomes and DNA, which is exothermic, rapid, and reversible and liposomes structure and shape is intact, and the second step is fusion and rearrangement of liposomes, which is slower, endothermic, and irreversible and occurs at approximately 1.2 molar lipid/DNA ratio [86]. During lipoplex formation, neutralization of charge of DNA and lipids affects the structure of both. DNA becomes more compact, resulting in less accessible to dyes or nucleases and lipid bilayer combines and/or aggregates resulting in fusion of liposomes and/or formation of multilamellar structure. However, which structure is most efficient for transfection is still unknown [87].

1.3.1.2 Complexes between cationic polymer and DNA (polyplexes)

Unlike cationic lipids, cationic polymers do not have a hydrophobic moiety and are completely soluble in aqueous medium (e.g., water). These polymers condense DNA more efficiently [88]. Cationic polymers mainly have different types of amines contributing to cationic charge, which can be protonated in physiological pH and binds with negatively charged DNA/RNA molecules. Hence, polymer/DNA polyplex formulation is denoted by the molar ratio of nitrogen (able to be protonated in physiological pH) of polycations to phosphates of nucleic acid (N/P ratio). Typically, a positive N/P ratio is used for *in vitro* gene delivery in order to have positively charged polyplexes, which prevents aggregation due to electrostatic repulsion and increases colloidal stability of polyplexes. Although particles of various sizes and shapes can be internalized *in vitro* [89], polyplex of optimum size and shape without any aggregation is mandatory for *in vivo* application [90]. The size of the polymer/DNA complex is dependent on type of cationic vehicles used, concentration of DNA, pH, type of buffer, and N/P ratio [5]. Cationic polymers strongly bind with DNA, resulting in complexes with multiple DNA molecules. The size of these complexes depends on the physical properties of cationic polymer.

The complexes of DNA and cationic polymers such as PLL, PEI, or dendrimers have been found toroidal to have structures of similar diameters but sizes varied with molecular weight of polymer. For example, Wolfert and Seymour found that polymer-DNA complexes with high molecular weight PLL (224 kDa) were 100 to 300 nm, while those with low molecular weight (~4 kDa) were 20 to 30 nm [91]. Moreover, polymer structure has an effect on the aggregation behavior of the complexes as less flexible intact dendrimers and polylysine complexes have been found to be aggregated [92]. These physiochemical characteristics of the complexes impact cellular uptake, endosomal escape, and nuclear targeting.

1.3.2 Binding and uptake

The route of non-viral DNA complexes depends on whether or not the targeting ligand is conjugated to complex. Several studies have demonstrated that cellular uptake of cationic complexes without any targeting ligand is guided through electrostatic interaction with anionic cell-surface proteoglycans [5]. Cationic liposome-DNA complexes were unable to transfect Raji cells that lack proteoglycans [93]. Transfection efficiency of PLL-DNA complexes was remarkably reduced by inhibition of proteoglycan sulfation, removal of cell surface glycosaminoglycans by glycosaminoglycan lysases, or addition of extracellular glycosaminoglycans to media [94]. However, another study showed that cellular uptake of DNA complexes is dependent on type of cell, carrier, and amount of glycosaminoglycan [50]. Uptake of PEI-DNA complexes by HeLa cells was reduced by inhibitor of protein kinase C (PKC), staurosporine, and cyclodextrin and following uptake, complexes were found in the actin filament [95]. Based on this evidence, the route of PEI-DNA complexes into cells is thought to be determined by a series of events. When PEI-DNA complexes bind with transmembrane heparin sulfate proteoglycans, called syndecan, clustering of proteoglycans into cholesterol-rich rafts on the cell surface occurs. This clustering stimulates phosphorylation of protein kinase C (PKC) and results in binding of syndecan into actin skeleton via linker proteins. Then, complexes are pulled into cells via phagocytosis [95]. In addition, several receptor ligands have conjugated to non-targeting vector to promote cells specific delivery. Some examples include asialoglycoprotein, epidermal growth factor, transferring, mannose, folate, etc. [5]. Complexes bound to cell surface receptors enter into cells via clathrin dependent endocytosis [5]. In general, the endocytic route can be divided into four different groups: phagocytosis, clathrin-mediated endocytosis (CME), caveolae-mediated endocytosis (CvME), and micropinocytosis (Figure 1.3).

1.3.2.1 Phagocytosis

Phagocytosis primarily exists in phagocytes such as macrophages, monocytes, neutrophils, and dendritic cells [96]. Large polyplexes and lipoplexes were suggested to be uptake via phagocytosis [95]. Complexes were first opsonized by opsonins in the bloodstream followed by binding to phagocytes and ingestion [97]. Following ingestion, phagosomes fuse with lysosomes to form phagolysosomes where they encounter acidic and enzymatic degradation [97]. This requires non-viral vectors to be designed in such a way that it protects the carrying gene from degradation. In addition, complexes can also be phagocytosed without opsonin, which is mediated by receptors on the cell surface. siRNA was targeted to beta-glycan receptor on macrophages by using ligand dectin-1 for treatment of inflammation by oral administration [98].

1.3.2.2 Clathrin-mediated endocytosis

Clathrin-mediated endocytosis is receptor dependent and often explored in the design of non-viral gene delivery. Transferrin, known to uptake by endocytosis, has been used to enhance endocytosis of non-viral complexes [99]. After binding to the cell

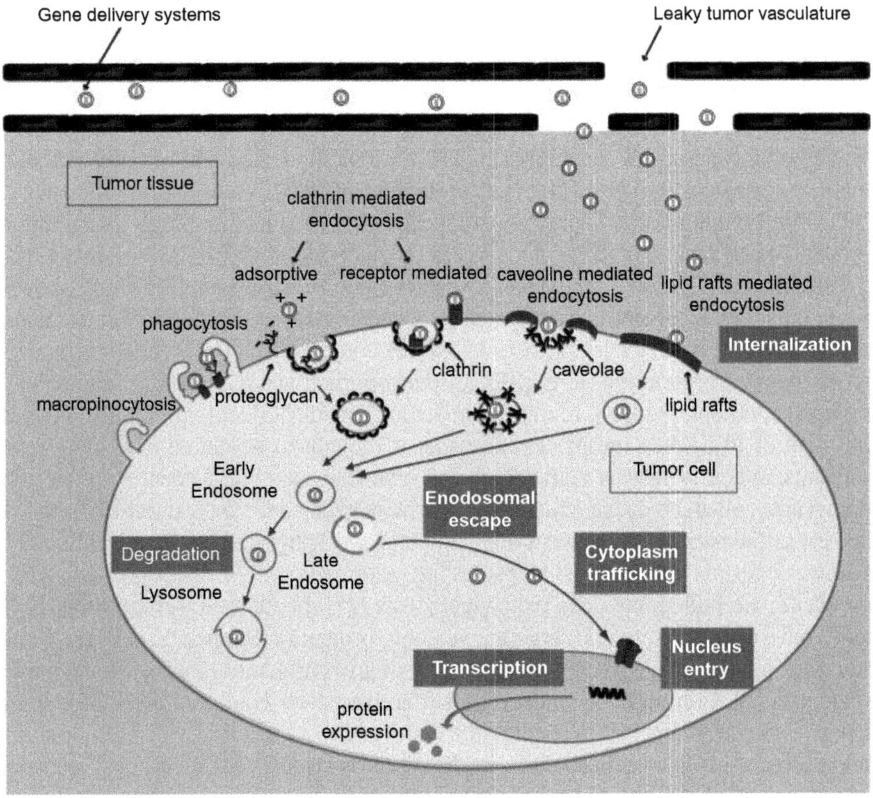

Figure 1.3 Schematic representation of uptake pathways of non-viral gene delivery systems and their fate in cytoplasm.
Figure from [51] with permission.

surface receptor, clathrins assemble together right on the cytosolic surface of membrane, which triggers deformation of the membrane into coated pits 100–150nm in size [100]. This process continues until vesicle fission occurs, which requires dynamin, a kind of GTPase [101], and cholesterol [102]. Following endocytosis, endosomes get maturated into late endosomes involving acidification by proton pumps of endosomal membrane and finally are fused into lysosomes (Figure 1.3) [99]. Some non-viral vectors such as PEI take advantage of the acidic environment of endosomes in order to release payload via a so-called proton sponge effect [103]. Nitrogen atoms of PEI are protonated using endosomal protons causing increase in chloride anion, which results in osmotic swelling of endosome, and finally endosomal membrane break-up and release complexes [103]. In the case of lipoplexes, payload is released via a flip-flop mechanism [104], where cationic lipids interact with anionic lipids from cytoplasm-facing monolayer of endosomes. This leads to release of payload into cytoplasm [104].

1.3.2.3 Caveolae-mediated endocytosis

Caveolae-mediated endocytosis initiate from caveolae, which is lined by dimeric protein caveolin and enriched with cholesterol and sphingolipid. They are a special flask-shaped structure with diameter of 50–100 nm located on cell membrane [97]. Caveolae-mediated endocytosis is receptor-mediated cholesterol and dynarrin-dependent pathway. Receptors present on caveolae such as insulin receptor and epidermal growth factor receptor can also mediate caveolae-mediated endocytosis [99]. During this process, complexes bind with the cell surface and move along the plasma membrane to caveolae invaginations. Then, GTPase dynamin mediates fission of caveolae from the membrane to generate cytosolic caveolar vesicles, which is generally considered as a non-acidic and non-digestive route of uptake [99]. Therefore, design of complexes exploiting caveolae-mediated endocytosis may by-pass the lysosomal degradation pathway. In contrast, Kiss et al. suggested that sometimes caveolar vesicle fuse with lysosomes [105]. Ligands such as folic acid, albumin, and cholesterol are known to be uptake by caveolae-mediated endocytosis [106].

1.3.2.4 Macropinocytosis

Macropinocytosis is also clathrin-independent endocytic pathway where a large amount of fluid-phase content is endocytosed non-specifically [97], and occurs via formation of actin-driven membrane extensions which collapse onto and fuse with plasma membrane and form vesicle, called macropinosome [107]. Fission from the cell membrane is mediated by small GTPase, Ras-related in brain (Rab) protein. Unlike other endocytosed vesicles, macropinosome does not have apparent coat structure and is of diameter larger than 0.2 um [99]. Macropinocytosis does not seem to have any selectivity and its intracellular fate depends on the cell type.

1.3.2.5 Non-endocytic pathways

In spite of its efficient cellular uptake pathways, endocytosis of genetic material causes poor intracellular availability due to the acidic and enzymatic environment of lysosomes. Therefore, non-endocytic pathways are beneficial for high intracellular availability of genetic materials. Generally, three technologies have been designed that by-pass endocytic pathways [99]: microinjection of genetic material into each cell using glass capillary pipettes; permeabilization mediated by pore-forming reagent such as streptolysin O or anionic peptides such as HA2 subunit of the influenza virus hemagglutinin; and electroporation using electric field to open pores in the cell. All of these techniques are invasive in nature and so are not suitable for *in vivo* application. In addition, penetration and fusion are naturally occurring non-endocytic pathways [99]. Penetration occurs via formation of holes in the cell membrane. Cationic peptides with a protein transduction domain such as TAT can directly penetrate cell membrane via receptor and energy-independent ways [108]. Polymeric vector, poly(amidoamine) (PAMAM) induced formation of hole in the fluid phase membrane [109]. In another study, protein transduction domain called Hph-1 conjugated PEI showed delivery siRNA though non-endocytic pathway and high transfection efficiency even at low temperature, suggesting an energy-independent pathway [110]. Another non-endocytic pathway, fusion, is mainly utilized by lipoplexes, which result in direct release of DNA to the cytoplasm. Although several studies suggest a minimum contribution of this pathway in cellular uptake of lipoplexes, it has a major role in endosomal release of lipoplexes [104].

1.3.3 Characteristics of non-viral gene delivery influencing uptake pathways

Since complexes of non-viral vector and DNA/RNA are a group of particles with heterogeneous size, shape, and surface charges, several pathways of cell uptake may be utilized by single cells. The uptake pathways of non-viral complexes depend on many factors including particle size, shape, and surface charge.

1.3.3.1 Size

Particle size of complexes is the most important factor in selecting the mode of endocytosis and uptake efficiency. For example, the uptake efficiency of 20 and 40nm particles via caveolae-mediated endocytosis was significantly higher (5–10 times) than that of 100nm [111]. Some studies revealed that particles of different sizes utilized different uptake pathways [112]. PEI/DNA complexes of smaller size (<500nm) are taken up by clathrin and caveolae endocytosis, whereas larger complexes (>500nm) are found to uptake via macropinocytosis pathway [99]. Similarly, uptake of PAMAM/DNA complexes is also dependent on particle size. Since PAMAM can form a hole with diameter of 15–40nm in cell membrane, complexes smaller than these holes can enter into cells by non-specific, non-endocytic pathways [99]. Non-targeting cationic DNA complexes between 70 and 90nm showed optimum gene transfer [113,114]. The transfection efficiency of PLGA-DNA complexes in COS-7 and HEK cells was higher while using complexes of size 70nm than complexes of size 200nm [114]. Similarly, the highest transfection was observed using PEI-DNA complexes with mean diameter 75 and 87nm [113]. Higher uptake of non-targeting complexes was observed with larger particles than smaller ones in some cases because larger complexes have more surface area for electrostatic interaction with cell membrane [5]. In the case of complexes with targeting ligands, uptake of complexes is strictly dependent on size of complexes [5,115]. Uptake of glycoside conjugated particles via a asialoglycoprotein receptor was found optimum for particles ≤70nm [115]. In other study, particles of 54–60nm showed the highest receptor-mediated endocytosis [116].

1.3.3.2 Surface charge

Another important factor for uptake is charge density of complexes. Cell membrane is negatively charged due to the presence of anionic membrane proteins, which results in better association and internalization rates of cationic complexes. However, if net positive charge is neutralized by anionic proteins present in blood circulation and in transfection medium, interaction of complexes and membrane protein will be reduced and complexes will aggregate, making it difficult to uptake [99].

1.3.3.3 Shape

Few studies have been done to relate shape and uptake pathway selection of non-viral complexes. Uptake of protein-coated spherical gold nanoparticles was found to be higher than rod-shaped ones in Hela cells [117]. Since DNA/non-viral vector complexes are heterogeneous in shape, its role in selection of uptake pathways is difficult to predict. The shapes of these complexes depend on the experimental conditions [99].

1.3.3.4 Cell type

Another important factor to be considered for transfection study is cell type. The uptake pathway of the complexes also depends on cell type. For example, PEI/DNA complexes enter into COS-7 cells mainly via the clathrin-dependent pathway [118]. In another study, macropinocytosis was found to be a more effective pathway than endocytosis for PEI/DNA complexes in CHO and Hela cells [119]. Caveolae, which is necessary for caveolae-mediated endocytosis, are rich in vessel wall lining of endothelial cells. Therefore, caveolae-mediated endocytosis is important for these endothelial cells [111]. Uptake of these complexes alone is not sufficient for transfection. It has been shown that PLL-DNA complexes enter into HepG2 cells through either clathrin-endocytosis or macropinocytosis, and only clathrin-dependent endocytosis afforded productive transfection in HepG2 cells [120].

1.3.4 Endosomal escape and dissociation of complexes

As compared to cytosol (pH 7.4), endosomal and lysosomal compartments of cells are acidic (pH 5.0–6.2) because of ATP-mediated proton accumulation [121]. This acidic pH is exploited by different non-viral vectors to prevent possible degradation. Chloroquine is a popular agent used in non-viral gene delivery to enhance transfection efficiency. Incorporation of chloroquine into DNA complexes prevents lysosomal degradation of the complexes since it raises pH endosomal and lysosomal compartments. Similarly, membrane-destabilizing peptides such as *N*-terminal peptides of Rhinovirus VP-1 or influenza virus HA-2 are used for endosomolysis since they interact with endosomal membrane in acidic pH and enhance the endosomal release of the complexes [5]. Some researchers have also used macromolecules with amine groups with low pKa values have buffering capacity, which causes endosomal swelling and lysis resulting in DNA release into cytoplasm via so-called "proton sponge" effects [5].

1.3.4.1 Lipoplexes

Lipoplexes are released from endosome by a flip-flap mechanism. In some studies, fusogenic lipids (e.g., DOPE) have been used to enhance transfection efficiency since it enables lipoplexes to fuse with endosomal membrane and release DNA molecules [122]. DOPE has the ability to catalyze the fusion process by undergoing from bilayer to inverted hexagonal structures [123]. Xu and Szoka demonstrated that lipoplexes destabilizes endosomal membrane inducing flip-flop movement of negatively charged phospholipids from cytoplasmic facing monolayer of endosomal membrane into inner monolayer, which causes formation of a neutral charge ion pair with a positively charged head group of cationic lipid. Consequently, electrostatic interaction between DNA and cationic lipid is reduced and ultimately DNA is released from lipoplexes into cytoplasm Figure 1.4 (a) [34]. Since cationic lipid covers large surface areas as compared to endosomal membrane surface of endosomal membrane, most of the cationic lipids cannot interact with anionic phospholipid and not all DNA can release into cytoplasm via this mechanism [124]. This mechanism is called flip-flop mechanism of lipid mixing. Another mechanism of endosomal escape for lipoplexes has been proposed (Figure 1.4), which disagrees

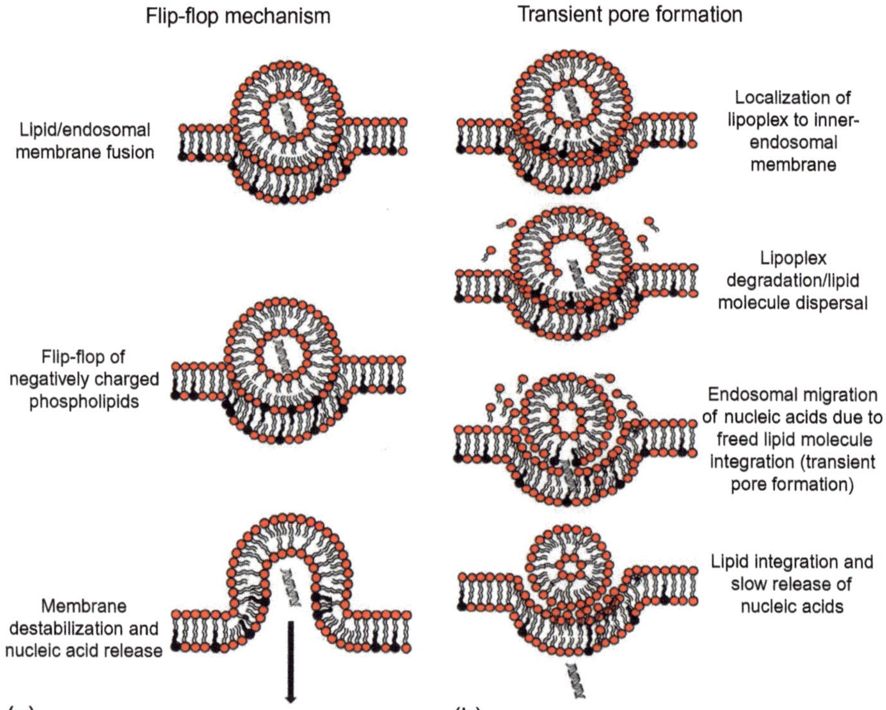

Figure 1.4 Schematic of endosomal escape by lipoplex reprinted with permission from [125]. Copyright (2013) American Chemical Society. The first approach is classical lipid mixing mechanism where endocytosed lipoplexes destabilized endosomal membrane. Destabilization induces flip-flop of anionic lipids from the cytoplasmic-facing monolayer to inner surface of endosomes and form a charge neutral pair with the cationic lipids. Finally, DNA is displaced from cationic lipid and released into cytoplasm. (b) In constrast, lipoplexes near the inner-endosomal membrane lose lipid molecules, resulting in passive release of nucleic acids.

the flip-flop mechanism of lipid mixing. According to this hypothesis, nucleic acids are moved to cytosol gradually, which is promoted by lipoplexes-facilitated multiple transient pores formation [125].

It has also been hypothesized that a detergent-like destabilization mechanism for endosomal escapes lipoplexes [87]. After releasing into cytoplasm, lipoplexes dissociate by interacting with a cytosolic membrane network such as the endoplasmic reticulum, golgi apparatus, mitochondria, and nuclear membrane [87].

1.3.4.2 Polyplexes

Unlike lipoplexes, cationic polymers cannot fuse or destabilize the endosomal membrane. The first-generation cationic polymers, polylysine or polyarginine, cannot escape from endosome and display insignificant transfection efficiency. Several strategies have been used to promote endosomal escape of polyplexes. Coincubation of adenovirus particles and polyplexes with cells is a common approach where adenovirus ruptures endosomal membrane [126].

However, this strategy raises serious concern regarding the toxicity and immune response of adenovirus particles. The common approach used in non-viral gene delivery is integration of fusogenic or endosome disrupting peptides. For example, amphipathic peptides, KALA and GALA, which undergo conformational change with pH change [127]. GALA contains repeating units of Glu-Ala-Leu-Ala, mimicking membrane-penetrating activity of viruses. At acidic pH, it changes from a random coil to a amphipathic alpha helix that has high membrane-binding capacity and aggregate that accumulate in the cell membrane, resulting in formation of pore [128]. Another cytolytic peptide of amino acid 26 from bee venom called melittin has also been used to promote endosomal escape. Ω-helix structure of melittin interacts with lipid membranes and forms toroidal pores (Figure 1.5). Melittin induce perturbations in membrane and bend it, resulting in continuous joining of the top and bottom layer of the bilayer. The presence of melittin in PEI complexes results in enhanced gene delivery up to 700-fold for luciferase reporter gene delivery and up to 20 fold for delivery of pEGFP [128].

Treatment of cells with drugs that prevent endosome acidification such as chloroquine reduces the transfection efficiency of PEI and explains the phenomenon of proton sponge hypothesis [129]. In addition, PEI/DNA polyplexes have been found to accumulate in lysosomes and get access to cytoplasm through local membrane damage [130]. A comparative study between PEI and PAM with PLL, at different pH, Cl^{-1} ion concentration, and volume of endosomes in CHO cells was performed to elucidate the effect of proton sponge in transfection. PEI and PAM significantly increased the Cl^{-1} ion concentration and volume of endosomes whereas PLL did not have any effect [73]. Although this mechanism of proton sponge hypothesis has been widely accepted, it is not fully explained. There is lack of direct evidence for polyplex-mediated compartmental escape by pH buffering. Recently, an alternative mechanism of proton sponge hypothesis was proposed [125]. According to this mechanism, polyplex tightly binds with the inner membrane of endosomes, resulting in time-dependent protonation-induced membrane permeabilization (Figure 1.6).

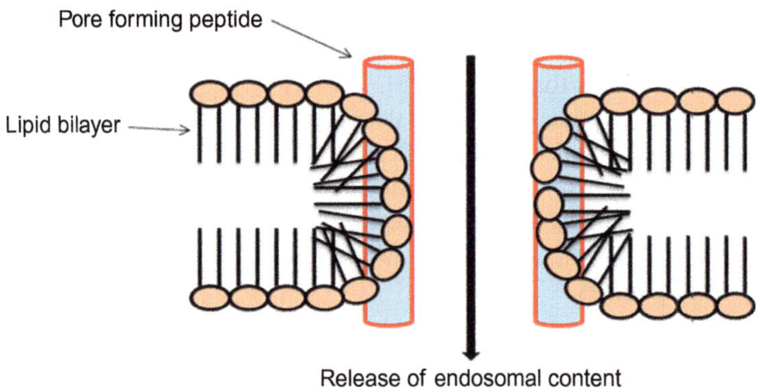

Figure 1.5 Formation of toroidal pore in bilayer by cytolytic peptides. These pore-forming peptides bind with bilayer, inducing bending and pore formation and thus releasing the endosomal contents. In the last few decades, cationic polymers (e.g., polyamidoamine dendrimers (PAM) and PEI) with intrinsic endosomolytic activity have also been developed [87]. High transfection efficiency of PEI-based polymer is believed to be the consequence of wide buffering capacity described as proton sponge hypothesis. Polymers including PEI and PAMAM dendrimers contain a large amount of secondary and tertiary amines and possess pKa values between physiological pH and lysosomal pH. At low pH of endosomes nitrogen of PEI are protonated and generate charge gradient inducing influx of Cl ion. As a result, water influxes and ultimately endosomes swell and rupture [87].

Since the size of polyplexes is much larger than nuclear pore, 25 nm polyplexes should release DNA once released into the cytosol to get access to the nucleus. Several studies have shown the role of dissociation in transfection, but the mechanism of intracellular dissociation is not understood clearly. Transfection efficiency of easily dissociable low molecular weight PLL was higher than high molecular weight indicating the effect of compactness [131]. Transfection efficiency was increased up to 187-fold by using reducible PLL polymers, which are easily degraded in intracellular environment [132].

1.3.5 Fate in cytoplasm and nuclear entry

Once released into cytoplasm, the complexes face additional hurdles to enter into the nucleus of host cell. In order to have access into transcriptional machinery, DNA must enter the nucleus from cytoplasm. Only 1 out of 1000 plasmids are estimated to enter into nucleus [133]. Cytoplasm possesses multiple barriers to nuclear entry of complexes, such as cytoskeletal elements that act as molecular sieves and prevent diffusion [134]. DNA fragmentation in cytoplasm by cytoplasmic nuclease is another hurdle. The half-life of DNA in cytoplasm of HeLa cells was estimated to be 90 minutes [135].

Some research groups have demonstrated microscopic study of microinjected lipoplexes. Microinjection of lipoplexes into cytoplasm dissociated within 1–2 hours and DNA disappeared with longer incubation time because of degradation [87]. To mimic the condition of endosomal released DNA, naked DNA was injected directly into cytoplasm and visualized for transfection, which was insignificant. But the cells incubation with lipoplexes under similar condition results significant transfection

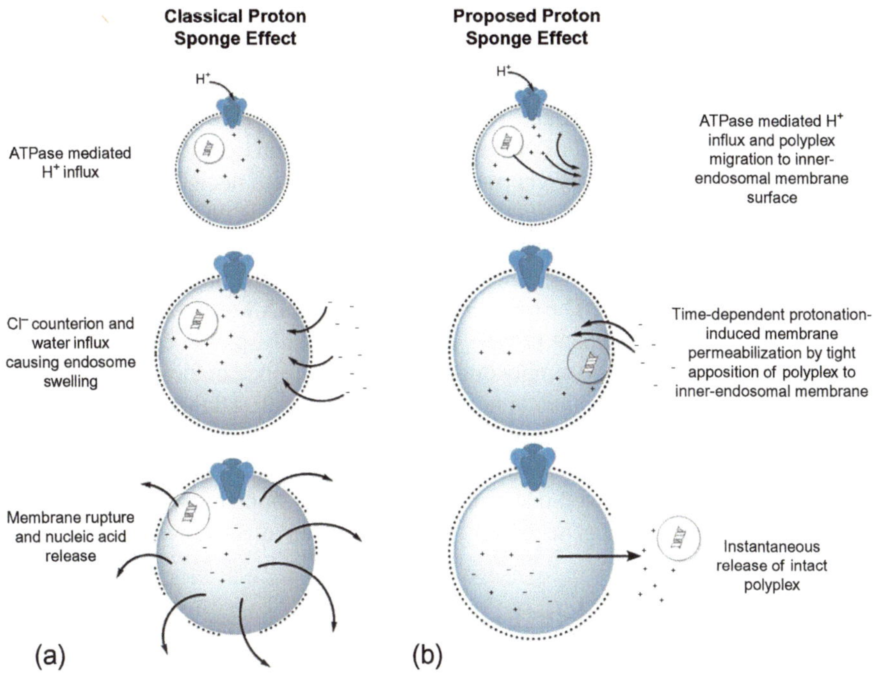

Figure 1.6 Schematic representation of endosomal escape by cationic polyplexes. Reprinted with permission from [125]. Copyright (2013) American Chemical Society. (a) Classical polyplex-induced proton sponge effect where protonation of amines in polyplexes causes influx of counterions (Cl^-) and water into endosomes, resulting in osmotic-induced lysis. (b) Alternative proposed mechanism of proton sponge effects based on recent studies where nucleic acid release depends on time-dependent protonation-induced membrane permeabilization by binding of the polyplexes and inner endosomal membrane.

[87]. This suggests the important role of endocytosis for efficient transfection, but the mechanism is still unknown. Some researchers believe the presence of cytoskeletal structure of cytoplasm results in slow diffusion of plasmid DNA, endocytosis may have important role to deliver DNA to perinuclear zone [134]. Cytoplasm of cell consists of proteins, microtubules and other organels which hinders the movement of large particles. For example, DNA of length more than 3000 base pairs are immobile in cytoplasm. However, cationic polyplexes can move either by interacting with anionic microtubules or motor proteins or by utilizing natural transport of endolysosome along the microtubules. Indeed, DNA injected near the nucleus has higher transgene expression as compared to far from the nucleus (60–90 um) [136]. It is believed that lipoplexes from endosomes fuse with endoplasmic reticulum and release DNA into the lumen of the reticulum [137]. Then DNA molecules enter the nucleus through the continuous network that exists between nuclear and reticulum membranes.

The delivery of plasmid into nucleus using polyplexes was studied by Pollard and colleagues [133]. PEI-DNA polyplexes were microinjected into the cytoplasm of cells, which resulted in higher transfection as compared to microinjection of naked DNA or

DOTAP/DNA lipoplexes. However, the role of PEI in trafficking DNA into nucleus was not found in this study. High transfection was due to another indirect mechanism such as better DNA protection or improved cytoplasmic mobility [133].

Nucleus is surrounded by double membrane composed of a 25 nm diameter nuclear pore complex (NPC). Entry of pDNA, which is released from lipoplex or polyplex, into nucleus can be processed either via passive diffusion through nuclear pore complex or via an ATP-dependent process. However, passive diffusion is possible only for compounds less than 9-11 nm in diameter [138], and an ATP-dependent process can be used for protein >20 kDa, which is obstructed by certain anti-nucleoprotein antibodies, and wheat gum agglutinin [139]. The transfer of exogenous DNA was reported to be similar as proteins, i.e., ATP-dependent process [136]. Additionally, pDNA can enter nucleus during nuclear envelope disassembly during cell division since higher transfection can be achieved in dividing cells than in non-dividing cells [140]. Higher transfection was observed when cells were transfected with lipoplexes or polyplexes at or near the M phases (mitosis) rather than at quiescent cell cycle phase such as G1 or early S phase [141]. The same study showed that unlike polyplexes or lipoplexes, transfection efficiency of adenovirus-based gene delivery systems was not affected by cell cycle phase.

1.4 Conclusion and future perspective

Varieties of non viral vectors have been discovered and used in gene delivery; however the clinical outcome is still insignificant. The fact behind this challenge is low efficiency. Gene delivery is a complete trajectory process where in each step's carrier vectors face specific barriers. The common barriers are, DNA binding and loading, stability in physiological environment, cellular communication and uptake, endosomal-lysosomal journey, cytosolic hurdle, nuclear entry, and transcription. The effect of physical and chemical properties of non viral vectors on these steps must be adequately understood. Current research should focus on the development of multifunctional and stimuli responsive carrier vectors which can have enough binding and loading capacity and undergoes rapid physical changes by intra cellular stimulus such that payload is efficiently released inside specific cells. On the other hand, intracellular molecular statistic (e.g. endosomal release, cytosolic migration and nuclear entry) should be addressed properly so that an appropriate carrier vector can be synthesized.

References

[1] Edelstein ML, Abedi MR, Wixon J, Edelstein RM. Gene therapy clinical trials worldwide 1989-2004-an overview. J Gene Med 2004;6:597–602.

[2] Macilwain C. World leaders heap praise on human genome landmark. Nature 2000;405:983–4.

[3] Guo X, Huang L. Recent advances in nonviral vectors for gene delivery. Acc Chem Res 2012;45:971–9.

[4] Pack DW, Hoffman AS, Pun S, Stayton PS. Design and development of polymers for gene delivery. Nat Rev Drug Discov 2005;4:581–93.

[5] Mintzer MA, Simanek EE. Nonviral vectors for gene delivery. Chem Rev 2009;109:259–302.
[6] Zhdanov RI, Podobed OV, Vlassov VV. Cationic lipid-DNA complexes-lipoplexes-for gene transfer and therapy. Bioelectrochemistry 2002;58:53–64.
[7] Scheule RK. The role of CpG motifs in immunostimulation and gene therapy. Adv Drug Deliv Rev 2000;44:119–34.
[8] Yew NS, Wang KX, Przybylska M, Bagley RG, Stedman M, Marshall J, et al. Contribution of plasmid DNA to inflammation in the lung after administration of cationic lipid:pDNA complexes. Hum Gene Ther 1999;10:223–34.
[9] Ruiz FE, Clancy JP, Perricone MA, Bebok Z, Hong JS, Cheng SH, et al. A clinical inflammatory syndrome attributable to aerosolized lipid-DNA administration in cystic fibrosis. Hum Gene Ther 2001;12:751–61.
[10] Yew NS, Zhao H, Przybylska M, Wu IH, Tousignant JD, Scheule RK, et al. CpG-depleted plasmid DNA vectors with enhanced safety and long-term gene expression *in vivo*. Mol Ther 2002;5:731–8.
[11] Yew NS, Zhao H, Wu IH, Song A, Tousignant JD, Przybylska M, et al. Reduced inflammatory response to plasmid DNA vectors by elimination and inhibition of immunostimulatory CpG motifs. Mol Ther 2000;1:255–62.
[12] Ginn SL, Alexander IE, Edelstein ML, Abedi MR, Wixon J. Gene therapy clinical trials worldwide to 2012 - an update. J Gene Med 2013;15:65–77.
[13] Marshall E. Gene therapy death prompts review of adenovirus vector. Science 1999;286:2244–5.
[14] Wolff JA, Malone RW, Williams P, Chong W, Acsadi G, Jani A, et al. Direct gene transfer into mouse muscle *in vivo*. Science 1990;247:1465–8.
[15] Ahmed M, Narain R. The effect of molecular weight, compositions and lectin type on the properties of hyperbranched glycopolymers as non-viral gene delivery systems. Biomaterials 2012;33:3990–4001.
[16] Song YK, Liu D. Free liposomes enhance the transfection activity of DNA/lipid complexes *in vivo* by intravenous administration. Biochim Biophys Acta 1998;1372:141–50.
[17] Guo W, Lee RJ. Efficient gene delivery using anionic liposome-complexed polyplexes (LPDII). Biosci Rep 2000;20:419–32.
[18] Kircheis R, Schuller S, Brunner S, Ogris M, Heider KH, Zauner W, et al. Polycation-based DNA complexes for tumor-targeted gene delivery *in vivo*. J Gene Med 1999;1:111–20.
[19] Brown MD, Schatzlein A, Brownlie A, Jack V, Wang W, Tetley L, et al. Preliminary characterization of novel amino acid based polymeric vesicles as gene and drug delivery agents. Bioconjug Chem 2000;11:880–91.
[20] Song YK, Liu F, Chu S, Liu D. Characterization of cationic liposome-mediated gene transfer *in vivo* by intravenous administration. Hum Gene Ther 1997;8:1585–94.
[21] Hickman MA, Malone RW, Lehmann-Bruinsma K, Sih TR, Knoell D, Szoka FC, et al. Gene expression following direct injection of DNA into liver. Hum Gene Ther 1994;5:1477–83.
[22] Yang JP, Huang L. Direct gene transfer to mouse melanoma by intratumor injection of free DNA. Gene Ther 1996;3:542–8.
[23] Yu WH, Kashani-Sabet M, Liggitt D, Moore D, Heath TD, Debs RJ. Topical gene delivery to murine skin. J Invest Dermatol 1999;112:370–5.
[24] Stechschulte SU, Joussen AM, von Recum HA, Poulaki V, Moromizato Y, Yuan J, et al. Rapid ocular angiogenic control via naked DNA delivery to cornea. Invest Ophthalmol Vis Sci 2001;42:1975–9.

[25] Isner JM. Arterial gene transfer of naked DNA for therapeutic angiogenesis: early clinical results. Adv Drug Deliv Rev 1998;30:185–97.

[26] Losordo DW, Vale PR, Symes JF, Dunnington CH, Esakof DD, Maysky M, et al. Gene therapy for myocardial angiogenesis: initial clinical results with direct myocardial injection of phVEGF165 as sole therapy for myocardial ischemia. Circulation 1998;98:2800–4.

[27] Smith BF, Baker HJ, Curiel DT, Jiang W, Conry RM. Humoral and cellular immune responses of dogs immunized with a nucleic acid vaccine encoding human carcinoembryonic antigen. Gene Ther 1998;5:865–8.

[28] Velaz-Faircloth M, Cobb AJ, Horstman AL, Henry SC, Frothingham R. Protection against Mycobacterium avium by DNA vaccines expressing mycobacterial antigens as fusion proteins with green fluorescent protein. Infect Immun 1999;67:4243–50.

[29] Meyer KB, Thompson MM, Levy MY, Barron LG, Szoka Jr FC. Intratracheal gene delivery to the mouse airway: characterization of plasmid DNA expression and pharmacokinetics. Gene Ther 1995;2:450–60.

[30] Felgner JH, Kumar R, Sridhar CN, Wheeler CJ, Tsai YJ, Border R, et al. Enhanced gene delivery and mechanism studies with a novel series of cationic lipid formulations. J Biol Chem 1994;269:2550–61.

[31] Niculescu-Duvaz D, Heyes J, Springer CJ. Structure-activity relationship in cationic lipid mediated gene transfection. Curr Med Chem 2003;10:1233–61.

[32] Zhi D, Zhang S, Wang B, Zhao Y, Yang B, Yu S. Transfection efficiency of cationic lipids with different hydrophobic domains in gene delivery. Bioconjug Chem 2010;21:563–77.

[33] Martin B, Sainlos M, Aissaoui A, Oudrhiri N, Hauchecorne M, Vigneron JP, et al. The design of cationic lipids for gene delivery. Curr Pharm Des 2005;11:375–94.

[34] Xu Y, Szoka Jr FC. Mechanism of DNA release from cationic liposome/DNA complexes used in cell transfection. Biochemistry 1996;35:5616–23.

[35] Heyes JA, Niculescu-Duvaz D, Cooper RG, Springer CJ. Synthesis of novel cationic lipids: effect of structural modification on the efficiency of gene transfer. J Med Chem 2002;45:99–114.

[36] Karmali PP, Kumar VV, Chaudhuri A. Design, syntheses and *in vitro* gene delivery efficacies of novel mono-, di- and trilysinated cationic lipids: a structure-activity investigation. J Med Chem 2004;47:2123–32.

[37] Koynova R, Tenchov B. Recent patents in cationic lipid carriers for delivery of nucleic acids. Recent Pat DNA Gene Seq 2011;5:8–27.

[38] Pedroso de Lima MC, Neves S, Filipe A, Duzgunes N, Simoes S. Cationic liposomes for gene delivery: from biophysics to biological applications. Curr Med Chem 2003;10:1221–31.

[39] Wheeler CJ, Sukhu L, Yang G, Tsai Y, Bustamente C, Felgner P, et al. Converting an alcohol to an amine in a cationic lipid dramatically alters the co-lipid requirement, cellular transfection activity and the ultrastructure of DNA-cytofectin complexes. Biochim Biophys Acta 1996;1280:1–11.

[40] Brown MD, Schatzlein AG, Uchegbu IF. Gene delivery with synthetic (non viral) carriers. Int J Pharm 2001;229:1–21.

[41] Woodle MC. Controlling liposome blood clearance by surface-grafted polymers. Adv Drug Deliv Rev 1998;32:139–52.

[42] Wasungu L, Hoekstra D. Cationic lipids, lipoplexes and intracellular delivery of genes. J Control Release 2006;116:255–64.

[43] Rehman Z, Zuhorn IS, Hoekstra D. How cationic lipids transfer nucleic acids into cells and across cellular membranes: recent advances. J Control Release 2013;166:46–56.

[44] Oberle V, Bakowsky U, Zuhorn IS, Hoekstra D. Lipoplex formation under equilibrium conditions reveals a three-step mechanism. Biophys J 2000;79:1447–54.
[45] Liu Y, Liggitt D, Zhong W, Tu G, Gaensler K, Debs R. Cationic liposome-mediated intravenous gene delivery. J Biol Chem 1995;270:24864–70.
[46] Monck MA, Mori A, Lee D, Tam P, Wheeler JJ, Cullis PR, et al. Stabilized plasmid-lipid particles: pharmacokinetics and plasmid delivery to distal tumors following intravenous injection. J Drug Target 2000;7:439–52.
[47] Han M, Chen JL, Hu Y, He CL, Shuai WP, Yu JH, et al. *In vitro* and *in vivo* tumor suppressive activity induced by human telomerase transcriptase-targeting antisense oligonucleotides mediated by cationic liposomes. J Biosci Bioeng 2008;106:243–7.
[48] He CX, Tabata Y, Gao JQ. Non-viral gene delivery carrier and its three-dimensional transfection system. Int J Pharm 2010;386:232–42.
[49] Alexis F, Pridgen E, Molnar LK, Farokhzad OC. Factors affecting the clearance and biodistribution of polymeric nanoparticles. Mol Pharm 2008;5:505–15.
[50] Ruponen M, Honkakoski P, Tammi M, Urtti A. Cell-surface glycosaminoglycans inhibit cation-mediated gene transfer. J Gene Med 2004;6:405–14.
[51] Morille M, Passirani C, Vonarbourg A, Clavreul A, Benoit JP. Progress in developing cationic vectors for non-viral systemic gene therapy against cancer. Biomaterials 2008;29:3477–96.
[52] Godbey WT, Mikos AG. Recent progress in gene delivery using non-viral transfer complexes. J Control Release 2001;72:115–25.
[53] Fischer D, Bieber T, Li Y, Elsasser HP, Kissel T. A novel non-viral vector for DNA delivery based on low molecular weight, branched polyethylenimine: effect of molecular weight on transfection efficiency and cytotoxicity. Pharm Res 1999;16:1273–9.
[54] Ogris M, Brunner S, Schuller S, Kircheis R, Wagner E. PEGylated DNA/transferrin-PEI complexes: reduced interaction with blood components, extended circulation in blood and potential for systemic gene delivery. Gene Ther 1999;6:595–605.
[55] Bettinger T, Remy JS, Erbacher P. Size reduction of galactosylated PEI/DNA complexes improves lectin-mediated gene transfer into hepatocytes. Bioconjug Chem 1999;10:558–61.
[56] Godbey WT, Wu KK, Mikos AG. Size matters: molecular weight affects the efficiency of poly(ethylenimine) as a gene delivery vehicle. J Biomed Mater Res 1999;45:268–75.
[57] Wightman L, Kircheis R, Rossler V, Carotta S, Ruzicka R, Kursa M, et al. Different behavior of branched and linear polyethylenimine for gene delivery *in vitro* and *in vivo*. J Gene Med 2001;3:362–72.
[58] Lungwitz U, Breunig M, Blunk T, Gopferich A. Polyethylenimine-based non-viral gene delivery systems. Eur J Pharm Biopharm 2005;60:247–66.
[59] Kunath K, von Harpe A, Fischer D, Petersen H, Bickel U, Voigt K, et al. Low-molecular-weight polyethylenimine as a non-viral vector for DNA delivery: comparison of physicochemical properties, transfection efficiency and *in vivo* distribution with high-molecular-weight polyethylenimine. J Control Release 2003;89:113–25.
[60] Godbey WT, Wu KK, Mikos AG. Poly(ethylenimine) and its role in gene delivery. J Control Release 1999;60:149–60.
[61] Zhang ZY, Smith BD. High-generation polycationic dendrimers are unusually effective at disrupting anionic vesicles: membrane bending model. Bioconjug Chem 2000;11:805–14.
[62] Park TG, Jeong JH, Kim SW. Current status of polymeric gene delivery systems. Adv Drug Deliv Rev 2006;58:467–86.
[63] Wolfert MA, Schacht EH, Toncheva V, Ulbrich K, Nazarova O, Seymour LW. Characterization of vectors for gene therapy formed by self-assembly of DNA with synthetic block co-polymers. Hum Gene Ther 1996;7:2123–33.

[64] van de Wetering P, Moret EE, Schuurmans-Nieuwenbroek NM, van Steenbergen MJ, Hennink WE. Structure-activity relationships of water-soluble cationic methacrylate/methacrylamide polymers for nonviral gene delivery. Bioconjug Chem 1999;10:589–97.
[65] van der Aa MA, Huth US, Hafele SY, Schubert R, Oosting RS, Mastrobattista E, et al. Cellular uptake of cationic polymer-DNA complexes via caveolae plays a pivotal role in gene transfection in COS-7 cells. Pharm Res 2007;24:1590–8.
[66] Funhoff AM, van Nostrum CF, Lok MC, Fretz MM, Crommelin DJ, Hennink WE. Poly(3-guanidinopropyl methacrylate): a novel cationic polymer for gene delivery. Bioconjug Chem 2004;15:1212–20.
[67] Funhoff AM, van Nostrum CF, Janssen AP, Fens MH, Crommelin DJ, Hennink WE. Polymer side-chain degradation as a tool to control the destabilization of polyplexes. Pharm Res 2004;21:170–6.
[68] Illum L, Jabbal-Gill I, Hinchcliffe M, Fisher AN, Davis SS. Chitosan as a novel nasal delivery system for vaccines. Adv Drug Deliv Rev 2001;51:81–96.
[69] Kiang T, Wen J, Lim HW, Leong KW. The effect of the degree of chitosan deacetylation on the efficiency of gene transfection. Biomaterials 2004;25:5293–301.
[70] Ji AM, Su D, Che O, Li WS, Sun L, Zhang ZY, et al. Functional gene silencing mediated by chitosan/siRNA nanocomplexes. Nanotechnology 2009;20:405103.
[71] Liu X, Howard KA, Dong M, Andersen MO, Rahbek UL, Johnsen MG, et al. The influence of polymeric properties on chitosan/siRNA nanoparticle formulation and gene silencing. Biomaterials 2007;28:1280–8.
[72] Haensler J, Szoka Jr FC. Polyamidoamine cascade polymers mediate efficient transfection of cells in culture. Bioconjug Chem 1993;4:372–9.
[73] Sonawane ND, Szoka Jr FC, Verkman AS. Chloride accumulation and swelling in endosomes enhances DNA transfer by polyamine-DNA polyplexes. J Biol Chem 2003;278:44826–31.
[74] Zinselmeyer BH, Mackay SP, Schatzlein AG, Uchegbu IF. The lower-generation polypropylenimine dendrimers are effective gene-transfer agents. Pharm Res 2002;19:960–7.
[75] Schatzlein AG, Zinselmeyer BH, Elouzi A, Dufes C, Chim YT, Roberts CJ, et al. Preferential liver gene expression with polypropylenimine dendrimers. J Control Release 2005;101:247–58.
[76] Ohsaki M, Okuda T, Wada A, Hirayama T, Niidome T, Aoyagi H. *In vitro* gene transfection using dendritic poly(L-lysine). Bioconjug Chem 2002;13:510–7.
[77] Eom KD, Park SM, Tran HD, Kim MS, Yu RN, Yoo H. Dendritic alpha, epsilon-poly(L-lysine)s as delivery agents for antisense oligonucleotides. Pharm Res 2007;24:1581–9.
[78] Inoue Y, Kurihara R, Tsuchida A, Hasegawa M, Nagashima T, Mori T, et al. Efficient delivery of siRNA using dendritic poly(L-lysine) for loss-of-function analysis. J Control Release 2008;126:59–66.
[79] Potter H. Electroporation in biology: methods, applications, and instrumentation. Anal Biochem 1988;174:361–73.
[80] Fynan EF, Webster RG, Fuller DH, Haynes JR, Santoro JC, Robinson HL. DNA vaccines: protective immunizations by parenteral, mucosal, and gene-gun inoculations. Proc Natl Acad Sci U S A 1993;90:11478–82.
[81] Prevette LE, Lynch ML, Kizjakina K, Reineke TM. Correlation of amine number and pDNA binding mechanism for trehalose-based polycations. Langmuir 2008;24:8090–101.
[82] Sakurai F, Inoue R, Nishino Y, Okuda A, Matsumoto O, Taga T, et al. Effect of DNA/liposome mixing ratio on the physicochemical characteristics, cellular uptake and intracellular trafficking of plasmid DNA/cationic liposome complexes and subsequent gene expression. J Control Release 2000;66:255–69.

[83] Ruysschaert JM, el Ouahabi A, Willeaume V, Huez G, Fuks R, Vandenbranden M, et al. A novel cationic amphiphile for transfection of mammalian cells. Biochem Biophys Res Commun 1994;203:1622–8.
[84] Xu Y, Hui SW, Frederik P, Szoka Jr FC. Physicochemical characterization and purification of cationic lipoplexes. Biophys J 1999;77:341–53.
[85] Radler JO, Koltover I, Salditt T, Safinya CR. Structure of DNA-cationic liposome complexes: DNA intercalation in multilamellar membranes in distinct interhelical packing regimes. Science 1997;275:810–4.
[86] Pector V, Backmann J, Maes D, Vandenbranden M, Ruysschaert JM. Biophysical and structural properties of DNA.diC(14)-amidine complexes. Influence of the DNA/lipid ratio. J Biol Chem 2000;275:29533–8.
[87] Elouahabi A, Ruysschaert JM. Formation and intracellular trafficking of lipoplexes and polyplexes. Mol Ther 2005;11:336–47.
[88] Ruponen M, Yla-Herttuala S, Urtti A. Interactions of polymeric and liposomal gene delivery systems with extracellular glycosaminoglycans: physicochemical and transfection studies. Biochim Biophys Acta 1999;1415:331–41.
[89] Gratton SE, Ropp PA, Pohlhaus PD, Luft JC, Madden VJ, Napier ME, et al. The effect of particle design on cellular internalization pathways. Proc Natl Acad Sci U S A 2008;105:11613–8.
[90] Davis ME, Chen ZG, Shin DM. Nanoparticle therapeutics: an emerging treatment modality for cancer. Nat Rev Drug Discov 2008;7:771–82.
[91] Wolfert MA, Seymour LW. Atomic force microscopic analysis of the influence of the molecular weight of poly(L)lysine on the size of polyelectrolyte complexes formed with DNA. Gene Ther 1996;3:269–73.
[92] Tang MX, Szoka FC. The influence of polymer structure on the interactions of cationic polymers with DNA and morphology of the resulting complexes. Gene Ther 1997;4:823–32.
[93] Mounkes LC, Zhong W, Cipres-Palacin G, Heath TD, Debs RJ. Proteoglycans mediate cationic liposome-DNA complex-based gene delivery *in vitro* and *in vivo*. J Biol Chem 1998;273:26164–70.
[94] Mislick KA, Baldeschwieler JD. Evidence for the role of proteoglycans in cation-mediated gene transfer. Proc Natl Acad Sci U S A 1996;93:12349–54.
[95] Kopatz I, Remy JS, Behr JP. A model for non-viral gene delivery: through syndecan adhesion molecules and powered by actin. J Gene Med 2004;6:769–76.
[96] Aderem A, Underhill DM. Mechanisms of phagocytosis in macrophages. Annu Rev Immunol 1999;17:593–623.
[97] Hillaireau H, Couvreur P. Nanocarriers' entry into the cell: relevance to drug delivery. Cell Mol Life Sci 2009;66:2873–96.
[98] Aouadi M, Tesz GJ, Nicoloro SM, Wang M, Chouinard M, Soto E, et al. Orally delivered siRNA targeting macrophage Map4k4 suppresses systemic inflammation. Nature 2009;458:1180–4.
[99] Xiang S, Tong H, Shi Q, Fernandes JC, Jin T, Dai K, et al. Uptake mechanisms of non-viral gene delivery. J Control Release 2012;158:371–8.
[100] Takei K, Haucke V. Clathrin-mediated endocytosis: membrane factors pull the trigger. Trends Cell Biol 2001;11:385–91.
[101] Wang G, Moniri NH, Ozawa K, Stamler JS, Daaka Y. Nitric oxide regulates endocytosis by S-nitrosylation of dynamin. Proc Natl Acad Sci U S A 2006;103:1295–300.
[102] Subtil A, Gaidarov I, Kobylarz K, Lampson MA, Keen JH, McGraw TE. Acute cholesterol depletion inhibits clathrin-coated pit budding. Proc Natl Acad Sci U S A 1999;96:6775–80.

[103] Martin ME, Rice KG. Peptide-guided gene delivery. AAPS J 2007;9:E18–29.
[104] Zelphati O, Szoka Jr. FC. Mechanism of oligonucleotide release from cationic liposomes. Proc Natl Acad Sci U S A 1996;93:11493–8.
[105] Kiss AL, Botos E. Endocytosis via caveolae: alternative pathway with distinct cellular compartments to avoid lysosomal degradation? J Cell Mol Med 2009;13:1228–37.
[106] Bareford LM, Swaan PW. Endocytic mechanisms for targeted drug delivery. Adv Drug Deliv Rev 2007;59:748–58.
[107] Conner SD, Schmid SL. Regulated portals of entry into the cell. Nature 2003;422:37–44.
[108] Gupta B, Levchenko TS, Torchilin VP. Intracellular delivery of large molecules and small particles by cell-penetrating proteins and peptides. Adv Drug Deliv Rev 2005;57:637–51.
[109] Hong S, Rattan R, Majoros IJ, Mullen DG, Peters JL, Shi X, et al. The role of ganglioside GM1 in cellular internalization mechanisms of poly(amidoamine) dendrimers. Bioconjug Chem 2009;20:1503–13.
[110] Lee H, Kim IK, Park TG. Intracellular trafficking and unpacking of siRNA/quantum dot-PEI complexes modified with and without cell penetrating peptide: confocal and flow cytometric FRET analysis. Bioconjug Chem 2010;21:289–95.
[111] Wang Z, Tiruppathi C, Minshall RD, Malik AB. Size and dynamics of caveolae studied using nanoparticles in living endothelial cells. ACS Nano 2009;3:4110–6.
[112] Rejman J, Oberle V, Zuhorn IS, Hoekstra D. Size-dependent internalization of particles via the pathways of clathrin- and caveolae-mediated endocytosis. Biochem J 2004;377:159–69.
[113] Xu DM, Yao SD, Liu YB, Sheng KL, Hong J, Gong PJ, et al. Size-dependent properties of M-PEIs nanogels for gene delivery in cancer cells. Int J Pharm 2007;338:291–6.
[114] Prabha S, Zhou WZ, Panyam J, Labhasetwar V. Size-dependency of nanoparticle-mediated gene transfection: studies with fractionated nanoparticles. Int J Pharm 2002;244:105–15.
[115] Rensen PC, Sliedregt LA, Ferns M, Kieviet E, van Rossenberg SM, van Leeuwen SH, et al. Determination of the upper size limit for uptake and processing of ligands by the asialoglycoprotein receptor on hepatocytes *in vitro* and *in vivo*. J Biol Chem 2001;276:37577–84.
[116] Gao H, Shi W, Freund LB. Mechanics of receptor-mediated endocytosis. Proc Natl Acad Sci U S A 2005;102:9469–74.
[117] Chithrani BD, Ghazani AA, Chan WC. Determining the size and shape dependence of gold nanoparticle uptake into mammalian cells. Nano Lett 2006;6:662–8.
[118] von Gersdorff K, Sanders NN, Vandenbroucke R, De Smedt SC, Wagner E, Ogris M. The internalization route resulting in successful gene expression depends on both cell line and polyethylenimine polyplex type. Mol Ther 2006;14:745–53.
[119] Hufnagel H, Hakim P, Lima A, Hollfelder F. Fluid phase endocytosis contributes to transfection of DNA by PEI-25. Mol Ther 2009;17:1411–7.
[120] Goncalves C, Mennesson E, Fuchs R, Gorvel JP, Midoux P, Pichon C. Macropinocytosis of polyplexes and recycling of plasmid via the clathrin-dependent pathway impair the transfection efficiency of human hepatocarcinoma cells. Mol Ther 2004;10:373–85.
[121] Ohkuma S, Poole B. Fluorescence probe measurement of the intralysosomal pH in living cells and the perturbation of pH by various agents. Proc Natl Acad Sci U S A 1978;75:3327–31.
[122] Felgner PL, Gadek TR, Holm M, Roman R, Chan HW, Wenz M, et al. Lipofection: a highly efficient, lipid-mediated DNA-transfection procedure. Proc Natl Acad Sci U S A 1987;84:7413–7.
[123] Koltover I, Salditt T, Radler JO, Safinya CR. An inverted hexagonal phase of cationic liposome-DNA complexes related to DNA release and delivery. Science 1998;281:78–81.

[124] El Ouahabi A, Thiry M, Pector V, Fuks R, Ruysschaert JM, Vandenbranden M. The role of endosome destabilizing activity in the gene transfer process mediated by cationic lipids. FEBS Lett 1997;414:187–92.
[125] Jones CH, Chen CK, Ravikrishnan A, Rane S, Pfeifer BA. Overcoming nonviral gene delivery barriers: perspective and future. Mol Pharm 2013;10:4082–98.
[126] Curiel DT, Agarwal S, Wagner E, Cotten M. Adenovirus enhancement of transferrin-polylysine-mediated gene delivery. Proc Natl Acad Sci U S A 1991;88:8850–4.
[127] Cho YW, Kim JD, Park K. Polycation gene delivery systems: escape from endosomes to cytosol. J Pharm Pharmacol 2003;55:721–34.
[128] Medina-Kauwe LK, Xie J, Hamm-Alvarez S. Intracellular trafficking of nonviral vectors. Gene Ther 2005;12:1734–51.
[129] Kichler A, Leborgne C, Coeytaux E, Danos O. Polyethylenimine-mediated gene delivery: a mechanistic study. J Gene Med 2001;3:135–44.
[130] Bieber T, Meissner W, Kostin S, Niemann A, Elsasser HP. Intracellular route and transcriptional competence of polyethylenimine-DNA complexes. J Control Release 2002;82:441–54.
[131] Schaffer DV, Fidelman NA, Dan N, Lauffenburger DA. Vector unpacking as a potential barrier for receptor-mediated polyplex gene delivery. Biotechnol Bioeng 2000;67:598–606.
[132] Read ML, Bremner KH, Oupicky D, Green NK, Searle PF, Seymour LW. Vectors based on reducible polycations facilitate intracellular release of nucleic acids. J Gene Med 2003;5:232–45.
[133] Pollard H, Remy JS, Loussouarn G, Demolombe S, Behr JP, Escande D. Polyethylenimine but not cationic lipids promotes transgene delivery to the nucleus in mammalian cells. J Biol Chem 1998;273:7507–11.
[134] Luby-Phelps K, Castle PE, Taylor DL, Lanni F. Hindered diffusion of inert tracer particles in the cytoplasm of mouse 3T3 cells. Proc Natl Acad Sci U S A 1987;84:4910–3.
[135] Lechardeur D, Sohn KJ, Haardt M, Joshi PB, Monck M, Graham RW, et al. Metabolic instability of plasmid DNA in the cytosol: a potential barrier to gene transfer. Gene Ther 1999;6:482–97.
[136] Dowty ME, Williams P, Zhang G, Hagstrom JE, Wolff JA. Proc Natl Acad Sci U S A 1995;92:4572–6.
[137] Kamiya H, Fujimura Y, Matsuoka I, Harashima H. Visualization of intracellular trafficking of exogenous DNA delivered by cationic liposomes. Biochem Biophys Res Commun 2002;298:591–7.
[138] Bonner WM. Protein migration into nuclei. I. Frog oocyte nuclei *in vivo* accumulate microinjected histones, allow entry to small proteins, and exclude large proteins. J Cell Biol 1975;64:421–30.
[139] Featherstone C, Darby MK, Gerace L. A monoclonal antibody against the nuclear pore complex inhibits nucleocytoplasmic transport of protein and RNA *in vivo*. J Cell Biol 1988;107:1289–97.
[140] Wilke M, Fortunati E, van den Broek M, Hoogeveen AT, Scholte BJ. Efficacy of a peptide-based gene delivery system depends on mitotic activity. Gene Ther 1996;3:1133–42.
[141] Brunner S, Sauer T, Carotta S, Cotten M, Saltik M, Wagner E. Cell cycle dependence of gene transfer by lipoplex, polyplex and recombinant adenovirus. Gene Ther 2000;7:401–7.

PEI and its derivatives for gene therapy

2

K.C. Remant Bahadur, H. Uludağ
University of Alberta, Edmonton, AB, Canada

Abbreviations

PEI	Polyethyleneimine
b-PEI	Branched polyethyleneimine
l-PEI	Linear polyethyleneimine
MW	Molecular weight
PEI25	25 kDa branched polyethyleneimine
PEI2	2 kDa branched polyethyleneimine
CS	Chitosan
DNA	Deoxyribonucleic acid
pDNA	Plasmid deoxyribonucleic acid
RNA	Ribonucleic acid
siRNA	Small interfering RNA
PLL	poly(L-lysine)
PDMAEM	poly[2-(dimethyamino)ethyl methacrylate]
PAMAM	poly(amidoamine)
-S-S-	Disulfide bonding
AuNPs	Gold nanoparticles
PBLG	Poly(γ-benzyl L-glutamate)
PCL	Polycaprolactone
PLA	Poly(L-lactide)
PEG	Polyethylene glycol
HA	hyaluronic acid
DTBP	dimethyl-3,3-dithiobispropionimidate
DSP	dithiobis(succinimidylpropionate)
SPDP	*N*-Succinimidyl 3-[2-pyridyldithio]-propionate
CBA	cystamine bisacrylamide
TAEI	tris[2-(acryloyloxy)ethyl]isocyanurate
Tf	transferrin
EGF	epidermal growth factor

Polymers and Nanomaterials for Gene Therapy. http://dx.doi.org/10.1016/B978-0-08-100520-0.00002-3
© 2016 Elsevier Ltd. All rights reserved.

2.1 Introduction

Gene delivery is pursued as a promising technology for the treatment of inherited and acquired diseases resulting from abnormal gene expression [1,2]. It is accomplished by delivery of exogenous genetic material (i.e., polynucleotides) to target cells using specific vectors since the delivery of polynucleotides on their own suffers from degradation in the physiological environment, and polynucleotides cannot be internalized by cells on their own. Two types of vectors, viral and non-viral, have been used in research activity and in clinical applications to date. Gene delivery efficiency of these vectors depends on their pharmacokinetic properties (e.g., biodistribution, circulation time, clearance, etc.) as well as conditions used at the time of gene transfer (e.g., serum effects, exposure time, post-transfection time for gene expression [3,4]). In spite of their low efficiency, non-viral vectors are used in a wide range of gene delivery applications due to their cost-effectiveness and especially superior safety profiles [5]. They also offer excellent molecular tunability of physicochemical properties (facile chemistry), stability for long-term storage and reconstitution, and unlimited capacity of gene sizes (cargo) for delivery [6,7]. Therefore, clinical attempts to treat inherited and acquired human diseases using non-viral vectors have been increasing [6]. Non-viral vectors were constructed from cationic polymers, dendrimers, lipids, and nanostructured particles such as nanotubes and nanorods, but cationic polymers have emerged as attractive candidate in recent years [1,5,8,9]. The studies in the last two decades have mainly used polyethylenimine (PEI), poly(L-lysine) (PLL), poly[2-(dimethyamino)ethyl methacrylate] (PDMAEM), poly(amidoamine) (PAMAM), and chitosan (CS) [10–13] in design of non-viral vectors. Despite intensive activity, however, concrete relationships among structure-property-functional performance remain incompletely described [14].

PEI has emerged as the most widely studied cationic polymer for design of non-viral vectors, as reviewed elsewhere [5,15–18]. It is available in linear (l-PEI) and branched (b-PEI) topology. The l-PEI is synthesized by cationic ring opening polymerization of 2-substituted 2-oxazoline followed by acid-catalyzed hydrolysis and b-PEI by acid catalyzed ring-opening polymerization of aziridine (Figure 2.1) [19–21]. The b-PEI possesses primary, secondary, and tertiary amine at a ratio close to 1:1:1, while the l-PEI possesses only secondary amines in polymer backbone and terminally capped with primary amines [22]. It has been reported that the gene delivery efficiency of PEIs depends on degree of polymerization (molecular weight), branching (topology), and the buffering capacity, which is a function of cationic charge density. Different types of PEIs have been studied to investigate structure-functional relationship, however, a concrete picture on such relationships has yet to emerge. In an earlier study, Godbey et al. reported that, under *in vitro* condition, the transfection efficiency of PEIs increases with molecular weight (MW) (70 kDa PEI > 10 kDa PEI > 1.8 kDa PEI) [23]. However, *in vivo* efficiency of the polymers decreases with the MW (25 kDa PEI > 50 kDa PEI > 800 kDa PEI) [24]. There have also been reported contradictory results on the influence of PEI branching and branching density [25,26]. Nevertheless, transfection efficiency of PEIs is comparatively higher than most non-viral carriers, but its non-biodegradability and intolerable toxicities (as well as non-specific interactions) pose problems for clinical applications [23,25,27–30].

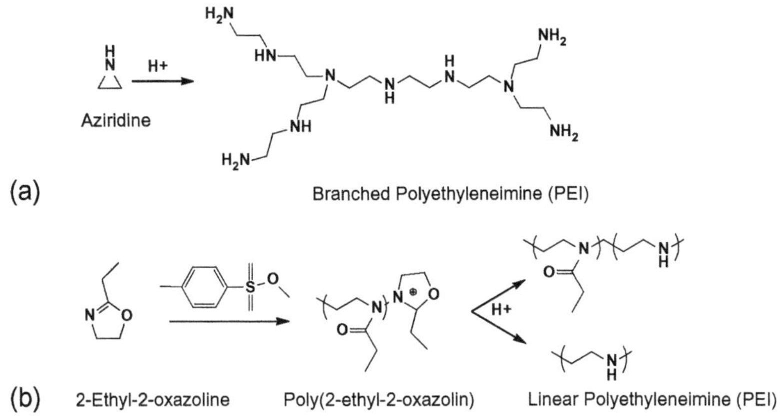

Figure 2.1 Synthesis of PEI polymers, (a) b-PEI synthesis by ring opening polymerization of aziridine, and (b) l-PEI synthesis by acid-catalyzed hydrolysis of poly(2-ethyl-2-oxazolin).

2.2 Criteria for non-viral vectors and general considerations

Polynucleotide complexes of cationic polymers, so-called polyplexes, exhibit some characteristic features that lie at the interface of molecular and bulk materials [31]. They display polydispersity across a given population, and it is a crucial challenge to formulate stable monodisperse polyplexes that possess well-defined surfaces properties and morphological geometries. Polynucleotides are either directly coupled onto PEIs (and modified PEIs) or electrostatically bound with PEIs for polyplex formation. There have been significant activities in formulating cell specific vectors by grafting targeting moieties that bind to desired cells [32–34]. This strategy enhances the efficiency through bio-recognition, but it can also shield the binding efficiency of the cationic polymers and may compact particles that inhibit proper release of the payloads. Attempts to functionalize the polyplexes with targeting ligands may pose a chemical challenge due to the small size of targeting ligands, which creates molecular crowing and restricts reactive functionalities into a narrow curvature, ultimately decreasing conjugation yields [35,36]. Furthermore, the charge density, electrostatic repulsion, nearest-neighbor interactions, and loss of colloidal stability during reaction may result in low conjugation yields [37]. Effective purification of bio-functionalized polyplexes is another complication since unbounded ligands will compete for binding with cell receptors and ultimately hinder cellular uptake of functionalized polyplexes. During *in-vivo* delivery, functionalized polyplexes are exposed to a complex mixture of endogenous molecules and opsonization may trigger aggregation and decrease the half-life of polyplexes in circulation and/or local tissue sites. However, continuous efforts in bioconjugate chemistry promise to address these challenges to develop multifunctional vectors. Therefore, ideal vectors to overcome extracellular and intracellular barriers should control: (i) the size/charge of polyplexes, (ii) variation in the features of polyplex population, (iii) efficiency of grafting for various moieties intended to improve PEI features, (iv) targeting ligands grafting to enhance cellular susceptibility via bio-recognition, and (v) stimuli response by generating labile linkages (e.g., disulfide -S-S-) that undergo intracellular cleavage and disassembly of polyplexes in the cytosolic environment. The overall goal is to engineer a vector to overcome the major physiological bottlenecks of gene delivery, namely cellular recognition in an organism, escape from endosomal-lysosomal network, cytoplasmic migration into the nucleus, and decondensation of the plasmid DNA (pDNA) necessary for transcriptional activity [8,37]. When PEI is used for short interfering RNA (siRNA) delivery to obtain gene silencing, the cargo does not need to reach the nucleus but rather released in the cytoplasm to reach the silencing machinery (i.e., RNA-induced silencing complex, RISC). In this chapter, we will focus on PEI and its derivatives, and present the current approaches to improve this promising vector to undertake polynucleotide delivery.

2.3 Rationale for the use of PEI for gene therapy

The basic foundation for the use of cationic polymers in gene delivery is the electrostatic interaction between the cationic primary amines of polymers and anionic phosphate groups of the polynucleotides that leads the formation of condensed polyionic complexes (polyplexes). Polyplexes protect the polynucleotides against enzymatic and non-enzymatic degradation, avoid the clearance through reticuloendothelial system (RES) and interaction with blood components, enhance cellular uptake via interactions with anionic cell surface proteoglycans and finally increase half-life in cytoplasm [38–40]. PEI was initially used in gene delivery in 1995 by Boussif et al., and it has since served as a "reference" non-viral vector for *in vitro* and *in vivo* applications [9,41]. The widely accepted benefit of PEIs is thought to be derived from their extraordinary cationic charge density and buffering capacity [41,42]. The secondary and tertiary amines of PEIs are responsible for generating the buffering capacity over a wide range of pH values that facilitates endosomal escape of carriers, which has been named the "proton-sponge" effect [43]. Chemical modification of PEIs converts primary and secondary amine to secondary and tertiary amines and ultimately increases the density of un-protonated amines at the physiological pH [27,28]. It has been reported that every fifth or sixth amine Ns of b-PEIs is protonated at the physiological pH, which is responsible for the polynucleotide complexation. Previous studies have shown that factors, such as size, surface charge, chemical composition, degradability, and stimuli sensitivity, affect cellular uptake and intracellular trafficking [44,45]. Delivery of smaller polynucleotides (e.g., 22-bp siRNA) is a more challenging task than larger polynucleotides (e.g., >2000-bp plasmid DNA) due to higher dissociation possibility of weakly packed complexes upon contact with polyanionic cell surfaces [15,46,47]. In general, b-PEIs are more effective in condensing/aggregating polynucleotides to form stabile complexes due to higher amine density and, accordingly, they display better efficiency [48]. In some studies, however, l-PEI has been reported to display superior efficiency *in vivo* compared to b-PEI [49]. Wightman et al. demonstrated the impact of PEI topology in gene delivery using different types of reporter genes [25].

Transfection efficiency of l-PEI (22 kDa; PEI22) was higher than b-PEI (25 kDa; PEI25), which was attributed to inherent instability of the resultant polyplexes under salt condition since the l-PEI complexes formed large aggregates at cell membranes, cytoplasm, and nucleus, whereas b-PEI complexes remained as small distinct particles even after cellular uptake and nuclear translocation [25,29,30]. In contrast, PEI25 displayed superior siRNA delivery efficacy compared to PEI22 and low MW b-PEIs [50]. Grayson et al. observed a remarkable siRNA delivery and gene silencing efficiency of PEI25 over PEI22 and low molecular weight b-PEI (0.8 kDa) [47]. The better performance of PEI25 was attributed to better cellular delivery, which was proportional to the branching density of the polymer [51]. However, the use of PEIs raises serious concern, either due to less than desirable efficiency or excessive cytotoxicity. Multiple studies in designing second-generation PEIs via chemical modification are actively pursued [15,44], as well as strategies to engineer prefabricated nano-carriers with PEI [52]. Chemical modification of PEI typically generates two changes: it alters the physicochemical properties of native PEI and it generates new chemical functionalities that alter the physical and biological properties of the native PEI. The buffering capacity of PEIs is usually altered since the chemical modification changes the degree of amines (i.e., primary to secondary and secondary to tertiary). This change finally impacts carrier efficiency of the polymers [27].

2.4 PEI Derivatives for gene delivery

2.4.1 Grafting targeting ligands on PEI

Gene delivery is a multistep process that involves transfer of an extracellular polynucleotide into a specific cell compartment suitable for the action of the polynucleotide. The efficiency of a carrier can be dramatically improved by increasing cell uptake by incorporating target-specific ligands [8,53]. Some of the common targeting ligands include RGD and HIV-1 Tat peptides, hyaluronic acid (HA), transferrin (Tf), epidermal growth factor (EGF), and folic acid (FA) [54–57]. Targeting ligands are conjugated onto polymer backbones either directly or via spacers where the ligand is separated from the backbone in a controlled fashion [58,59]. While designing a target specific delivery vehicle, the unbound ligands need to be removed, bound ligands should be well exposed on the surface of polyplexes for effective cell binding, and the density of bound ligand should be sufficient for effective target (receptor) interactions [60]. In one of the first studies on targeting ligands, Kunath et al. reported a promising approach by designing target-specific PEIs using RGD peptides for integrin-positive cells [58]. The integrin targeting RGD peptides were grafted onto PEIs via SPDP and PEG for polyplex formation with pDNA. The carriers composed with RGD grafted via SPDP displayed effective gene expression in integrin-positive Mewo cells, but not in integrin-negative A549 cells. The higher gene expression was due to higher cell binding of the polyplexes. Conjugation with PEG, however, showed insignificant targeting due to PEG shielding the cationic charges of PEIs. In another approach, polyplexes

(PEI/DNA) were coated with PEGylated PEI through which targeting ligands were grafted [59,61,62]. Suk et al. grafted RGD or HIV-1 Tat peptides onto PEI/DNA complexes via coating technology using PEGylated PEIs [61]. The cellular uptake of the complexes was significantly improved as well as subsequent transfection efficiency in neuronal cells. The synergistic effect of targeting ligand and PEG spacer was investigated *in vivo* gene delivery in another study. Ogris et al. reported efficient gene transfer to subcutaneous Neuro2A neuroblastoma tumors of syngeneic A/J mice upon intravenous injection of Tf-PEG coated PEI/DNA complexes [62]. In the same study, the delivery of EGF-PEG coated PEI/DNA complexes was achieved via intravenous injection to target human hepatocellular carcinoma xenografts in SCID mice. In these studies, the levels of gene expression in tumors were significantly higher than other tissues. Systemic administrated Tf-PEG-PEI/DNA complexes encoding TNF induced efficient tumor necrosis and inhibited tumor growth. Some studies have reported grafting of targeting ligands via electrostatic interaction by using polyanionic molecules such as HA [63]. This strategy is simple and straightforward and generates duel benefit in gene delivery, by neutralizing the cationic charge density of polyplexes and enhancing specificity toward the desired receptor (i.e., CD$$ in the case of HA) and cell type.

2.4.2 Chemical modification to shield cationic charge density

Although native PEIs possess excellent capacity to complex and condense polynucleotides for cellular delivery, serious concerns regarding aggregation on cell surfaces, endosomal stability, and induction of necrosis [44,64] led to chemical modifications of PEIs to enhance functional performance. Anionic moieties are grafted onto PEIs to neutralize cationic charge density of the polyplexes and decrease cellular toxicity. Zintchenko et al. synthesized a library of PEI25 modified with hydrophobic and corresponding carboxylated moieties [15]. Such a modification decreased the toxicity of the polymers, where the carboxylated polymers were almost non-toxic. Modified polymers were more potent for siRNA delivery as compared to the parent polymers. Minimally carboxylated polymers (9% succinic anhydride) displayed better silencing efficacy, probably indicating the adverse effect of the introduced anionic charges on binding efficiency. The effects of anionic charges and alkyl chain length on PEI25 have been also explored by carboxylation using ω-bromoalkylcarboxylic acid (Figure 2.2) [65–67]. Binding efficiency was inversely affected by the degree of substitution, but polyplex stability was improved with higher hydrophobic chain lengths. Cytotoxicity generally decreased with degree of substitution regardless of the alkycarboxylate chain length, indicating the effect of charge neutralization. Carboxylation improved the siRNA delivery efficiency, but the optimum efficiency was observed only at 6-9% of primary amine modification regardless of the alkyl chain length.

2.4.3 Hydrophobic modifications

Hydrophobic amino acids and poly(amino acids) were grafted onto PEI25 to improve biocompatibility and carrier efficiency [67,68]. The PEI25 was modified with different amino acids to generate variable hydrophobicity and *in vitro* gene delivery

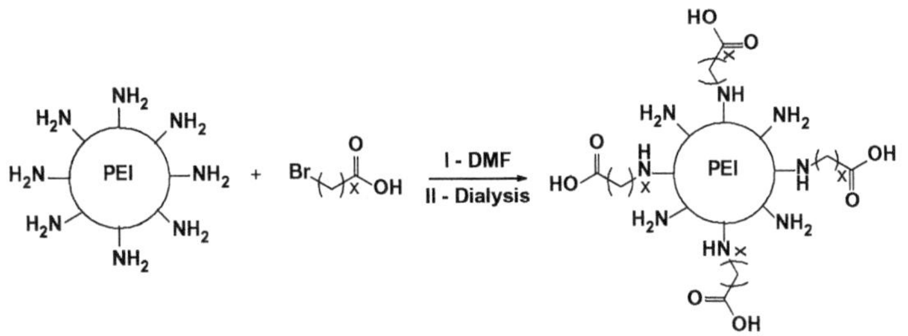

Figure 2.2 Chemical modification of PEI25 with alkycarboxylate of different chain length (x=2, 6, 10, and 16) [66].

efficiency was systematically investigated [43]. The polymers modified with moderate hydrophobic amino acids (e.g., alanine) showed improved transfection efficiency even in the presence of serum, whereas the effect of stronger hydrophobic amino acid (e.g., leucine) was insignificant. This observation indicates that a proper balance between the hydrophobicity and the hydrophilicity in PEI25 derivatives was an essential factor for achieving optimum efficacy. Poly(γ-benzyl L-glutamate) (PBLG) has been also used as a poly(amino acid) for PEI25 grafting. Chen et al. observed an effective down-regulation of luciferase gene expression in HeLa, CT26, and 4T1 cells using PBLG-grafted PEI25 [69]. More recently, poly(phenylalanine) was grafted onto PEI25, and the polymer was used to mediate delivery of a therapeutic gene, rev-casp-3, to induce tumor apoptosis and inhibition of tumor growth with insignificant side effect in a xenograft model [70,71].

Linear aliphatic acids and steroids were grafted onto PEI25 to generate water-soluble lipopolymers [72,73] to enhance membrane interaction and trafficking into cells [74]. Hydrophobic modification is generally expected to enhance the gene delivery efficacy of PEI25, but some studies have observed an optimum efficiency at a certain degree of substitution (e.g., ~25 mol%) beyond which the efficiency was decreased [75]. Alshamsan et al. evaluated a library of PEI25 grafted with different aliphatic lipids (C8 to C18) and reported promising results on structure-function relationship with oleic and stearic acid grafted PEI25 (Figure 2.3) [76]. The modified polymers displayed better siRNA binding capacity and formed more serum compatible complexes, resulting in superior siRNA delivery compared to parent polymer and some commercial transfecting agents. Among the aliphatic acids, stearic acid modification demonstrated better effect, although it was not clear if this lipid was the optimal among a large number of feasible lipid substituents. Stearic acid conjugated PEIs were also explored for STAT3 down-regulation in B16 melanoma cells [77]. Upon administration, siRNA complexes of stearic acid-conjugated PEI25 suppressed tumor growth by STAT3 inhibition, and secondarily by increased IL-6 secretion and Caspase 3 activity, while decreasing VEGF secretion. Since chemical modifications alter the buffering capacity of cationic PEI25 due to decreased primary amine density, cationic hydrophobic moieties were grafted to generate hydrophobicity while conserving cationic charge density and buffering capacity [78]. Mahato et al. successfully conjugated 6-(*N,N,N,N*-tetramethylguanidinium chloride)-hexanoic acid onto PEI25 and

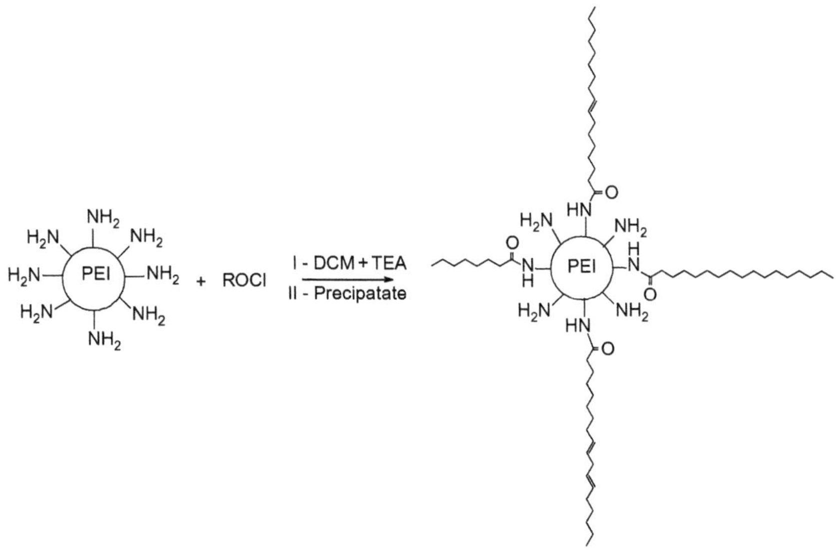

Figure 2.3 Hydrophobic modification of PEIs with different types of aliphatic lipids [76].

resulting polymers displayed reduced toxicity and improved siRNA delivery (GFP and GAPDH-specific) to different cell lines, which was superior to parent polymer and commercial transfecting agent Lipofectamine™ [78].

Chemical modification of low MW polymers is an alternative approach to overcoming the inherent challenges of high MW PEIs [72,79]. Cholesterol was initially grafted onto low MW PEIs to prepare water-soluble lipopolymers for gene delivery. The efficiency of cholesterol derivatized low MW PEIs was remarkably higher than the parent polymers, PEI1.8, PEI10, and PEI25 [73,80]. Cholesterol derivetized low MW PEIs were also used for siRNA delivery. Cholesterol modified b-PEI (1.8 kDa) significantly reduced (40%) the production of VEGF in PC-3 cells and tumor volume by 55% (21 days) by using VEGF specific siRNA [81]. The related deoxycholic acid (DA) was also integrated into cationic polymer to improve cell penetration efficiency. Kim et al. reported down-regulation of matrix metalloproteinase-2 activity in smooth muscles cells by using siRNA complexes with DA-grafted b-PEI (2 kDa) [82]. Alternatively, low MW PEIs were used for modification with aliphatic lipids and such a modification was generally found to enhance the efficiency of the polymers irrespective of grafting materials. In an earlier study, Thomas et al. modified low MW PEI with multiple aliphatic hydrocarbons, but only dodecyl- and hexadecyl-derivatives showed positive effect for *in vitro* transfection efficiency [43]. Recently, our group performed multiple studies using lipid-modified low MW PEIs (aliphatic lipids with different chain lengths and degree of unsaturation) [72] (Figure 2.4). These substitutions generate relatively non-toxic PEI derivatives (but more toxic than native low MW PEI) and displayed superior efficiency compared to parent polymers. The linoleic acid derivative in particular showed extraordinary performance compared to other substitutions. In one study, Teng et al. reported that oleic acid grafted b-PEI

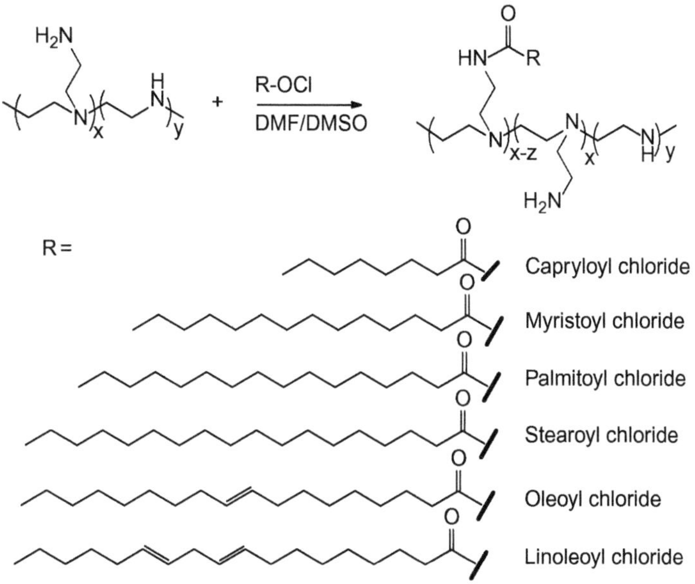

Figure 2.4 Hydrophobic modification of low MW (2 kDa) PEI with aliphatic lipids of variable chain lengths and degree of unsaturation [72].

(2.0 kDa) significantly (~50%) down-regulated luciferase gene expression, where free siRNA and parent polymers were not effective [76,83]. The studies on structure-function relationship of aliphatic lipid grafted onto low MW PEIs indicated that binding efficiency of polymer decreased with grafting density, which could be beneficial for complex dissociation at the appropriate site. Schroeder et al. reported that a higher grafting density of aliphatic lipid on 0.6 kDa PEI showed better siRNA delivery and the improved efficiency was attributed to consequences of polymer/siRNA dissociation after cellular internalization [84].

The siRNA delivery with lipid-substituted PEIs has been used to sensitize tumor cells against drugs by targeting drug efflux proteins that are associated with multidrug resistance in cancer cells, which limit the outcome of chemotherapy. Breast cancer resistance proteins (BCRP, ABCG2) and P-glycoprotein are common efflux proteins associated with human cancers. Modified PEIs have been used to target efflux protein genes [77]. Aliphatic lipid (C8 to C18) conjugation enhanced the siRNA delivery efficacy of low MW PEI (2 kDa) and consequently down-regulated efflux proteins [84–86]. The effective siRNA delivery sensitized the human breast cancer cells (MDA-MB-231) against several anti-cancer drugs (e.g., doxorubicin, paclitaxel, and mitoxantrone) through down-regulation of anti-apoptotic gene survivin [86]. The lipid-grafted polymers have been used to deliver multiple siRNAs for better control of malignant cells. Aliabadi et al. observed a significant apoptosis in drug-resistant breast cancer cells after dual silencing of Mcl-1 and P-glycoprotein via linoleic acid-modified PEI mediated siRNA delivery [84]. The therapeutic efficiency after dual silencing was significantly higher than when targeting a single mediator. Chen et al.

also reported a similar approach in an *in vivo* study targeting VEGFR2 and EGFR expression using PEI/siRNA complexes. The anti-tumor activity of cisplatin was significantly increased in non-small cell lung cancer xenografts and subsequently tumor growth was inhibited and survival time of the mice was increased in this way [87].

2.4.4 Stimuli-sensitive construct

Electrostatic interaction between cationic polymers and polynucleotides is the foundation of non-viral gene delivery, but some studies revealed that it could inhibit the appropriate release of the polynucleotides, which could be the rate-limiting step in gene delivery [40]. To this end, stimuli-sensitive biodegradable polymers have been explored to trigger intracellular disassembly of complexes. Small molecular weight cationic polymers have been used as a template to design these vectors that transform their physicochemical properties in response to extracellular and, intracellular stimuli [88]. The common physiological stimuli include pH, redox-potential, temperature, and enzyme-sensitivity, along with artificial stimuli that include light, ultrasound, and magnetic fields [89,90]. Stimuli-sensitive vectors were designed using high MW cationic polymers composed of biodegradable linkages such as disulfide (-S-S-), hydazone (C=N-N-), ketal (R_1R_2-C-(OR_3)2), and ester bonds (R-(CO)-OR) [91–93]. An additional important benefit of this strategy is lower toxicity on cells in the long term since induction of degradation significantly avoids any possible accumulation of higher MW polymers in cells and tissues.

2.4.4.1 Redox-sensitive construct

The disulfide (-S-S-) linkage is one of the most studied strategies to exploit the large difference between intracellular (1-11 mM) and extracellular (~1.5 μM) glutathione concentration [94,95]. This linkage has sufficient stability in oxidizing extracellular environment but gets cleaved in reducing intracellular compartment via thiol-disulfide exchange [96]. Disulfide cross-linked cationic polymers were designed either by thiolation followed by cross-linking, or by using -S-S- linked cross-linkers (e.g., dimethyl-3,3'-dithiobispropionimidate (DTBP), *N*-succinimidyl 3-[2-pyridyldithio]-propionate (SPDP), dithiobis(succinimidyl propionate) (DSP), cystamine bisacrylamide (CBA), etc.) [97–99] (Figure 2.5).

Neu et al. synthesized -S-S- cross-linked PEI25/DNA complexes using the homo-functional cross-linker SPDP and reported beneficial effect in both *in vitro* and *in vivo* gene delivery [99,100]. These complexes showed minimum interaction with major blood components like albumin and erythrocytes, and enhanced DNA protection. Transfection efficiency was significantly increased as a result of redox-triggered DNA release. Upon systemic administration, -S-S- linked PEI25/DNA complexes increased the half-life of the pDNA in blood, and the complexes were mainly accumulated in lungs and livers with prominent transfection in liver. Disulfide cross-linking of low MW PEIs was also explored as an alternative approach to designing less toxic and redox-sensitive vectors [79,98,101]. The oxidation of thio-capped low MW PEIs generated redox-sensitive cationic polymers with strong DNA binding capacity [92]. The *in vitro* gene

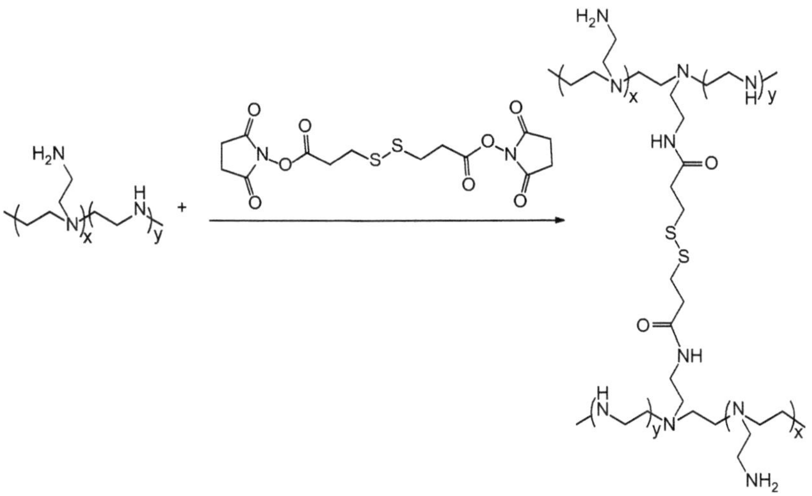

Figure 2.5 Disulfide cross-linking of PEI using SPDP as a cross-linker.

delivery efficiency of -S-S- linked polymers was significantly increased, even at higher serum concentration. Many studies reported on the design of -S-S- linked low MW PEIs [98]. Gosselin et al. synthesized -S-S- linked 800 Da PEI using the homo-functional cross-linkers DSP and DTBP [97]. Cross-linking remarkably increased the transfection efficiency in CHO cells, which was comparable to PEI25. The increased transfection efficiency was attributed to intracellular reductive cleavage and disassembly of the complexes. Recently, Breunig et al. reported an extensive study using -S-S- cross-linked low MW PEIs [98], where the transfection efficiency was studied in different cell lines (CHOK1, COS-7, NIH/3T3, HepG2, HCT116, HeLa, and HEK-293). The efficiency was remarkably higher than commercial transfection reagents including SuperFect™, Lipofectamine™, and JetPEI™. Others have also combined redox-sensitive polymers with functional ligands. Park K et al., for example, synthesized -S-S- linked polymer using PEI2 and CBA, followed by grafting with cell-specific targeting ligand HA [79]. The efficiency of the polymers was high for *in vitro* gene delivery even in high serum concentration (50 %), and it was further increased by the HA grafting.

In the literature, some studies showed cellular uptake as a primary limitation for effective transgene expression and some intracellular unpacking of the complexes. Breunig et al. suggested that intracellular unpacking is the major limitation for effective gene expression. They explored siRNA delivery in CHO-K1/EGFP cells and revealed a direct correlation between polymer branching and cellular uptake. Cell uptake efficiency increased with polymer branching: l-PEI < l-PEI cross-linked via -S-S- bonds (ssPEI) < 25 kDa b-PEI]. However, -S-S- cross-linked polymers performed better for down-regulating a model protein (EGFP) as compared to PEI25, even though the uptake of PEI25 complexes was prominent [51]. Hong et al. developed another approach in which thiol-terminated siRNA at both 3'-ends were complexes with l-PEI (2.0 kDa) (with/without –SH capping) and then the polyplexes were allowed to develop inter- and intra-molecular cross-linking via aerial oxidation. These polyplexes displayed extraordinary stability in polyanionic environment, were less toxic, and very effective in down-regulating GFP protein expression [92]. The -S-S- cross-linking was reported to further enhance the efficacy of hydrophobically modified low MW PEIs. As an example, Lee et al. prepared PEI/siRNA complexes using -S-S- cross-linked DC-grafted 2 kDa PEI. These complexes showed higher stability in polyionic environment, and better efficacy in down-regulation of MMP-2 expression as compared to the regular complexes [102]. However, our studies did not yield such an improvement. The transfection efficiency of PEI2 was significantly improved by grafting aliphatic lipids in our hands but -S-S- cross-linking of these polymers had not showed any beneficial effect [72,103]. The -S-S- cross-linked PEI (8.4 kDa) was also used for delivery of the human telomerase reverse transcriptase (hTERT) gene *in vitro* and *in vivo* [104]. The transfection of HepG2 cells by SS-PEI/hTERT siRNA showed significant silencing of hTERT mRNA and reduced hTERT protein level, telomerase activity, cell growth, and enhanced cell apoptosis. Upon systemic administration, these complexes displayed inhibition of subcutaneous HepG2 tumors in a mouse model with an insignificant effect on liver and kidney functioning.

2.4.5 pH-sensitive PEI constructs

The pH difference between normal tissues, bloodstream (pH ~ 7.4), cancerous tissues (pH ~ 5.7 to 7.8), and endosomal/lysosomal compartment (pH ~ 5.5) has been exploited in designing pH-sensitive vectors [52,105–107]. These vectors are used to target acidic extracellular matrix or intracellular organelles. Generally these vectors are designed using the polymers either grafted with pH-sensitive pendent groups (e.g., sulfonamide) or composed with pH-labile chemical bonding [108,109]. The pH-sensitive pendent groups display a prominent physical deformation with a slight change in pH value. Insertion of pH-sensitivity into cationic polymer generate significant shielding/de-shielding effect. In one study, Sethuraman et al. designed a pH-responsive ternary complexes using PEI and poly(methacyloyl sulfadimethoxine)-block-PEG (PSD-b-PEG) as a labile fragment [52]. Coating with the PEGylated polymer significantly inhibited the adsorption of biomolecules, but also decreased the transfection efficiency (at pH 7.4) and cellular toxicity due to PEI shielding. However, at low pH (~6.6), the polyplexes displayed significantly higher transfection efficiency and toxicity due to detachment of PSD-b-PEG that ultimately exposed the PEI to cellular interactions. Alternatively, PEIs could be either cross-linked or grafted onto polymer backbones via acid-cleavable bonding, such as esters, ketals, and imines, which undergo enhanced hydrolysis in acidic milieu [109–112]. Upon hydrolysis, the increased osmotic pressure enables endosomal disruption, dislodges the polynucleotide, and thereby enhances the cytosolic release [111]. The pH-sensitive polycations prepared by cross-linking low MW PEIs using ester-and/or amide linkers significantly boosted carrier efficiency both *in vitro* and *in vivo* models with acceptable toxicity profiles [19]. The enhanced efficiency of these polymers was due to intracellular disassembly of the complexes from acidic hydrolysis of ester linkages. Recently, a hyper-branched cationic polymer, poly(ester amine)s (PEAs), was prepared using tris[2-(acryloyloxy)ethyl]isocyanurate(TAEI) and low MW PEIs [113]. This approach yielded a promising *in vitro* and *in vivo* delivery system, presumably governed by intracellular polymeric degradation through ester linkages. Similarly, imine cross-linking of cationic polymers was reported; in one study, Kim et al. demonstrated a significant transfection efficiency of imine cross-liked PEI, which had enough stability at higher pH but underwent rapid hydrolysis at the endosomal pH (~5.5) [114].

2.4.6 Temperature-sensitive PEI constructs

Polynucleotide delivery has been attempted with thermo-responsive PEIs that were obtained by grafting with temperature-sensitive polymer segments (e.g., Pluronics™ and N-isopropylacrylamide, NIPAM) [115–117]. These constructs were intended to disrupt the endosomal/lysosomal compartment by using cold-shock treatments and releasing the entrapped polyplexes. In one study, Choi et al. demonstrated a promising process to generate temperature-sensitive hydrogels via emulsification/solvent evaporation method using a PEI-cross-linked Pluronic [118]. It was reported that endocytosed Pluronic/PEI hydrogels with temperature-dependent swelling/shrinking property (swelling upon temperature decrease from 37 °C to 20 °C) disrupted the endosomal compartment and enhanced cytoplasmic release after a cold-shock at 15 °C, consequently increasing *in vitro* gene silencing [119].

2.5 PEGylation of PEI

The PEG is the most widely substituted polymer on cationic polymers to generate a hydrophilic corona, decrease adsorption of plasma proteins and cellular elements, and increase residence time in circulation [30,120]. Hydrophilic PEG is expected to reduce polymer/DNA binding strength, undesirable aggregation, opsonization by plasma proteins and systemic toxicity of cationic polymers, improve solubility of DNA complexes in physiological condition, and help to introduce degradable bonds [62,121,122]. However, PEGylation of cationic polymer also shields charges, decrease binding efficiency by steric hindrance, and reduce cellular uptake of polyplexes [30,123]. Therefore, in some studies, targeting ligands are incorporated with PEG to compensate the adverse effect of PEGylation [61,62]. Schiffelers et al. reported a significant inhibition of tumor growth and angiogenesis in mice bearing subcutaneous N2A neuroblastomas by systemic administration of PEI-PEG-RGD/VEGF-siRNA [124]. The specific effects of PEGylation exclusively depended on the degree of substitution and its MW [125,126]. Starting with low MW PEI and PEG, multiple groups designed water-soluble, biodegradable, biocompatible cationic copolymers for gene delivery [127]. The copolymer prepared with 2.0 kDa PEGylated PEI formed nano-sized almost neutral and non-toxic polyplexes with three times higher *in vitro* transfection efficiency than the starting PEIs. Some studies have explored the integration of high MW PEGs into PEIs under controlled grafting density so that it can generate polymers just enough to neutralize the toxicity while maintaining DNA binding efficiency [128]. Neu et al. prepared minimally (one or two/PEIs) PEGylated PEI25 using 20 and 30 kDa PEGs followed by -S-S- cross-linking using DSP. DNA binding efficiency of PEI25 was not altered even after PEGylation but cytotoxicity and hemolytic activity was significantly reduced. *In vitro* transfection efficiency was increased by 10-fold in this case as compared to PEI25. The combined modification (PEGylation followed by -S-S- cross-linking) of PEI25 synergistically prolonged navigation time of DNA in the bloodstream. Unlike DNA delivery, steric effect of PEGylation is prominent in siRNA delivery, and studies show that the efficiency is enhanced by grafting a lower density of longer (5.0 kDa) PEG moieties onto PEI25 [126]. PEGylation of PEI25 increased the stability of PEI/siRNA complexes against heparin displacement and RNase digestion, and enhanced silencing of β-galactosidase. Weber et al. reported an effective silencing of GAPDH and inhibition of HIV replication using PEG grafted PEI25 as a carrier vector [129]. The overall findings show that PEGylation improves the biocompatibility of PEIs, but the influence of MW and degree of substitution is still contradictory. Both MW and substitution degree had a direct impact on carrier efficiency of PEGylated PEIs [130], and the optimal effect is a function of substitution level as well as N/P ratio used to make complexes.

2.6 Polyester conjugation on PEI

Amphiphilic polymers with hydrophilic PEG and a flexible hydrophobic segments, such as polycaprolactone (PCL) and poly(L-lactide) (PLA), are attractive materials in drug delivery applications [131]. These polymers self-assemble into a nano-sized micellar core-shell structure in an aqueous environment that can be used as an effective depot for hydrophobic drugs for systemic delivery [131,132]. These systems have been modified for functional delivery of multiple therapeutic agents including polynucleotides [133,134] by grafting PLL, PEI, and chitosan-like cationic polymers [135,136]. Liu et al. observed a beneficial effect of PEG-co-PCL grafting onto PEI25 in buffering capacity, stability, and surface charge of polyplexes [136]. These polymers showed negligible toxicity with significant siRNA-mediated gene silencing compared to parent polymer (PEI25) [137]. The grafting density had a direct effect on protection and stability of polynucleotides, with higher grafting giving better stability. The cationic block of the polymers served as an anchoring site for polynucleotides and penetrated into the drug containing hydrophobic core [138]. These constructs reduced the inherent toxicity of cationic polymers by virtue of surface charge neutralization [22,139,140].

2.7 Chitosan conjugation on PEI

PEIs have been grafted with chitosan to improve solubility, biocompatibility, endosomal escape, and unpacking of the complexes [141–143]. The common and most studied reaction for this grafting is imine reaction (Figure 2.6) [144]. The overall idea is to combine the biological properties of chitosan with gene transfection efficiency of PEIs into new generation polymer with significant synergistic activity.

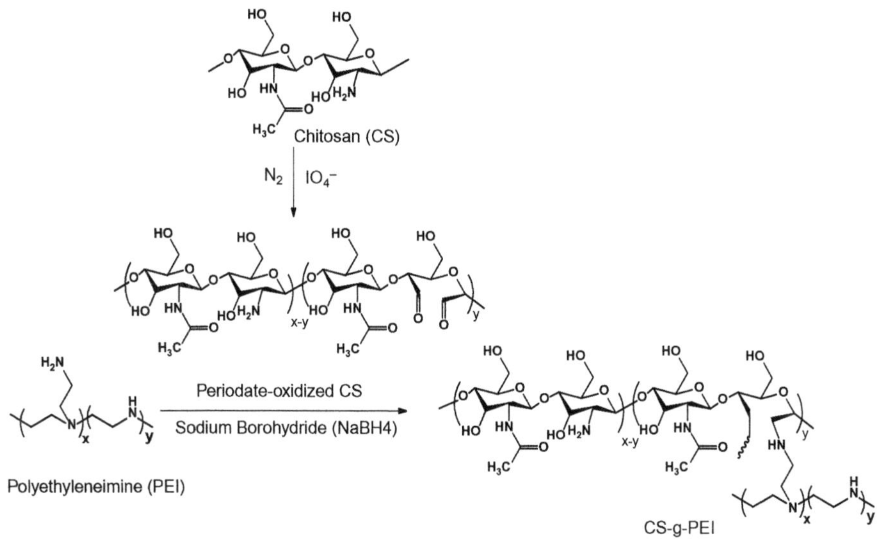

Figure 2.6 Synthesis of chitosan grafted PEI polymers

Triphati et al. reported that l-PEI grafted chitosan (PEI-g-CS) displayed significant transfection efficiency, endosomal escape, and DNA release kinetics compared to the parent polymers (l-PEI and CS) and commercial transfecting agents (Lipofectamine™) [145]. These polymers have also shown remarkable gene (EGFP and onco-protein AKt1) silencing and inhibited survival and proliferation of A549 cells and decrease malignancy and metastasis in lung xenograft model [144]. Low MW PEIs have been also grafted with chitosan [142,146]. These polymers showed higher transfection efficiency, which was attributed to the improved biocompatibility, cellular uptake, and buffering capacity [147]. The *in vivo* gene delivery investigation of these polymers showed a significant gene expression in spleen of male Balb/c mice. The efficiency of PEI-g-CS polymers has been increased by further incorporation of a targeting ligand [147,148]. Jiang et al. reported a promising DNA delivery to liver cells with galactosylated poly(ethylene glycol)-chitosan-graft-PEI [149,150].

2.8 PLL conjugation on PEI

While b-PEIs and PLL are the most explored cationic polymers in gene delivery [3,151], both polymers individually display some critical drawbacks; PEI is non-degradable and relatively cytotoxic and PLL has no buffering effect and its efficacy is limited [47,152]. These two polymers were integrated together physically or chemically to generate an improved cationic polymer. With chemical integration, PLL was grafted onto PEIs to improve biodegradability and biocompatibility [21,153–155] and gene delivery efficiency of PEI was generally increased with PLL grafting. These constructs were reported to enhance anti-tumor treatment in a xenograft model [154], and the efficiency could be further increased by grafting targeting ligands [155].

2.9 Metallic nanoparticles with PEIs

Gold nanoparticles (AuNPs) have recently emerged as a promising material due to facial synthesis, highly tunable optical properties that can be utilized in sensing, imaging, and therapy for different diseases. PEIs were grafted onto metal AuNPs to import its beneficial gene delivery features [156,157]. Gene delivery using AuNPs was accompanied in early 1990s using DNA- coated AuNPs introduced into target cells by particle bombardment [158,159]. Cationic polymers were grafted onto AuNPs to generate a cationic monolayer for polynucleotide binding by electrostatic adsorption [160–163]. Common cationic ligands exploited in AuNPs are cationic amino acids, PEI, chitosan, quaternary ammonium chains, 2-aminoethanethiol, and cationic lipids [156,164–166], but we will only focus on PEI modification. Thomas et al. prepared 2 kDa PEI grafted AuNPs (PEI-AuNPs) for pDNA delivery to COS-7 cells [156]. Transfection efficiency of the nanoparticles was slightly increased (25%) with PEI grafting while the efficiency was more significantly increased (50 %) by adding *N*-dodecyl–modified 2 kDa PEI. The beneficial effect of *N*-dodecyl–PEI2 was due to increased cellular uptake of the carriers. In some studies, high MW PEIs (PEI25) was used to stabilize AuNPs since its tertiary amines act as a reductant as well as a stabilizer [167,168]. Song et al. formulated mono-dispersed PEI25-AuNPs by incubating aqueous $HAuCl_4$ with PEI25 at room temperature, which was subsequently complexed with siRNA. A remarkable cellular uptake of PEI25-AuNPs/siRNA was seen in MDA-MB435 cells *in vitro*. These complexes display significant knock-down of GFP and endogenous cell-cycle kinase (PLK1) expression, and induce apoptosis [169]. In some studies, intermediate molecules were used to reduce and graft AuNPs onto polymeric materials. This strategy improves the size distribution and functionalization of nanoparticles. Catechols are common hydrophobic molecules that reduce counter molecules by donating lone-pair electrons and convert itself into a highly reactive quinone form that has high affinity toward amines and thiols via Michael-types addition reaction [15,169]. Lee et al. reported facile synthesis of AuNPs using catechol, which was then functionalized with PEI25/PEG [170]. These nanoparticles exhibited low toxicity and displayed excellent GFP silencing in MDA-MB435 human breast cancer cells.

2.10 Pharmacokinetics and bio-distribution of PEI delivery vectors

Pharmacokinetic aspects of non-viral vectors are believed to be one area that limits the efficiency of the carriers. Therefore, research into developing stable, non-toxic vectors that can encapsulate sufficient genetic materials (DNA and RNA), deliver to specific cells, and display significant expression will be beneficial. Systematic studies on pharmacokinetics and bio-distribution of PEI-based gene delivery has been poorly reported. The stability of the complexes in the bloodstream and their clearance, improper protection of polynucleotides upon systemic administration (e.g., intravenous or intramuscular), and poor nuclear targeting are the major hurdles of *in vivo* gene delivery [171,172]. Stability of the complexes upon systemic administration is the first hurdle since various poly-anionic blood components will interact instantly with polyplexes [173]. This could lead to unexpected aggregation, clearance through the reticuloendothelial system (RES), or dissociation of complexes. The half-life of higher MW PEIs complexes in bloodstream is generally shorter than low MW counterparts [22,174]. Many studies proposed PEGylation as the best strategy since it generates a hydrated sheath around the complexes that increase retention in bloodstream [121]. In addition, it neutralizes the cationic surface charge, thereby inhibiting the possible adsorption and non-specific interaction with cell organelles [175].

In an early study, *in vivo* DNA delivery using l-PEI (22 kDa) was studied in mice and bio-distribution of the complexes was systematically investigated [49]. Gene expression was prominently observed in different organs but gene expression was significantly higher in lungs compared to heart, liver, spleen, and kidney. In contrast, *in vivo* gene delivery efficiency of b-PEI25 was relatively low [44], while displaying intolerable toxicity. Kunath et al. performed a pharmacokinetics and bio-distribution study of different PEI/DNA complexes in balb/c mice using ^{125}I-PEIs [22]. This study showed that blood concentration of complexes was nearly identical irrespective of the type of complexes. However, organ accumulation was significantly different; e.g., accumulation of low MW PEI complexes in kidney was significantly higher than large MW PEI complexes. The accumulation in other organs (e.g., liver, spleen, lungs) was not different. PEIs derivatized with hydrophilic polymers are frequently used for *in vivo* gene delivery due to better compatibility. The efficiency of PEI grafted chitosan was significantly higher than PEI25 and gene expression was prominently higher in spleen, followed by heart and brain [147]. Chitosan was believed to increase the stability/biocompatibility of the complexes to enhance delivery into target cells compared to PEI25. Many researchers have focused on tumor- targeted gene delivery/therapy for siRNA delivery specifically. In one study, Schiffelers et al. observed superior siRNA delivery by cRGD grafted PEG-PEIs compared to PEI itself [124]. The injection of siRNA and PEIs/siRNA complexes did not show any kind of accumulation in tumor as well as in other organs but the accumulation of cRGD-grafted complexes was distinctly observed in tumor tissues. These observations indicate that design of hydrophilic derivatives of PEIs and tailoring with targeting ligand significantly improves the target specific gene delivery.

2.11 Concluding remarks

Considerable progress has been achieved in recent years in delivery of gene medicines using PEI-based vectors. However, efficacy of the vectors has not improved sufficiently to overtake the viral vectors. The drawbacks of the non-viral carriers are due to the lack of target specificity, instability in physiological environment, and significant safety concerns (e.g., immunogenicity). Gene delivery, being a multistep process, will need vectors that can overcome each hurdle of delivery trajectory. The advancements in bio-conjugate chemistry and proper functionalization of the PEI polymers promise multi-functional PEI derivatives that can generate a new avenue for diagnosis and therapy of human diseases. Stimulus-responsive PEI derivatives in particular that can deliver multiple therapeutic agents will benefit the field enormously.

Acknowledgements

The gene delivery research in the authors' lab was financially supported by operating grants from Alberta Innovates Health Sciences (AIHS), Natural Sciences and Engineering Council of Canada (NSERC), Canadian Institutes of Health Research (CIHR), and Canadian Breast Cancer Foundation (CBCF). We also appreciate equipment support from Alberta Heritage Foundation for Medical Research (AHFMR) and NSERC. We are grateful to numerous past trainees in the Uludag lab who contributed to development of non-viral delivery systems, in particular Dr. Vanessa Incani, Dr. Meysam Abbasi, Dr. Hamidreza M. Aliabadi, Dr. Laura Rose, and Dr. Charlie Hsu.

References

[1] Putnam D. Polymers for gene delivery across length scales. Nat Mater 2006;5(6):439–51.
[2] Schroeder A, et al. Alkane-modified short polyethyleneimine for siRNA delivery. J Control Release 2012;160(2):172–6.
[3] Waehler R, Russell SJ, Curiel DT. Engineering targeted viral vectors for gene therapy. Nat Rev Genet 2007;8(8):573–87.
[4] Nishikawa M, Huang L. Nonviral vectors in the new millennium: delivery barriers in gene transfer. Hum Gene Ther 2001;12(8):861–70.
[5] Mintzer MA, Simanek EE. Nonviral vectors for gene delivery. Chem Rev 2009;109(2):259–302.
[6] Han S, et al. Development of biomaterials for gene therapy. Mol Ther 2000;2(4): 302–17.
[7] Godbey WT, Mikos AG. Recent progress in gene delivery using non-viral transfer complexes. J Control Release 2001;72(1–3):115–25.
[8] Pack DW, et al. Design and development of polymers for gene delivery. Nat Rev Drug Discov 2005;4(7):581–93.
[9] Morille M, et al. Progress in developing cationic vectors for non-viral systemic gene therapy against cancer. Biomaterials 2008;29(24–25):3477–96.
[10] Mansouri S, et al. Chitosan-DNA nanoparticles as non-viral vectors in gene therapy: strategies to improve transfection efficacy. Eur J Pharm Biopharm 2004;57(1):1–8.
[11] Kumar VV, Singh RS, Chaudhuri A. Cationic transfection lipids in gene therapy: successes, set-backs, challenges and promises. Curr Med Chem 2003;10(14):1297–306.
[12] Davis ME. Non-viral gene delivery systems. Curr Opin Biotechnol 2002;13(2):128–31.
[13] Park TG, Jeong JH, Kim SW. Current status of polymeric gene delivery systems. Adv Drug Deliv Rev 2006;58(4):467–86.
[14] Gebhart CL, Kabanov AV. Evaluation of polyplexes as gene transfer agents. J Control Release 2001;73(2-3):401–16.
[15] Zintchenko A, et al. Simple modifications of branched PEI lead to highly efficient siRNA carriers with low toxicity. Bioconjug Chem 2008;19(7):1448–55.

[16] Lungwitz U, et al. Polyethylenimine-based non-viral gene delivery systems. Eur J Pharm Biopharm 2005;60(2):247–66.
[17] Iwai M, et al. Polyethylenimine-mediated suicide gene transfer induces a therapeutic effect for hepatocellular carcinoma in vivo by using an Epstein-Barr virus-based plasmid vector. Biochem Biophys Res Commun 2002;291(1):48–54.
[18] Aigner A, et al. Delivery of unmodified bioactive ribozymes by an RNA-stabilizing polyethylenimine (LMW-PEI) efficiently down-regulates gene expression. Gene Ther 2002;9(24):1700–7.
[19] Thomas M, et al. Full deacylation of polyethylenimine dramatically boosts its gene delivery efficiency and specificity to mouse lung. Proc Natl Acad Sci U S A 2005;102(16):5679–84.
[20] Jeong JH, et al. DNA transfection using linear poly(ethylenimine) prepared by controlled acid hydrolysis of poly(2-ethyl-2-oxazoline). J Control Release 2001;73(2–3):391–9.
[21] Fischer D, et al. A novel non-viral vector for DNA delivery based on low molecular weight, branched polyethylenimine: effect of molecular weight on transfection efficiency and cytotoxicity. Pharm Res 1999;16(8):1273–9.
[22] Kunath K, et al. Low-molecular-weight polyethylenimine as a non-viral vector for DNA delivery: comparison of physicochemical properties, transfection efficiency and in vivo distribution with high-molecular-weight polyethylenimine. J Control Release 2003;89(1):113–25.
[23] Godbey WT, Wu KK, Mikos AG. Size matters: molecular weight affects the efficiency of poly(ethylenimine) as a gene delivery vehicle. J Biomed Mater Res 1999;45(3):268–75.
[24] Abdallah B, et al. A powerful nonviral vector for in vivo gene transfer into the adult mammalian brain: polyethylenimine. Hum Gene Ther 1996;7(16):1947–54.
[25] Wightman L, et al. Different behavior of branched and linear polyethylenimine for gene delivery in vitro and in vivo. J Gene Med 2001;3(4):362–72.
[26] Boletta A, et al. Nonviral gene delivery to the rat kidney with polyethylenimine. Hum Gene Ther 1997;8(10):1243–51.
[27] Forrest ML, et al. Partial acetylation of polyethylenimine enhances in vitro gene delivery. Pharm Res 2004;21(2):365–71.
[28] Akinc A, et al. Exploring polyethylenimine-mediated DNA transfection and the proton sponge hypothesis. J Gene Med 2005;7(5):657–63.
[29] Godbey WT, et al. Poly(ethylenimine)-mediated transfection: a new paradigm for gene delivery. J Biomed Mater Res 2000;51(3):321–8.
[30] Godbey WT, Wu KK, Mikos AG. Tracking the intracellular path of poly(ethylenimine)/DNA complexes for gene delivery. Proc Natl Acad Sci U S A 1999;96(9):5177–81.
[31] Saha K, et al. Gold nanoparticles in chemical and biological sensing. Chem Rev 2012;112(5):2739–79.
[32] Radwan SH, Azzazy HM. Gold nanoparticles for molecular diagnostics. Expert Rev Mol Diagn 2009;9(5):511–24.
[33] Zayats M, et al. Biocatalytic growth of Au nanoparticles: from mechanistic aspects to biosensors design. Nano Lett 2005;5(1):21–5.
[34] Daniel MC, Astruc D. Gold nanoparticles: assembly, supramolecular chemistry, quantum-size-related properties, and applications toward biology, catalysis, and nanotechnology. Chem Rev 2004;104(1):293–346.
[35] Algar WR, et al. The controlled display of biomolecules on nanoparticles: a challenge suited to bioorthogonal chemistry. Bioconjug Chem 2011;22(5):825–58.
[36] Bhattacharya R, Mukherjee P. Biological properties of "naked" metal nanoparticles. Adv Drug Deliv Rev 2008;60(11):1289–306.

[37] Nel AE, et al. Understanding biophysicochemical interactions at the nano-bio interface. Nat Mater 2009;8(7):543–57.
[38] Alexis F, et al. Factors affecting the clearance and biodistribution of polymeric nanoparticles. Mol Pharm 2008;5(4):505–15.
[39] Dykxhoorn DM, Palliser D, Lieberman J. The silent treatment: siRNAs as small molecule drugs. Gene Ther 2006;13(6):541–52.
[40] Schaffer DV, et al. Vector unpacking as a potential barrier for receptor-mediated polyplex gene delivery. Biotechnol Bioeng 2000;67(5):598–606.
[41] Boussif O, et al. A versatile vector for gene and oligonucleotide transfer into cells in culture and in vivo: polyethylenimine. Proc Natl Acad Sci U S A 1995;92(16):7297–301.
[42] Funhoff AM, et al. Endosomal escape of polymeric gene delivery complexes is not always enhanced by polymers buffering at low pH. Biomacromolecules 2004;5(1):32–9.
[43] Thomas M, Klibanov AM. Enhancing polyethylenimine's delivery of plasmid DNA into mammalian cells. Proc Natl Acad Sci U S A 2002;99(23):14640–5.
[44] Kircheis R, Wightman L, Wagner E. Design and gene delivery activity of modified polyethylenimines. Adv Drug Deliv Rev 2001;53(3):341–58.
[45] Chen DJ, et al. Structure-function relationships of gene delivery vectors in a limited polycation library. J Control Release 2005;103(1):273–83.
[46] Bolcato-Bellemin AL, et al. Sticky overhangs enhance siRNA-mediated gene silencing. Proc Natl Acad Sci U S A 2007;104(41):16050–5.
[47] Grayson AC, Doody AM, Putnam D. Biophysical and structural characterization of polyethylenimine-mediated siRNA delivery in vitro. Pharm Res 2006;23(8):1868–76.
[48] Kwok A, Hart SL. Comparative structural and functional studies of nanoparticle formulations for DNA and siRNA delivery. Nanomedicine 2011;7(2):210–9.
[49] Goula D, et al. Polyethylenimine-based intravenous delivery of transgenes to mouse lung. Gene Ther 1998;5(9):1291–5.
[50] Hassani Z, et al. Lipid-mediated siRNA delivery down-regulates exogenous gene expression in the mouse brain at picomolar levels. J Gene Med 2005;7(2):198–207.
[51] Breunig M, et al. Mechanistic investigation of poly(ethylene imine)-based siRNA delivery: disulfide bonds boost intracellular release of the cargo. J Control Release 2008;130(1):57–63.
[52] Sethuraman VA, Na K, Bae YH. pH-responsive sulfonamide/PEI system for tumor specific gene delivery: an in vitro study. Biomacromolecules 2006;7(1):64–70.
[53] Sakhalkar HS, et al. Leukocyte-inspired biodegradable particles that selectively and avidly adhere to inflamed endothelium in vitro and in vivo. Proc Natl Acad Sci U S A 2003;100(26):15895–900.
[54] Brooks H, Lebleu B, Vives E. Tat peptide-mediated cellular delivery: back to basics. Adv Drug Deliv Rev 2005;57(4):559–77.
[55] Lundberg P, Langel U. A brief introduction to cell-penetrating peptides. J Mol Recognit 2003;16(5):227–33.
[56] Jeong JH, Kim SW, Park TG. Novel intracellular delivery system of antisense oligonucleotide by self-assembled hybrid micelles composed of DNA/PEG conjugate and cationic fusogenic peptide. Bioconjug Chem 2003;14(2):473–9.
[57] Reddy JA, et al. Optimization of folate-conjugated liposomal vectors for folate receptor-mediated gene therapy. J Pharm Sci 1999;88(11):1112–8.
[58] Kunath K, et al. Integrin targeting using RGD-PEI conjugates for in vitro gene transfer. J Gene Med 2003;5(7):588–99.
[59] Kursa M, et al. Novel shielded transferrin-polyethylene glycol-polyethylenimine/DNA complexes for systemic tumor-targeted gene transfer. Bioconjug Chem 2003;14(1):222–31.

[60] Schaffer DV, Lauffenburger DA. Optimization of cell surface binding enhances efficiency and specificity of molecular conjugate gene delivery. J Biol Chem 1998;273(43):28004–9.
[61] Suk JS, et al. Gene delivery to differentiated neurotypic cells with RGD and HIV Tat peptide functionalized polymeric nanoparticles. Biomaterials 2006;27(29):5143–50.
[62] Ogris M, et al. Tumor-targeted gene therapy: strategies for the preparation of ligand-polyethylene glycol-polyethylenimine/DNA complexes. J Control Release 2003;91(1-2):173–81.
[63] Bahadur KCR, Thapa B, Xu P. Design of serum compatible tetrary complexes for gene delivery. Macromol Biosci 2012;12(5):637–46.
[64] Moghimi SM, et al. A two-stage poly(ethylenimine)-mediated cytotoxicity: implications for gene transfer/therapy. Mol Ther 2005;11(6):990–5.
[65] Oskuee RK, et al. Alkylcarboxylate grafting to polyethylenimine: a simple approach to producing a DNA nanocarrier with low toxicity. J Gene Med 2009;11(10):921–32.
[66] Oskuee RK, et al. The impact of carboxyalkylation of branched polyethylenimine on effectiveness in small interfering RNA delivery. J Gene Med 2010;12(9):729–38.
[67] Tian H, et al. Gene transfection of hyperbranched PEI grafted by hydrophobic amino acid segment PBLG. Biomaterials 2007;28(18):2899–907.
[68] Tian H, et al. Micellization and reversible pH-sensitive phase transfer of the hyperbranched multiarm PEI-PBLG Copolymer. Chemistry 2006;12(16):4305–12.
[69] Chen J, et al. A highly efficient siRNA carrier of PBLG modified hyperbranched PEI. Macromol Biosci 2009;9(12):1247–53.
[70] Fu C, et al. Hydrophobic poly (amino acid) modified PEI mediated delivery of rev-casp-3 for cancer therapy. Biomaterials 2012;33(18):4589–96.
[71] Xia J, et al. Hydrophobic polyphenylalanine-grafted hyperbranched polyethylenimine and its in vitro gene transfection. Macromol Biosci 2011;11(2):211–8.
[72] Neamnark A, et al. Aliphatic lipid substitution on 2 kDa polyethylenimine improves plasmid delivery and transgene expression. Mol Pharm 2009;6(6):1798–815.
[73] Mahato RI, et al. Intratumoral delivery of p2CMVmIL-12 using water-soluble lipopolymers. Mol Ther 2001;4(2):130–8.
[74] Linder ME, Deschenes RJ. New insights into the mechanisms of protein palmitoylation. Biochemistry 2003;42(15):4311–20.
[75] Doody AM, et al. Characterizing the structure/function parameter space of hydrocarbon-conjugated branched polyethylenimine for DNA delivery in vitro. J Control Release 2006;116(2):227–37.
[76] Alshamsan A, et al. Formulation and delivery of siRNA by oleic acid and stearic acid modified polyethylenimine. Mol Pharm 2009;6(1):121–33.
[77] Alshamsan A, et al. The induction of tumor apoptosis in B16 melanoma following STAT3 siRNA delivery with a lipid-substituted polyethylenimine. Biomaterials 2010;31(6):1420–8.
[78] Mahato M, Kumar P, Sharma AK. Amphiphilic polyethylenimine polymers mediate efficient delivery of DNA and siRNA in mammalian cells. Mol Biosyst 2013;9(4):780–91.
[79] Park K, et al. Target specific systemic delivery of TGF-beta siRNA/(PEI-SS)-g-HA complex for the treatment of liver cirrhosis. Biomaterials 2011;32(21):4951–8.
[80] Bajaj A, Kondaiah P, Bhattacharya S. Synthesis and gene transfection efficacies of PEI-cholesterol-based lipopolymers. Bioconjug Chem 2008;19(8):1640–51.
[81] Kim WJ, et al. Efficient siRNA delivery using water soluble lipopolymer for anti-angiogenic gene therapy. J Control Release 2007;118(3):357–63.
[82] Kim D, et al. Facial amphipathic deoxycholic acid-modified polyethyleneimine for efficient MMP-2 siRNA delivery in vascular smooth muscle cells. Eur J Pharm Biopharm 2012;81(1):14–23.

[83] Teng L, Xie J, Lee RJ. Enhanced siRNA delivery using oleic acid derivative of polyethylenimine. Anticancer Res 2012;32(4):1267–71.

[84] Aliabadi HM, Mahdipoor P, Uludag H. Polymeric delivery of siRNA for dual silencing of Mcl-1 and P-glycoprotein and apoptosis induction in drug-resistant breast cancer cells. Cancer Gene Ther 2013;20(3):169–77.

[85] Aliabadi HM, et al. Effective down-regulation of breast cancer resistance protein (BCRP) by siRNA delivery using lipid-substituted aliphatic polymers. Eur J Pharm Biopharm 2012;81(1):33–42.

[86] Montazeri Aliabadi H, et al. Induction of apoptosis by survivin silencing through siRNA delivery in a human breast cancer cell line. Mol Pharm 2011;8(5):1821–30.

[87] Grzelinski M, et al. Critical role of G(alpha)12 and G(alpha)13 for human small cell lung cancer cell proliferation in vitro and tumor growth in vivo. Clin Cancer Res 2010;16(5):1402–15.

[88] Bauhuber S, et al. Delivery of nucleic acids via disulfide-based carrier systems. Adv Mater 2009;21(32–33):3286–306.

[89] Yokoyama M. Gene delivery using temperature-responsive polymeric carriers. Drug Discov Today 2002;7(7):426–32.

[90] Hatakeyama H, et al. Development of a novel systemic gene delivery system for cancer therapy with a tumor-specific cleavable PEG-lipid. Gene Ther 2007;14(1):68–77.

[91] Lin YL, et al. Degradable, pH-sensitive, membrane-destabilizing, comb-like polymers for intracellular delivery of nucleic acids. Biomaterials 2010;31(27):7150–66.

[92] Peng Q, Zhong Z, Zhuo R. Disulfide cross-linked polyethylenimines (PEI) prepared via thiolation of low molecular weight PEI as highly efficient gene vectors. Bioconjug Chem 2008;19(2):499–506.

[93] Shim MS, Kwon YJ. Stimuli-responsive polymers and nanomaterials for gene delivery and imaging applications. Adv Drug Deliv Rev 2012;64(11):1046–59.

[94] Yang J, et al. Evaluation of disulfide reduction during receptor-mediated endocytosis by using FRET imaging. Proc Natl Acad Sci U S A 2006;103(37):13872–7.

[95] Wu G, et al. Glutathione metabolism and its implications for health. J Nutr 2004;134(3):489–92.

[96] Meng F, Hennink WE, Zhong Z. Reduction-sensitive polymers and bioconjugates for biomedical applications. Biomaterials 2009;30(12):2180–98.

[97] Gosselin MA, Guo W, Lee RJ. Efficient gene transfer using reversibly cross-linked low molecular weight polyethylenimine. Bioconjug Chem 2001;12(6):989–94.

[98] Breunig M, et al. Breaking up the correlation between efficacy and toxicity for nonviral gene delivery. Proc Natl Acad Sci U S A 2007;104(36):14454–9.

[99] Neu M, et al. Crosslinked nanocarriers based upon poly(ethylene imine) for systemic plasmid delivery: in vitro characterization and in vivo studies in mice. J Control Release 2007;118(3):370–80.

[100] Neu M, et al. Stabilized nanocarriers for plasmids based upon cross-linked poly (ethylene imine). Biomacromolecules 2006;7(12):3428–38.

[101] Wang Y, Chen P, Shen J. The development and characterization of a glutathione-sensitive cross-linked polyethylenimine gene vector. Biomaterials 2006;27(30):5292–8.

[102] Lee D, et al. Bioreducible crosslinked polyelectrolyte complexes for MMP-2 siRNA delivery into human vascular smooth muscle cells. Pharm Res 2012;29(8):2213–24.

[103] Remant Bahadur KC, Uludag H. A comparative evaluation of disulfide-linked and hydrophobically-modified PEI for plasmid delivery. J Biomater Sci Polym Ed 2011;22(7):873–92.

[104] Xia W, et al. Bioreducible polyethylenimine-delivered siRNA targeting human telomerase reverse transcriptase inhibits HepG2 cell growth in vitro and in vivo. J Control Release 2012;157(3):427–36.
[105] Stubbs M, McSheehy PM, Griffiths JR. Causes and consequences of acidic pH in tumors: a magnetic resonance study. Adv Enzyme Regul 1999;39:13–30.
[106] Mok H, et al. pH-Sensitive siRNA nanovector for targeted gene silencing and cytotoxic effect in cancer cells. Mol Pharm 2010;7(6):1930–9.
[107] Venugopalan P, et al. pH-sensitive liposomes: mechanism of triggered release to drug and gene delivery prospects. Pharmazie 2002;57(10):659–71.
[108] Na K, Lee ES, Bae YH. Adriamycin loaded pullulan acetate/sulfonamide conjugate nanoparticles responding to tumor pH: pH-dependent cell interaction, internalization and cytotoxicity in vitro. J Control Release 2003;87(1–3):3–13.
[109] Shim MS, Kwon YJ. Acid-responsive linear polyethylenimine for efficient, specific, and biocompatible siRNA delivery. Bioconjug Chem 2009;20(3):488–99.
[110] Shim MS, Kwon YJ. Controlled delivery of plasmid DNA and siRNA to intracellular targets using ketalized polyethylenimine. Biomacromolecules 2008;9(2):444–55.
[111] Shim MS, Kwon YJ. Controlled cytoplasmic and nuclear localization of plasmid DNA and siRNA by differentially tailored polyethylenimine. J Control Release 2009;133(3):206–13.
[112] Forrest ML, Koerber JT, Pack DW. A degradable polyethylenimine derivative with low toxicity for highly efficient gene delivery. Bioconjug Chem 2003;14(5):934–40.
[113] Wang M, et al. Tris[2-(acryloyloxy)ethyl]isocyanurate cross-linked low-molecular-weight polyethylenimine as gene delivery carriers in cell culture and dystrophic mdx mice. Bioconjug Chem 2012;23(4):837–45.
[114] Kim YH, et al. Polyethylenimine with acid-labile linkages as a biodegradable gene carrier. J Control Release 2005;103(1):209–19.
[115] Zintchenko A, Ogris M, Wagner E. Temperature dependent gene expression induced by PNIPAM-based copolymers: potential of hyperthermia in gene transfer. Bioconjug Chem 2006;17(3):766–72.
[116] Schwerdt A, et al. Hyperthermia-induced targeting of thermosensitive gene carriers to tumors. Hum Gene Ther 2008;19(11):1283–92.
[117] Wang JJ, et al. Recent advances of chitosan nanoparticles as drug carriers. Int J Nanomedicine 2011;6:765–74.
[118] Choi SH, Lee SH, Park TG. Temperature-sensitive pluronic/poly(ethylenimine) nanocapsules for thermally triggered disruption of intracellular endosomal compartment. Biomacromolecules 2006;7(6):1864–70.
[119] Lee SH, et al. Thermally sensitive cationic polymer nanocapsules for specific cytosolic delivery and efficient gene silencing of siRNA: swelling induced physical disruption of endosome by cold shock. J Control Release 2008;125(1):25–32.
[120] Malhotra M, Tomaro-Duchesneau C, Prakash S. Synthesis of TAT peptide-tagged PEGylated chitosan nanoparticles for siRNA delivery targeting neurodegenerative diseases. Biomaterials 2013;34(4):1270–80.
[121] Ogris M, et al. PEGylated DNA/transferrin-PEI complexes: reduced interaction with blood components, extended circulation in blood and potential for systemic gene delivery. Gene Ther 1999;6(4):595–605.
[122] Fu J, et al. New polymeric carriers for controlled drug delivery following inhalation or injection. Biomaterials 2002;23(22):4425–33.
[123] Walker GF, et al. Toward synthetic viruses: endosomal pH-triggered deshielding of targeted polyplexes greatly enhances gene transfer in vitro and in vivo. Mol Ther 2005;11(3):418–25.

[124] Schiffelers RM, et al. Cancer siRNA therapy by tumor selective delivery with ligand-targeted sterically stabilized nanoparticle. Nucleic Acids Res 2004;32(19):e149.

[125] Brus C, et al. Physicochemical and biological characterization of polyethylenimine-graft-poly(ethylene glycol) block copolymers as a delivery system for oligonucleotides and ribozymes. Bioconjug Chem 2004;15(4):677–84.

[126] Mao S, et al. Influence of polyethylene glycol chain length on the physicochemical and biological properties of poly(ethylene imine)-graft-poly(ethylene glycol) block copolymer/SiRNA polyplexes. Bioconjug Chem 2006;17(5):1209–18.

[127] Ahn CH, et al. Biodegradable poly(ethylenimine) for plasmid DNA delivery. J Control Release 2002;80(1–3):273–82.

[128] Neu M, et al. Bioreversibly crosslinked polyplexes of PEI and high molecular weight PEG show extended circulation times in vivo. J Control Release 2007;124(1–2):69–80.

[129] Weber ND, et al. PEGylated poly(ethylene imine) copolymer-delivered siRNA inhibits HIV replication in vitro. J Control Release 2012;157(1):55–63.

[130] Fitzsimmons RE, Uludag H. Specific effects of PEGylation on gene delivery efficacy of polyethylenimine: interplay between PEG substitution and N/P ratio. Acta Biomater 2012;8(11):3941–55.

[131] Duncan R. The dawning era of polymer therapeutics. Nat Rev Drug Discov 2003;2(5):347–60.

[132] Whitehead KA, Langer R, Anderson DG. Knocking down barriers: advances in siRNA delivery. Nat Rev Drug Discov 2009;8(2):129–38.

[133] Bahadur KCR, Xu P. Multicompartment intracellular self-expanding nanogel for targeted delivery of drug cocktail. Adv Mater 2012;24(48):6479–83.

[134] Merkel OM, et al. In vivo SPECT and real-time gamma camera imaging of biodistribution and pharmacokinetics of siRNA delivery using an optimized radiolabeling and purification procedure. Bioconjug Chem 2009;20(1):174–82.

[135] Liu XX, et al. PAMAM dendrimers mediate siRNA delivery to target Hsp27 and produce potent antiproliferative effects on prostate cancer cells. ChemMedChem 2009;4(8):1302–10.

[136] Zheng M, et al. Enhancing in vivo circulation and siRNA delivery with biodegradable polyethylenimine-graft-polycaprolactone-block-poly(ethylene glycol) copolymers. Biomaterials 2012;33(27):6551–8.

[137] Liu Y, et al. Biophysical characterization of hyper-branched polyethylenimine-graft-polycaprolactone-block-mono-methoxyl-poly(ethylene glycol) copolymers (hy-PEI-PCL-mPEG) for siRNA delivery. J Control Release 2011;153(3):262–8.

[138] Liu P, et al. A mPEG-PLGA-b-PLL copolymer carrier for adriamycin and siRNA delivery. Biomaterials 2012;33(17):4403–12.

[139] Grimm D, Kay MA. Therapeutic application of RNAi: is mRNA targeting finally ready for prime time? J Clin Invest 2007;117(12):3633–41.

[140] Merdan T, et al. PEGylation of poly(ethylene imine) affects stability of complexes with plasmid DNA under in vivo conditions in a dose-dependent manner after intravenous injection into mice. Bioconjug Chem 2005;16(4):785–92.

[141] Lavertu M, et al. High efficiency gene transfer using chitosan/DNA nanoparticles with specific combinations of molecular weight and degree of deacetylation. Biomaterials 2006;27(27):4815–24.

[142] Jiang HL, et al. Chitosan-graft-polyethylenimine as a gene carrier. J Control Release 2007;117(2):273–80.

[143] Lou YL, et al. Poly(ethylene imine)-g-chitosan using EX-810 as a spacer for nonviral gene delivery vectors. J Biomed Mater Res A 2009;88(4):1058–68.

[144] Jere D, et al. Chitosan-graft-polyethylenimine for Akt1 siRNA delivery to lung cancer cells. Int J Pharm 2009;378(1–2):194–200.

[145] Tripathi SK, et al. Linear polyethylenimine-graft-chitosan copolymers as efficient DNA/siRNA delivery vectors in vitro and in vivo. Nanomedicine 2012;8(3):337–45.
[146] Wong K, et al. PEI-g-chitosan, a novel gene delivery system with transfection efficiency comparable to polyethylenimine in vitro and after liver administration in vivo. Bioconjug Chem 2006;17(1):152–8.
[147] Tripathi SK, et al. Depolymerized chitosans functionalized with bPEI as carriers of nucleic acids and tuftsin-tethered conjugate for macrophage targeting. Biomaterials 2012;33(16):4204–19.
[148] Park SC, et al. Branched polyethylenimine-grafted-carboxymethyl chitosan copolymer enhances the delivery of pDNA or siRNA in vitro and in vivo. Int J Nanomedicine 2013;8:3663–77.
[149] Jiang HL, et al. Galactosylated poly(ethylene glycol)-chitosan-graft-polyethylenimine as a gene carrier for hepatocyte-targeting. J Control Release 2008;131(2):150–7.
[150] Jiang HL, et al. Mannosylated chitosan-graft-polyethylenimine as a gene carrier for Raw 264.7 cell targeting. Int J Pharm 2009;375(1–2):133–9.
[151] Kadlecova Z, et al. Hyperbranched polylysine: a versatile, biodegradable transfection agent for the production of recombinant proteins by transient gene expression and the transfection of primary cells. Macromol Biosci 2012;12(6):794–804.
[152] Putnam D, et al. Polymer-based gene delivery with low cytotoxicity by a unique balance of side-chain termini. Proc Natl Acad Sci U S A 2001;98(3):1200–5.
[153] Fischer D, et al. In vitro cytotoxicity testing of polycations: influence of polymer structure on cell viability and hemolysis. Biomaterials 2003;24(7):1121–31.
[154] Tian H, et al. Polylysine-modified polyethylenimine inducing tumor apoptosis as an efficient gene carrier. J Control Release 2013;172(2):410–8.
[155] Dai J, et al. Polyethylenimine-grafted copolymer of poly(l-lysine) and poly(ethylene glycol) for gene delivery. Biomaterials 2011;32(6):1694–705.
[156] Thomas M, Klibanov AM. Conjugation to gold nanoparticles enhances polyethylenimine's transfer of plasmid DNA into mammalian cells. Proc Natl Acad Sci U S A 2003;100(16):9138–43.
[157] Hu C, et al. Low molecular weight polyethylenimine conjugated gold nanoparticles as efficient gene vectors. Bioconjug Chem 2010;21(5):836–43.
[158] Yang NS, et al. In vivo and in vitro gene transfer to mammalian somatic cells by particle bombardment. Proc Natl Acad Sci U S A 1990;87(24):9568–72.
[159] Larregina AT, et al. Direct transfection and activation of human cutaneous dendritic cells. Gene Ther 2001;8(8):608–17.
[160] Lee SK, et al. Effective gene silencing by multilayered siRNA-coated gold nanoparticles. Small 2011;7(3):364–70.
[161] Elbakry A, et al. Layer-by-layer assembled gold nanoparticles for siRNA delivery. Nano Lett 2009;9(5):2059–64.
[162] Hurst SJ, Lytton-Jean AK, Mirkin CA. Maximizing DNA loading on a range of gold nanoparticle sizes. Anal Chem 2006;78(24):8313–8.
[163] Han G, Martin CT, Rotello VM. Stability of gold nanoparticle-bound DNA toward biological, physical, and chemical agents. Chem Biol Drug Des 2006;67(1):78–82.
[164] Sandhu KK, et al. Gold nanoparticle-mediated transfection of mammalian cells. Bioconjug Chem 2002;13(1):3–6.
[165] Li P, et al. Cationic lipid bilayer coated gold nanoparticles-mediated transfection of mammalian cells. Biomaterials 2008;29(26):3617–24.
[166] Ghosh PS, et al. Efficient gene delivery vectors by tuning the surface charge density of amino acid-functionalized gold nanoparticles. ACS Nano 2008;2(11):2213–8.

[167] Sun X, Dong S, Wang E. One-step preparation of highly concentrated well-stable gold colloids by direct mix of polyelectrolyte and HAuCl4 aqueous solutions at room temperature. J Colloid Interface Sci 2005;288(1):301–3.
[168] Kim K, et al. Self-assembly of poly(ethylenimine)-capped Au nanoparticles at a toluene-water interface for efficient surface-enhanced raman scattering. Langmuir 2008;24(14):7178–83.
[169] Song WJ, et al. Gold nanoparticles capped with polyethyleneimine for enhanced siRNA delivery. Small 2010;6(2):239–46.
[170] Lee Y, et al. Controlled synthesis of PEI-coated gold nanoparticles using reductive catechol chemistry for siRNA delivery. J Control Release 2011;155(1):3–10.
[171] Kawabata K, Takakura Y, Hashida M. The fate of plasmid DNA after intravenous injection in mice: involvement of scavenger receptors in its hepatic uptake. Pharm Res 1995;12(6):825–30.
[172] Mumper RJ, et al. Polyvinyl derivatives as novel interactive polymers for controlled gene delivery to muscle. Pharm Res 1996;13(5):701–9.
[173] Plank C, et al. Activation of the complement system by synthetic DNA complexes: a potential barrier for intravenous gene delivery. Hum Gene Ther 1996;7(12):1437–46.
[174] Kunath K, et al. The structure of PEG-modified poly(ethylene imines) influences biodistribution and pharmacokinetics of their complexes with NF-kappaB decoy in mice. Pharm Res 2002;19(6):810–7.
[175] Kircheis R, et al. Polycation-based DNA complexes for tumor-targeted gene delivery in vivo. J Gene Med 1999;1(2):111–20.

Natural polymers in nucleic acid delivery

3

H. Montazeri Aliabadi
Chapman University School of Pharmacy, Irvine, CA, USA

3.1 Introducing cells to exogenous nucleic acids

3.1.1 Nucleic acids as therapeutics

The ability to replace a defected or missing gene in target cells as well as the addition of new exogenous genes that could create or enhance a specific intracellular mechanism is without a doubt one of the most fascinating ideas in health sciences. The idea, however, is not a brand new one anymore. The first clinical trials for gene therapy were undertaken in the 1980s, and it seemed that it would soon take clinical settings by storm [1]. There are more than 1800 monogenic hereditary disorders that would benefit greatly of the ability to replace a missing or defective gene. The first attempt to correct a hereditary disorder (in this case, adenosine deaminase deficiency) was undertaken in 1990 by Blaese and Anderson using a gene transfer method. Since then, there have been 95 approved gene-transfer trials worldwide directed at treating monogenic disorders [2]. In a clinical trial in 1999 for the treatment of partial deficiency in ornithine transcarbamylase (OTC), Jesse Gelsinger and one other patient received high doses of an adenoviral vector. A massive inflammatory response, liver injury, and intravascular coagulations claimed his life in four days. This incident brought a wrongful death lawsuit against the University of Pennsylvania and its researchers, and triggered still ongoing concerns about the safety of this approach [3]. In addition to the challenges that scientists have faced for the safe and efficient delivery of genetic material to specific cells, and to induce stable expression of the gene of interest, unresolved ethical issues have been a constant subject of controversy, and headlines in scientific journals (e.g., "Gene therapy: cursed, or inching towards credibility?" [4] and even pop culture (e.g., popular movies including *I am Legend* and *Splice*). The distinctive ethical challenges include the question of who (and at what stage of their condition) should be included in human testing, how to balance potential risks versus benefits, application to germline tissue, and cosmetic applications (e.g., genetic enhancement) [5].

Over the past decade, advances in vector technologies (including improved vector production methods, refined vectors, and enhanced transduction rates) have made the process simpler and more efficient [6]. Among different strategies used in gene therapy, the addition of a missing or defected gene is the most extensively studied. The genetic material is internalized into the target cell, and enters the nucleus where it is

Polymers and Nanomaterials for Gene Therapy. http://dx.doi.org/10.1016/B978-0-08-100520-0.00003-5
© 2016 Elsevier Ltd. All rights reserved.

further processed. The internalized exogenous DNA can be integrated into the host chromosome by covalent binding, or it can exist as an episome, which could be lost during cell division, but could persist in relatively quiescent tissues, due to slow turnover [7]. The integrated DNA would therefore persist longer in cells with rapid turnover (e.g., haematopoietic cells); however, they could potentially activate or disrupt nearby genes by insertional mutagenesis [1]. On the other hand, gene correction has also attracted scientists recently. In this method, genomic sequence is either corrected for an existing mutation, or altered by creating a new mutation (e.g., to make the cells resistant to HIV infection), which is performed using zinc finger nucleases and DNA recombination technologies [8].

The other equally fascinating idea is the ability to knock down a specific gene (or a group of functionally related genes) permanently, or for a limited time. Post-transcriptional gene silencing is a younger idea that was born on the verge of the twenty-first century. The possibility of interfering with the protein expression via endogenous pathways was first recognized in the late 1990s when Napoli et al. reported their attempt to over-express chalcone synthase in pigmented petunia petals by introducing a chimeric petunia CHS gene; unexpectedly, they observed a block in anthocyanin biosynthesis and a 50-fold decrease in mRNA levels of the protein of interest [9]. Soon after, it was shown that exogenous double-stranded RNA can interfere with protein expression at mRNA level, and RNA interference (RNAi) was born. This is a general term for various strategies for degrading the mRNA responsible for the synthesis of target protein, which all depend on the RNA-induced silencing complex (RISC) that contains Argonaute 2 (Ago2) enzyme. A double-stranded RNA (either endogenous or introduced exogenously) is incorporated into this complex, and only one strand is retained in the complex (guide strand) as it matches the target mRNA sequence. Upon identification and binding to complementary region of mRNA, Ago2 cleaves the mRNA, and the mRNA fragments are further degraded by the intracellular RNases [10]. Protein synthesis could also be inhibited by blocking the translation (rather than mRNA degradation) depending on the degree of matching sequences between the guide strand and the target mRNA [11]. RNAi strategies are categorized as microRNAs (miRNA), short hairpin RNAs (shRNA), or small interfering RNA (siRNA). While miRNAs are an endogenous group of non-coding hairpin RNAs designed for post-transcriptional regulation of protein expression, they do not require a perfect match with target mRNA, and are therefore capable of silencing a group of proteins [12]. shRNAs are another hairpin structure that is transferred to cytoplasm for silencing [13]. Transgenes encoding miRNA and shRNA are delivered to a cell by expression vectors, similar to the protocol for gene therapy, where they are integrated into the cell genome and lead to a stable expression of the hairpin structures. Unlike miRNAs, shRNAs target only one specific protein. Synthetic siRNAs are the third RNAi method, which provide the cell with the final mediator of miRNA- or shRNA-based silencing (20-25 pairs of double-stranded RNA) [14]. Silencing achieved via siRNA is temporary and limited by the molecules delivered with each dose.

3.1.2 Applications in cancer

Cancer is a heterogeneous disease, which means many different mechanisms are involved in carcinogenesis, and cancer cell enhanced proliferation and survival [15,16]. Therefore, there are plenty of opportunities to manipulate the cancerous cells to modify the fate of the cells in our favor. These tactics could be categorized as direct manipulation of the cancer cells to initiate apoptosis and/or suppress the cell proliferation, induction of specific enzyme expression to convert a prodrug, enhancement of the immune response to tumor antigens, silencing proteins involved in cell proliferation and survival, and protecting body against chemotherapy [17].

The first strategy is based on assumption that a defective gene with a crucial role can be replaced by gene delivery to the affected cells, which in turn, may supress the enhanced growth of the malignant cells. Many studies have shown the relative efficiency of this approach *in vitro*, and even *in vivo*, in animal models. For example, transfection of breast or ovarian cancer cells with BRCA1 has shown significant growth arrest and apoptosis induction [18,19]. However, this strategy was not efficient in Phase I and Phase II clinical studies in ovarian cancer patients, which might be due to unstable expression of the gene [20]. Other examples for this strategy include replacing the mutated genes that have been shown to play an important role in carcinogenesis, e.g., P53 [21] and adenomatous polyposis coli (APC; mutated in majority of colorectal cancers; [22,23]). Transfecting tumor cells with these genes have also shown significant effect on inducing apoptosis and causing growth arrest *in vivo* [17], which is somewhat surprising, considering that not all cells start consistent expression of these proteins after treatment; however, a "bystander effect" have been seen in non-transfected cells, which could be due to the effect of these proteins on other mechanisms involved in general growth and survival of tumor cells, including angiogenesis [24–26]. *In vitro* and *in vivo* studies have demonstrated antiproliferative effect for delivering p53 to tumor bearing mice, which has significantly increased the survival rate in this animal model [27]. Similarly, transfection of cancer cells with PTEN (alteration of which has been associated with different types of cancer, including bladder and prostate cancer as well as glioblastoma) has been shown to not only inhibit cell growth but also sensitize the cells to chemotherapeutic agents *in vitro* [28,29], and *in vivo* [30].

The other approach in cancer gene therapy has been delivering the gene responsible for encoding an enzyme that could potentially transform a non-toxic prodrug to a cytotoxic drug. This cytotoxic agent could potentially affect neighboring cells as well, which again would create a strong "bystander effect." The other advantage of this approach is the non-toxic nature of the prodrug, which eliminates the risk of toxicity for non-target healthy cells. Herpes simplex virus thymidine kinase (HSV-tk) has been used to phosphorylate ganciclovir to phosphogancyclovir, which in turn is metabolized by cellular enzymes to triphosphate that could inhibit DNA synthesis through competition with dGTP [31,32]. Conversion of 5-fluorocytosine to 5-fluorouracil (used mainly in colorectal cancer) by cytosine deaminase [33–35] and nitroreductases (e.g., nitroimidazole reductase) that converts anthracyclines (e.g., 5-(aziridin-1-YL)-2, 4-dinitrobenzamide) [36–38] are other examples of this approach.

Immunogenetic strategies are based on the idea of enhancing a specific immune response against the tumor cells or their components. This strategy is unique in the sense that unlike most anticancer strategies, it does not depend on targeting tumor cells, but on a systemic and wide-spread response via immune system effectors [39]. However, immunogenetic strategies have not been as promising as expected. A cell surface glycoprotein known as carcinoembryonic antigen (CEA) that is over-expressed in the majority of colorectal cancer patients, and its mutated form has been used as a vaccine in multiple clinical trials [40]. While the vaccine is well tolerated, clinical outcomes do not indicate clear efficiency in advanced colorectal cancer [41–43]. Another clinically evaluated approach has been *ex vivo* transduction of interlukin-2 (IL-2) in autologous fibroblasts [44] or direct intratumoral injection of IL-2 encoding plasmids [45]. While immune response could be detected after both methods, tumor regression has not been observed. On the other hand, down-regulation of Major Histocompatibility Complex (MHC) class I is one of the tumor cell defensive mechanisms to avoid immune response. However, while delivering plasmids encoding HLA-B7 to melanoma patients has shown promising results [46], the same strategy did not cause tumor regression in colorectal cancers [47].

The toxic effect of anticancer agents on the bone marrow is one of the major concerns and complications of chemotherapy, which could be a dose-limiting factor. Therefore, it is not surprising that several approaches based on gene therapy have been studied based on the idea that transferring drug-resistance genes to haematopoietic stem cells can protect them against myelosupression [48]. P-glycoprotein (P-gp), encoded by multidrug resistance 1 (MDR1) gene, was one of the most intensely studied proteins in the turn of twenty-first century and was considered the key to resistance against cytotoxicity of many anticancer agents. In fact, studies in Mdr1 transgenic mice indicated the possibility of protecting hematopoietic cells against a variety of chemotherapeutics [49]. However, further studies have demonstrated potential complications due to involvement of P-gp in other intracellular pathways [50,51]. It has also been shown that transferring vectors expressing specific variants of dihydrofolate reductase (DHFR) could create a similar protection against antifolate drugs [52]. However, antifolate agents are often used in combination with other anticancer drugs, and therefore, studies have been designed to develop vectors expressing DHFR and other resistance-creating proteins [53]. Expression of DNA repair proteins, e.g., methylguanine methyltransferase (MGMT), have also been used for this purpose [54].

Finally, silencing proteins involved in cell proliferation and/or survival as well as multidrug resistance have been studied along with small molecule drugs. All different RNAi strategies have been extensively evaluated for their efficacy in tumor suppression along with, and independent of chemotherapy. For silencing via siRNA see our previous review paper [14].

3.1.3 Applications in tissue engineering

Initiating or enhancing the expression of certain proteins could also have significant impact on therapeutic strategies other than cancer therapy. Delivering genetic material for tissue regeneration has been the most studied approach in this category, and has been mostly focused on bone and spinal cord injuries, as well as implanting scaffolds for extended release of nucleic acids. In addition to augmentation of healing process of wounds and bone fractures, gene therapy has emerged as a possible solution to limited neural regeneration through administration of factors promoting axonal growth [55]. Scaffolds, on the other hand, are implants or injectable hydrogels that provide a matrix for gene, drug, or cell delivery [56].

Bone is one of the few organs that have the regeneration potential even in adulthood, and the gene therapy approach offers the opportunity for direct delivery of genes encoding bone morphogenetic proteins (BMPs) [57]. In this approach, unlike cancer treatment or genetic disorders, the goal is not to replace a defective gene, but to deliver transcription or growth factors [58]. A variety of osteogenic factors have been studied in this approach, which include at least 15 different BMPs [59] (mostly BMP-2 and BMP-7), platelet-derived growth factor (PDGF), fibroblast growth factor (FGF), and vascular endothelial growth factor (VEGF), to name a few. Different neurotrophic factors have also been used as a gene therapy-based strategy in spinal cord injuries. For example, delivering genetic material encoding brain-derived neurotrophic factor (BDNF) increased neuronal and oligodendrocyte survival after spinal cord injury in an animal model [60]. Also, delivering vectors expressing glial cell line-derived neurotrophic factor (GDNF) has been reported to reduce tissue damage after such injuries while improving locomotor and behavioral recovery for periods of up to 5 weeks following injury in a similar animal model [61]. Vectors encoding VEGF, as well as interleukin 10 (IL-10; to reduce the inflammatory cascade and increase neuronal survival [62]) have also been studied for similar purposes. RNAi-based approaches have also been used in spinal cord injuries. Silencing NgR (a common receptor for most important myelin-associated inhibitors) via lentiviral siRNA vector of NgR1 has shown effective promotion of neurite outgrowth in vitro [63]. Using photomechanical waves (PMWs) for cellular internalization of GFAP and vimentin (proteins involved in neurogenic bladder) siRNAs has also demonstrated functional recovery in a rat spinal cord injury model [64].

Scaffolds have been used both to deliver cells (e.g., stem cells) or genetic material to enhance the expression of proteins involved in tissue regeneration (e.g., growth factors). Scaffolds could provide a matrix to direct the growth of the cells that it delivers, or the cells that migrate from the surrounding tissue. For a thorough review of scaffolds and genetic materials see review [56].

3.2 Delivering nucleic acids to target cells

3.2.1 Routes of delivery

Nucleic acid delivery is usually undertaken via two major strategies: *in vivo* and *ex vivo*. The basic idea for the *ex vivo* route is to isolate stem, progenitor, or differentiated cells from the patient, perform the genetic modification, and then return the cell to the body, and therefore, it is usually performed only when removal of the cells of interest is feasible [65]. Also, the modified cells that are returned to the body should be able to survive, proliferate, and preferably, have a selective advantage over unmodified cells [2]. The *ex vivo* approach has been studied in hereditary disorders, as a source of a specific protein, or in infectious and neoplastic conditions. In this approach, the patient is not directly exposed to the gene or the carrier, and the cells can be selected (to eliminate non-transduced cells), expanded, and/or differentiated, before being re-administered to the patient [66]. The best example of a successful *ex vivo* approach is haematopoietic stem cell transplantation (HST) [67]. With years of experience in the field, multiple clinical trials have been recently conducted on HST-based *ex vivo* gene therapy for several monogenic diseases of the blood and the immune system [66]. Despite relative success, there are, of course, challenges associated with this approach. After isolation of the cells, the first step is to culture the cells *ex vivo*. A major concern at this stage is the risk of mutations (in case of prolonged *in vitro* growth) and the possibility of irreversible changes in phenotype, viability, and proliferation capacity of the cells [68,69]. Also, multiple transfections might be required to assure transduction of the majority of the cells, which could again prolong the culture stage and enhance the issues associated with the maintenance of the cells and their properties [70]. Patient "conditioning" might also be required before re-administration of modified cells, which has been well-studied in HST (by irradiation or chemotherapy) [67] and T-cell therapies (by lymphoid-depletion regiment) [71]. *Ex vivo* approach has also been used in RNAi silencing [72].

The *in vivo* approach could be undertaken by delivering the vector directly to the target tissue (also known as "in situ") or as a systemic injection. While direct delivery seems to pose fewer challenges, the target tissue is not always readily accessible. Systemic delivery has been performed in different animal models using different approaches; however, targeting specific tissue or even cells for gene transfer remains a major hurdle. A variety of approaches have been evaluated for systemic delivery of nucleic acids, which could be categorized as viral and non-viral approaches, and they are both discussed in the following sections.

3.2.2 Viral carriers

Viruses have been naturally designed for effective transfection. Therefore, it is not surprising that several viral vectors, including adenoviral, adeno-associated viral, and retroviral vectors, have been studied extensively for gene delivery, and in fact, almost 70% of clinical trials for gene therapy have used viral vectors [73]. Limiting gene transfer to target cells, however, is a major challenge, since the natural tropism of the viral vector does not necessarily match the therapeutic plan (with rare exceptions, e.g., herpes virus, which is for neuronal gene delivery) [74]. Two main strategies to enhance the selectivity for the targeted cells are transcriptional and transductional targeting. While transcriptional targeting is based on using promotors that are only active in the cells of interest [75], transductional targeting is an attempt to modify the natural tropism of the viral vector toward the components of target cells [76]. Vector development has mostly progressed using adenovirus, adeno-associated virus, and vectors that are derived from retroviruses, particularly lentiviruses [77]. Transductional targeting in these viral vectors has mostly been accomplished using one of these strategies: (i) Pseudotyping: where a viral attachment protein from a different viral strain (or even family) is transferred to the selected vector [78–80]; (ii) Adaptors: which are moieties (e.g., antibodies, ligands, etc.) with two specific binding sites, one for cell surface receptor and one for viral attachment protein [81–84]; and (iii) Genetic incorporation: where ligand encoding gene is incorporated into viral vector [85,86]. Engineering adeno-associated viruses for gene delivery are reviewed in detail in [87] and [88].

Despite significant improvement of the viral vectors engineered for gene delivery, there are still major concerns to be addressed. An ideal virus-based vector would take advantage of virus highly evolved transduction machine, but minimizes the replication and toxicity by deleting parts of the viral genome [6]. However, the immune response to viral infection is still a considerable hurdle. The disaster of OTC trial in 1999 demonstrated the possibility of a severe immune response to a seemingly safe dose for other patients, and the National Institutes of Health (NIH) recommends that "all research participants enrolled in gene-transfer clinical trials should be monitored for several types of acute toxicities before and after vector administration ... routinely include a research participant's immune status, cytokine profile, and predisposing or underlying conditions that might elevate an individual's sensitivity to a particular vector" [89]. Adenoviral vectors seem to pose the most risk of immunogenicity. On the other hand, insertional mutations have also raised concerns with using viral vectors, which could even disrupt the expression of tumor suppressor genes or activate an oncogene mechanism [90]. It has been shown that gene insertion by viral vectors is not entirely random, and certain retroviral vectors (including lentivirus) demonstrate tendency for insertion in transcriptionally active regions of genome, which further increases the risk of insertional mutation [91,92].

3.2.3 Non-viral carriers

In addition to a long-lasting expression of the transduced gene, the safety profile of the delivery system is an important deciding factor in selecting the gene carrier. Non-viral delivery systems, while not efficient as viral vectors, do offer some distinct advantages including less pathogenicity and lower risk of insertional mutagenesis, as well as low cost and ease of production [93]. The category of non-viral carriers for gene delivery is a diverse group. It includes approaches known as "physical" methods, as well as multicomponent supramolecular carriers designed to enhance the efficacy of gene delivery. Physical methods could be categorized as: (a) Electroporation: This method uses electric pulse to create temporary "pores" in the cell membrane, as an entrance for the genetic material [94]. It has recently been used "in situ" for localized gene delivery to specific tissue (e.g., tumors) [95]. (b) Gene "gun" (ballistic bombardment): Metal microparticles coated with genetic material are "shot" towards the target cells to penetrate into cytoplasm. It has limited use *in vivo* (dermal delivery in animal models) [96,97]; (c) Hydrodynamic pressure: High volumes of the DNA-containing solution are injected IV, which creates enough hydrodynamic pressure to force DNA into tissue [98]. This technique could pose safety issues for clinical use, unless utilized under monitored pressure to a limb [99]; and (d) Microinjection: DNA is injected into a single cell using a microsyringe, which indicates limited use *in vitro* or *ex vivo* [100].

Lipids, proteins, and polymers have been used in different capacities and serve a variety of purposes in gene delivery systems. Liposomes, which encapsulate nucleic acids in a lipid bilayer envelope, seem to mimic a viral vector to some extent, and are probably the oldest nanocarrier used for gene delivery (delivering SV40 DNA to monkey kidney cells,

1980) [101]. Cationic lipids have also been used to form lipoplexes with genetic material based on ionic interactions with negatively charged nucleic acids [102]. Cationic polymers, including poly(l-lysine) (PLL) and polyethyleneimine (PEI), have also been used for delivery of nucleic acids (often known as polyplexes) based on the same principle [103–105]. PEI is considered by many as a gold standard for non-viral delivery of nucleic acids, due to its ability to condense nucleic acids and to enhance endosomal escape based on "proton sponge" effect [106]. Polymerosomes [107], cell-penetrating peptides [108], and inorganic nanoparticles [109] have also been used for nucleic acid delivery. The focus of the next section will be on the natural polymers used for this purpose.

3.3 Polymers as nucleic acid carriers

Different types of polymers have been utilized in delivery systems in a variety of capacities. The main advantage of polymers as delivery systems is their structural flexibility, which affords the researcher convenient manipulation of the physicochemical characteristics of the delivery system; polymer properties such as molecular weight, charge density, solubility, and hydrophobicity could be designed and engineered at will; as well as addition of desired chemical groups or targeting moieties for further functionalization. This flexibility is reflected in the overwhelming volume of studies on nucleic acid delivery using polymeric carriers. Both naturally occurring and synthetic polymers have been explored for this purpose [14].

3.3.1 Natural polymers

Naturally occurring polymers have been applied to a wide range of gene therapeutics, from nanoparticles to three-dimensional scaffolds, and have been applied to different routes of administration successfully as non-viral gene therapy systems. There are distinctive advantages associated with natural polymers, which make them more suitable for certain applications. One of the general advantages of natural polymers is their biodegradability in biological systems, which prevents accumulation of the polymer in the body, or inside the cells. This usually means a lower toxicity of the delivery system, which as mentioned, is a major concern in gene delivery [110]. Some of the natural polymers used as delivery systems include alginate, collagens, gelatin, fibrins, albumin, elsinan, pectin, galactan, curdlan, gellan, levan, emulsan, dextran, pullulan, gluten, elastin, fibroin, hyaluronic acid, cellulose, starch, chitosan (chitin), scleroglucan, heparin, chondroitin 6-sulfate, and polyhydroxyalkanoates [56].

Polymer-based nanoparticles for gene delivery serve different purposes: (a) they are designed to form a complex with the nucleic acid (usually based on ionic interaction), in order to protect their cargo from early degradation by DNases or RNases that are abundant in biological settings (including blood circulation, extracellular fluids, and cytoplasm) [111,112]. In fact, it has been shown that the half-life of degradation of naked plasmid DNA in serum is less than 10 minutes [113]; (b) even without degradation, nucleic acid molecules are cleared from blood circulation quite rapidly. While plasmid DNA is mostly cleared by liver (making the half-life of naked DNA less than 1 minute) [114], siRNA is rapidly cleared through glomerular filtration [115]. Polyplexes could change the pharmacokinetic profile and biodistribution pattern of plasmid DNA drastically with appropriate size and optimum surface properties (surface charge and stealth properties); (c) nucleic acids are negatively charged due to the phosphate groups on their backbone, and so is the cell membrane phospholipid bilayer. Therefore, it is not surprising that naked nucleic acid have negligible cellular uptake [14,114]. Cationic polymers neutralize the negative charge of nucleic acid (and even in most cases, create an overall positive charge), and increase the chance of interaction of the nanoparticle with cell membrane. However, while it is essential for the polymer to form a stable complex with nucleic acid, it is equally important for the complex to dissociate in cytoplasm in a timely manner; as long as it is associated with polymer, nucleic acid is not available for its intended action.

Cell-less scaffolds that carry genetic material and transfection reagents are known as gene-activated matrix (GAM) [116]. A specific type of GAMs is designed based on natural polymers (e.g., collagen) that carry naked DNA, and provide a localized source of gene transfer *in vivo*. Therefore, it is important for the scaffold to allow cells to infiltrate inside, which is usually accomplished by natural polymers and their biocompatible characteristics. Also, the matrix based on natural polymers, could degrade gradually, which releases the DNA to the surrounding tissue [110]. Some of the most studied natural polymers used in gene delivery will be discussed in more detail in the following sections.

3.3.2 Chitosan and derivatives

Chitosan, the result of alkaline deacetylation of chitin, is a natural-based, biocompatible, and biodegradable polymer [117], and is composed of randomly distributed α-(1-4)-linked d-glucosamine (deacetylated unit) and N-acetyl-d-glucosamine (acetylated unit). The attractive safety profile and low immunogenicity [118], as well as polycationic character of chitosan [119], make it a suitable candidate for nucleic acid protection against pre-mature degradation and intracellular delivery. Effective cellular internalization of DNA after complexation with chitosan has been reported in a variety of cells including human epithelial colorectal adenocarcinoma cells (Caco-2) [120], human embryonic kidney cells (HEK293) [121], human osteosarcoma cell line (MG63) [122], human hepatoma cells (HepG2) [123], human epithelial cervical cancer cells (HeLa) [124], and primary chondrocytes [125]. Different factors can affect the transfection efficiency of chitosan-delivered DNA, including the degree of deacetylation, molecular weight and positive charge of chitosan, and the pH of media [126–129]. The degree of polymerization is an important deciding factor, and a chain of 24 monomers in chitosan has been shown to enhance gene delivery characteristics of this polymer [130]. Two methods have been used for gene delivery using chitosan: simple complexation due to ionic interaction, and gelation, where plasmid is adsorbed on the surface of nanoparticles forming the gel or encapsulated in the gel matrix as nanoparticles. A recent study has shown that nanoparticles formed by these methods are in the 100–230 nm in size, and 37–48 mV in zeta potential, and can achieve 100% loading efficiency with 100 μg/mL, regardless of the method of preparation [131]. Selected studies on delivery of nucleic acids using chitosan or chitosan-derived polymers have been summarized in Table 3.1.

Table 3.1 **Selected studies for nucleic acid delivery using chitosan and derivatives**

Polymer	Nucleic Acid	Target Protein	Design	Cells/Animals (Route)	Outcome	Ref.
UC & DC	Plasmid	Luc & GFP	*In Vivo*	Balb/c mice (IV)	↑ Efficiency with ↓ DD	[132]
UC	Plasmid	Luc	*In Vitro*	HEK 293	↑ pH = ↓ Transfection	[133]
UC	AsODN	β-Gal	*In Vivo*	SP Rats	Efficient silencing	[135]
UC & TMC	Plasmid	Luc	*In Vitro*	Caco-2 cells	Transfection $\geq$ PEI	[136]
UC (Hydrogel)	siRNA	TG2	*In Vivo*	Athymic nude mice	↓ Tumor growth	[142]
UC	Plasmid	FVIII	*In Vitro / Vivo*	Cos-7 cells/FVIII knockout mice (oral)	Transfection/FVIII expression for 30 days	[137]
UC	Plasmid	IFN-γ	*In Vivo*	STAT4 -/- Blab/c mice	↓ Airway inflammation	[138]
UC	Plasmid	GFP	*In Vitro/Vivo*	HEK 293/CD-1 mice (In utero)	Postnatal gene expression	[139]
UC	Plasmid	β-Gal/Luc	*In Vivo*	Balb/c mice (IM)	Transfection	[140]
UC	siRNA	COX-2	*In Vitro/Vivo*	Macrophages (RAW 264.7)/ C57BL mice	↓ COX-2 expression/↓ Kidney damage	[141]
TMC-Cys	Plasmid	GFP	*In Vitro/Vivo*	HEK293/Balb/c mice (IM)	Enhanced Transfection	[147]
UC & TMC	siRNA	GFP	*In Vitro*	HEK293	↑ Efficiency with TMC	[146]
HTFP	Plasmid	Luc	*In Vitro*	KB Cells	Efficiency $\approx$ PEI	[148]
C-PEI-PEG	Plasmid	GFP	*In Vivo*	Nu/nu mice	Efficient transfection	[151]
HPC-FA	AsODN	P-gp	*In Vivo*	Nu/nu mice (IP)	↓ Tumor growth	[149]

GC-g-PEI	Plasmid	Luc	*In Vitro/Vivo*	HepG2/Balb/c mice (IP)	↑ Transfection than PEI/Target hepatocytes	[153]
HA/C-g-PEG	DNA/siRNA	GFP	*In Vitro*	HEK 293	Transfection/silencing	[155]
C-imi-HC-PEG	Plasmid	B-Gal/ GFP/ Luc	*In Vitro*	ND7/23, NIH 3T3 & DRG	Transfection specificity for neuronal cells	[152]
C-IAA-PEG	siRNA	GAPDH	*In Vivo*	Balb/c & C57BL mice (IV & IN)	Effective silencing of the reporter enzyme	[156]
Arg-C	Plasmid	GFP	*In Vitro*	HEK 293 & Cos-7	Transfection > UC	[157]
TM-BZ-CS	Plasmid	GFP	*In Vitro*	Huh7 hepatoma cells	Transfection > UC	[158]
CS/CG	Plasmid	PDGF	*In Vitro*	PDLCs	Transfection & proliferation	[144]
HA/CS	Plasmid	TGF-β1	*In Vitro*	Chondrocytes	Cell proliferation	[159]
MC	Plasmid	IL-12	*In Vitro/Vivo*	Dendritic cells/Balb/c mice (IT)	Transfection > UC/↓ Tumor growth	[150]

Arg-C: Arginine - Chitosan; **AsODN**: Antisense Oligonucleotides; **β-Gal**: Beta-Galactosidase; **C-IAA-PEG**: Chitosan - imidazole - 4 - Acetic Acid - Polyethylene Glycol; **C-imi-HC-PEG**: Chitosan - Imidazole - HC - Polyethylene Glycol; **COX-2**: Cyclooxygenase Type 2; **C-PEI-PEG**: Chitosan - polyethyleneimine - Polyethylene glycol conjugates; **CS/CG**: Chitosan/Collagen; **DC**: Deacetylated Chitosan; **DD**: Degree of Deacetylation; **DRG**: Dorsal Root Ganglia; **FVIII**: Factor VIII; **GAPDH**: Glyceraldehyde 3-Phosphate Dehydrogenase; **GC-g-PEI**: Galactosylated Chitosan-graft-Polyethyleneimine; **GFP**: Green Fluorescence Protein; **HA/C-g-PEG**: Hyaluronic acid/chitosan-grafted-Polyethylene Glycol; **HA/CS**: Hyaluronic acid/Chitosan; **HPC-FA**: Hydroxypropyl-chitosan conjugated with folic acid; **HTFP**: Histidine-conjugated Trimethylated chitosan - Folate - PEG conjugate; **IFN-γ**: Interferon gamma; **IL-12**: Interleukin - 12; **IM**: Intramuscular; **IN**: Intranasal; **IP**: Intraperitoneal; **IT**: Intratumoral; **IV**: Intravenous; **Luc**: Luciferase; **MC**: Mannosylated Chitosan; **PDGF**: Platelet Derived Growth Factor; **PDLCs**: Periodontal Ligament Cells; **PEG**: Polyethylene Glycol; **PEI**: Polyethyleneimine; **P-gp**: P-glycoprotein; **TG2**: Transglutaminase; **TGF-β1**: Transforming Growth Factor - beta 1; **siRNA**: Small Interfering RNA; **SP**: Sprague-Dawley; **TM-BZ-CS**: N-(4-N,N-dimethylaminobenzyl) chitosan; **TMC-Cys**: Trimethyl chitosan - cysteine conjugate; **UC**: Unmodified Chitosan

The importance of the degree of deacetylation of chitosan has been studied *in vivo*. Centelles et al. recently reported an *in vivo* study in an animal model using hydrodynamic IV injection of nanoparticles prepared with two variations of chitosan with degrees of acetylation of 85% and 75% [132]. In this study, a decrease in deacetylation degree (using acetic anhydride) increased the transgene expression of luciferase and green fluorescence protein (GFP; two reporter genes), and accelerated the DNA release rate. The presence of serum, salts, and a variation in pH could also affect the transfection efficacy of complexes prepared with chitosan. In 2010, Nimesh et al. reported a systematic study designed to evaluate the effects of these factors in the *in vitro* transfection efficacy of low molecular weight chitosan [133]. Preparing the complexes in PBS (pH = 7.4) instead of water increased the hydrodynamic diameter of the complexes from 243 to 1244 nanometers. Also, the presence of serum seemed to cause aggregation of the complexes. And as expected, increasing the pH and serum percentage had a negative effect on the transfection efficiency of the complexes in HEK 293 cells. Molecular weight of chitosan has a significant impact on the size of chitosan-nucleic acid complex. It has been reported that a decrease in chitosan molecular weight decreases chitosan-plasmid complex diameter. This might seem unexpected, as condensation of DNA should be more effectively accomplished with higher molecular weights. However, at high molecular weight (>150 kDa), chitosan is less soluble and more susceptible to aggregation [134].

Chitosan has been studied for nucleic acid delivery via different routes of administration. Topical application of chitosan/antisense oligonucleotides targeting the expression of beta-galactosidase has demonstrated up to 85% silencing of the targeted protein in adult and Sprague Dawley rats skin samples [135]. It has also been shown that due to its muco-adhesive properties, chitosan is a proper candidate for oral DNA delivery. In a recent study reported by Plapied et al., complexes of a plasmid-encoding luciferase and chitosan produced from a fungal source showed significant transfection efficiency *in vitro* using Caco-2 cells [136]. Interestingly, complexes prepared with fungal trimethylchitosan did not induce significant transduction, which the authors concluded to be due to over-strong interaction between DNA and polymer. Oral delivery of plasmids expressing Factor VIII using chitosan in FVIII knockout mice has shown FVIII detection for up to 30 days, which indicates a potential therapeutic benefit in hemophilia treatment [137]. Intranasal administration of interferon gamma (IFN-γ) and chitosan nanoparticles to wild type and STAT4 knockout mice has demonstrated efficiency in reducing established allergen-induced airway inflammation and hyper-responsiveness [138]. Yang et al. recently reported *in vivo* transfection via injection of chitosan/plasmid complexes into amniotic sac [139]. In this study, chitosan was evaluated as a non-viral carrier for in utero gene transfer in the treatment of monogenic disorders diagnosed in the fetus. Expression of GFP was confirmed in lung and intestine of the mice pups. Intramuscular injection of chitosan microspheres carrying two different plasmids (beta-galactosidase and luciferase) has also been reported in an animal model. Significant production of both proteins indicated the possibility of encapsulating two different plasmids without affecting their structural and functional integrity [140].

In 2014, Yang et al. reported siRNA delivery via intraperitoneal injection of chitosan complexes, in order to silence cyclooxygenase type 2 (COX-2) [141]. COX-2 plays an important role in kidney injury in obstructive nephropathy, and in this study performed on a mice model of unilateral ureteral obstruction (UUO), COX-2 silencing diminished tubular damage, parenchymal inflammation, and tumor necrosis factor-alpha (TNF-α), as well as heme oxygenase I and cleaved caspase 3, indicating lower oxidative stress and apoptosis, respectively. On the other hand, chitosan has also been used to form thermosensitive hydrogels for localized delivery of siRNAs. In an *in vivo* study performed in athymic nude mice, chitosan hydrogel showed temperature-dependent liquid/solid phase transition, and formed endothermic hydrogel after intratumoral injection in A375SM and MDA-MB-231 xenograft models [142]. Alexa555-labeled siRNA delivery showed localization of siRNAs inside tumor cells, and infiltration of tumor cells into hydrogel. Silencing transglutaminase (TG2) using this delivery system resulted in 72% and 92% tumor growth reduction in A375SM and MDA-MB-231 xenografts, respectively.

Chitosan could also form porous scaffolds with interconnected pores by lyophilisation [143]. Chitosan has been studied for providing gene-activated matrix that both serves as therapeutic gene expression, as well as a structural template for cell adhesion, proliferation, and synthesis of extracellular matrix. Chitosan nanoparticles carrying plasmids encoding platelet-derived growth factor (PDGF) in a chitosan-collagen scaffold have been used *in vitro* to assess a cell-seeded GAM for periodontal tissue engineering. Growth factor was released by the periodontal ligament cells (PDLCs) for over 6 weeks, and the cells retained their fibroblast-like shape and filled the pores of the scaffold [144]. The delivery of RNAi-based GAMs is a newly emerging area of research that has been reviewed elsewhere recently, and chitosan may display considerable potential in this field [145].

Structural modification of chitosan has been shown to enhance the efficacy of this polymer in nucleic acid delivery. In 2010, Dehousse et al. compared the efficacy of trimethylchitosan (TMC) in delivering siRNA to HEK293 cells to unmodified chitosan. Under pH condition of cell culture medium, the siRNA condensation by polymer was significantly affected, which in turn diminished the efficacy of the polyplex. TMC/siRNA complexes, however, were not affected, and showed a reliable stability in physiological conditions [146]. Cysteine-conjugated TMC has also demonstrated superior efficacy *in vitro* and *in vivo*, where GFP expression was significantly higher than commercially available Lipofectamine2000® in HEK293 cells as well as animal models after intramuscular injection [147]. Conjugation of histidine to chitosan backbone is another structural modification that has been evaluated for nucleic acid delivery. Nanoparticles formed by complexation of plasmid DNA with histidine-conjugated chitosan decorated with folate as targeting moiety and polyethylene glycol (PEG) grafts (HTFP) has demonstrated similar *in vitro* transfection efficacy to PEI [148]. Using folic acid-conjugated hydroxypropyl-chitosan has also been studied to target cancer cells that over-express folic acid receptor. In an *in vivo* study in a KB-A-1 xenograft model in Balb/c nu/nu mice, the nanoparticles decorated with the targeting moiety showed a significantly higher efficiency in targeting the tumor with oligonucleotides targeting P-glycoprotein (P-gp) expression. The tumor growth suppression with the targeted delivery was significantly higher than the same nanoparticles without folic acid moiety [149]. Mannose receptors are overexpressed on antigen-presenting cells (APCs), such as macrophages and immature dendritic cells, and therefore mannosylated chitosan (MC) has been studied for targeting dendritic cells. In a CT-26 carcinoma cells xenograft model, MC complexes with interleukin - 12 (IL-12) plasmid significantly induced cell cycle arrest and apoptosis [150]. PEGylation is a well-studied method to create stealth properties and enhance blood circulation time for polyplexes that would be otherwise removed from systemic circulation rapidly. A PEG-conjugated chitosan-PEI polyplex has shown efficient transfection in a C6 xenograft model in nude mice after intravenous injection [151].

Targeting peripheral sensorial neurons using modified chitosan has also been reported [152]. In this study, Oliveria et al. reported modification of chitosan structure with imidazole and thiol grafts to enhance transfection efficiency and to conjugate targeting ligands, respectively. Non-toxic fragments of tetanus toxin (HC) that demonstrated specific interaction with peripheral neurons was then grafted to the thiol moiety, and targeted transfection was confirmed in dorsal root ganglion dissociated primary cultures using GFP encoding plasmids. Targeting specific cells has also been achieved with localized delivery using hydrogels. Galactosylated chitosan grafted with PEI (GC-g-PEI) has been loaded in thermosensitive biodegradable hydrogels prepared from poly(organophosphazene), and studied for specificity for hepatocytes [153]. *In vitro*, the complexes of GC-g-PEI and luciferase plasmid showed a higher transfection efficacy than PEI in a hepatoblastoma cell line (HepG2) with lower cytotoxicity, whereas *in vivo*, intraperitoneal injection of radio-labeled complex-loaded hydrogel showed liver specificity.

Structural modification of chitosan might have other indications as well. Zhu et al. reported a strong interaction between O-carboxylmethylchitosan (OCMCS) and diplamitoyl-sn-glycero-3-phosphocholine (DPPC), a component of pulmonary cell membrane, which induced the fusion of small DPPC multilamellar vesicles (MLV) to form larger vesicles [154]. This strong interaction could potentially increase the effectiveness of OCMCS for gene delivery. The structures of some of the natural polymers discussed in this chapter are shown in Figure 3.1.

3.3.3 Poly (amino acids)

Poly (amino acids) have a relatively long history in nucleic acid delivery, and poly (L-lysine) (PLL) is the first (and still mostly studied) natural polymer used for nucleic acid in this category [160]. Although efficiently internalized into the cell, unlike PEI, PLL does not display a significant "proton sponge" effect, and therefore show lower levels of endosomal escape, which is reflected as poor transfection efficiency [161]. However, in transfection efficiency, while ranked lower than polyornithine (PLO), PLL is still more efficient than polyarginine in delivering DNA [162]. Many studies have evaluated strategies to improve the endosomal escape, and adding grafts of L-tryptophan and poly (histidine) have indicated promising results, which in case of

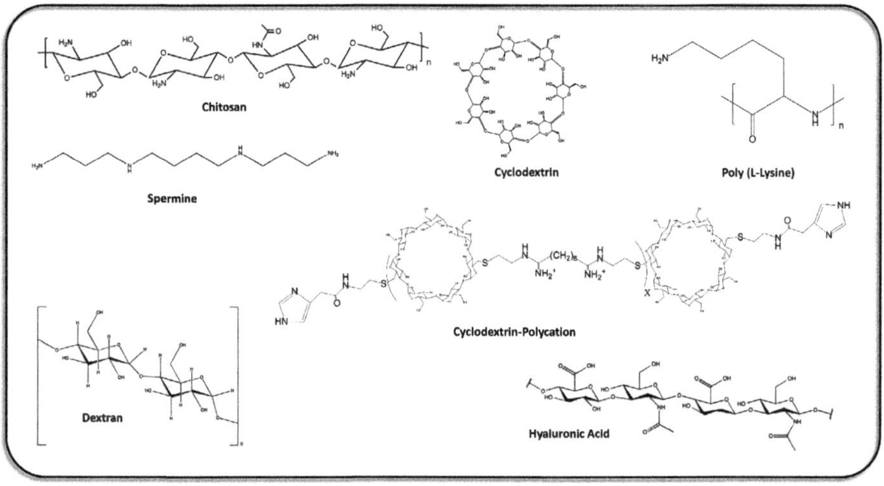

Figure 3.1 Chemical structure of selected natural polymers used for nucleic acid delivery.

histidine was explained by protonation of histidine in pH < 6, and therefore, induction of a proton sponge effect in endosome [163,164]. PLL/DNA polyplexes have also been coated with a lipid bilayer using a lipid hydration method, which enhanced the transfection efficiency 10-folds compared to uncoated particles. Interestingly, adding an octaarginine component to the delivery system showed a significant 1000-fold improvement in transfection [165]. Despite the biodegradable nature of PLL, similar to other natural polymers, toxicity is another problem with this polymer. In addition to hydrophilic barrier created by PEG, which creates stealth properties for nanoparticles formed with nucleic acid, PEG grafts on PLL backbone [166] or adding blocks of PEG [167] have shown an improvement of the toxicity profile of polyplexes. Other modifications on PLL structure to improve gene transfer capability have been reviewed by Tiera et al. [168].

3.3.4 Other natural polymers

Cyclodextrins (CD) are naturally occurring cyclic oligosaccharides that consist of (α-1,4)-linked α-D-glucopyranose, and have a basket-like form with "inner/outer" amphiphilic characteristics. It has been known since the nineteenth century, but the ring structure was revealed by Freudenberg et al. in the late 1930s [169]. The common CDs are α-, β-, and γ- with six, seven, and eight α-D-glucopyranose units, respectively, among which β-cyclodextrin has the lowest solubility, which could be due to internal hydrogen bonds [170]. Due to instable complexation with DNA, unmodified CD has limited application in gene delivery [171], and therefore, CD derivatives have been mostly used for this purpose. CDs have been functionalized to form a linkage between polymers to form a larger molecule for gene delivery. One example for this approach is linkage of low molecular weight PEI blocks with CD, which has shown a significant improvement in transfection efficiency compared to naïve PEI [172]. However, this requires proper optimization of the grafting ratio of CDs, since a variety of PEI modifications with CD has shown a drop in transfection efficiency, which might be due to lowering of the proton sponge effect and endosomal escape [173]. Also, β-CD functionalized with carboxymethyl has been combined with quarternized chitosan and achieved 40% transfection efficacy for Math1 expression in mammalian cochlear hair cells and lower toxicity than liposomes [174]. Lai recently reported a comprehensive review of different approaches in gene delivery using CDs [169].

Dextran is a branched, high molecular weight polysaccharide that has been modified for gene delivery. Polycations based on dextran-spermine have demonstrated activity in transfecting cells *in vitro* [175]. While intramuscular injection of dextran-spermine-pSV-LacZ complex in mice induced significant local gene expression, IV injection of the same complexes did not produce any expression, which was improved after partial PEGylation of the complexes, mostly in liver and due to binding to the galactose receptors of parenchymal cells in liver (and not the mannose receptors of nonparenchymal cells) [176]. Diethylaminoethyl (DEAE)-functionalized dextran has also been studied for gene delivery to macrophages; however, its use is usually limited to *in vitro* experiments due to toxicity and low transfection efficiency *in vivo* [177]. Galactosylated-chitosan has been grafted with dextran (to increase the stability of the complex), and plasmid DNA-containing complexes were efficient in transfection of Chang liver cells [178]. PEI-graft-dextran has been studied to increase the biocompatibility of the gene delivery system [179].

Human serum albumin is the most well-studied plasma protein and an attractive carrier due to its availability in pure form, in addition to biocompatibility and low immunogenicity, which is common in natural polymeric delivery systems. Cationized albumin could serve as a potential non-viral vector system for gene delivery [180]. It has been shown that cationized albumin could form complexes with plasmid DNA through spontaneous self-assembly, which could induce transfection *in vitro* with help of chloroquine for endosomal escape [181]. PEI-albumin conjugates have also been used for genetic immunization, and intravenous injection of the complexes resulted in both pulmonary mucosal and systemic immune responses [182]. On the other hand, many studies have evaluated albumin-coated microbubble destruction gene delivery, which is based on the cavitation of microbubbles induced by ultrasound application [183]. In 2002, Taniyama et al. showed effective transfection of wild-type p53 plasmid DNA to endothelial and vascular smooth muscle cells with albumin-coated microbubbles and ultrasound *in vivo*, which resulted in significant inhibition of the ratio of neointimal-to-medial area [184].

Nucleic acids can be loaded onto gelatin nanoparticles via physical encapsulation, electrostatic attraction, or complexation with surface modifying groups [185]. Type B gelatin is negatively charged at neutral pH of 7.0; however, it can physically encapsulate nucleic acids without electrostatically condensing DNA. The physically encapsulated plasmid DNA is trapped in a hydrogel-type matrix and is protected in the systemic circulation and after cellular internalization [186]. Another strategy is to modify the surface of gelatin particles with a quaternary amine such as cholamine, which increases ionic interaction and absorbs the DNA to gelatin particle [187]. Oral gene delivery using type B gelatin has been studied for several years, and LacZ plasmid transfection in the tibialis anterior muscle of 6-week-old BALB/c mice was demonstrated by Leong et al. after intramuscular injection [188]. Bhavsar et al. designed DNA-loaded type B gelatin nanoparticles in poly (ε-caprolactone) (PCL) microspheres using a "double emulsion-like" technique they called "nanoparticles-in-microsphere oral system." When PCL matrix is degraded by lipases in the intestinal tract after oral administration, the released DNA-containing gelatin nanoparticles can be internalized by GI lumen for transfection [189–191]. Collagen, gelatin, fibroin, fibrin, elastin, and alginate have also been used in creating scaffolds for gene delivery in tissue regeneration [56].

3.4 Conclusion

Non-viral delivery systems have improved significantly in the past few decades in delivering nucleic acids to the target cells. However, they are still inferior to viral vectors in terms of permanent and consistent gene expression at therapeutically relevant concentrations. They do, however, offer distinct advantages over viral vectors, as well as other non-viral gene transferring vehicles, including biodegradability, biocompatibility, and low immunogenicity. Extended studies have been devoted to taking advantage of this safe profile, while minimizing the shortcomings of natural polymers, and the accomplishments in the last few years suggest the feasibility of this approach, and the long way ahead to achieve a safe and efficient gene delivery system based on natural polymers.

References

[1] Kay MA. State-of-the-art gene-based therapies: the road ahead. Nat Rev Genet 2011; 12(5):316–28.

[2] O'Connor TP, Crystal RG. Genetic medicines: treatment strategies for hereditary disorders. Nat Rev Genet 2006;7(4):261–76.

[3] Smith L, Byers JF. Gene therapy in the post-Gelsinger era. JONAS Healthc Law Ethics Regul 2002;4(4):104–10.

[4] Branca MA. Gene therapy: cursed or inching towards credibility? Nat Biotechnol 2005;23(5):519–21.

[5] Kimmelman J. The ethics of human gene transfer. Nat Rev Genet 2008;9(3):239–44.

[6] Thomas CE, Ehrhardt A, Kay MA. Progress and problems with the use of viral vectors for gene therapy. Nat Rev Genet 2003;4(5):346–58.

[7] Kennedy S, et al. Experiments in transgenic mice show that hepatocytes are the source for postnatal liver growth and do not stream. Hepatology 1995;22(1):160–8.

[8] Urnov FD, et al. Genome editing with engineered zinc finger nucleases. Nat Rev Genet 2010;11(9):636–46.

[9] Napoli C, Lemieux C, Jorgensen R. Introduction of a chimeric chalcone synthase gene into petunia results in reversible co-suppression of homologous genes in trans. Plant Cell 1990;2(4):279–89.

[10] Raemdonck K, et al. Maintaining the silence: reflections on long-term RNAi. Drug Discov Today 2008;13(21-22):917–31.

[11] Guo H, et al. Mammalian microRNAs predominantly act to decrease target mRNA levels. Nature 2010;466(7308):835–40.

[12] Zhang Y, Wang Z, Gemeinhart RA. Progress in microRNA delivery. J Control Release 2013;172(3):962–74.

[13] Sliva K, Schnierle BS. Selective gene silencing by viral delivery of short hairpin RNA. Virol J 2010;7:248.

[14] Aliabadi HM, et al. Supramolecular assemblies in functional siRNA delivery: where do we stand? Biomaterials 2012;33(8):2546–69.

[15] Burrell RA, et al. The causes and consequences of genetic heterogeneity in cancer evolution. Nature 2013;501(7467):338–45.

[16] Meacham CE, Morrison SJ. Tumour heterogeneity and cancer cell plasticity. Nature 2013;501(7467):328–37.

[17] McCormick F. Cancer gene therapy: fringe or cutting edge? Nat Rev Cancer 2001; 1(2):130–41.

[18] Holt JT, et al. Growth retardation and tumour inhibition by BRCA1. Nat Genet 1996;12(3):298–302.

[19] Shao N, et al. Induction of apoptosis by the tumor suppressor protein BRCA1. Oncogene 1996;13(1):1–7.

[20] Tait DL, et al. A phase I trial of retroviral BRCA1sv gene therapy in ovarian cancer. Clin Cancer Res 1997;3(11):1959–68.

[21] Vogelstein B, Lane D, Levine AJ. Surfing the p53 network. Nature 2000; 408(6810):307–10.

[22] Kaplan KB, et al. A role for the Adenomatous Polyposis Coli protein in chromosome segregation. Nat Cell Biol 2001;3(4):429–32.

[23] Fodde R, et al. Mutations in the APC tumour suppressor gene cause chromosomal instability. Nat Cell Biol 2001;3(4):433–8.

[24] Nishizaki M, et al. Recombinant adenovirus expressing wild-type p53 is antiangiogenic: a proposed mechanism for bystander effect. Clin Cancer Res 1999;5(5):1015–23.

[25] Bouvet M, et al. Adenovirus-mediated wild-type p53 gene transfer down-regulates vascular endothelial growth factor expression and inhibits angiogenesis in human colon cancer. Cancer Res 1998;58(11):2288–92.

[26] Dameron KM, et al. The p53 tumor suppressor gene inhibits angiogenesis by stimulating the production of thrombospondin. Cold Spring Harb Symp Quant Biol 1994;59:483–9.

[27] Harris MP, et al. Adenovirus-mediated p53 gene transfer inhibits growth of human tumor cells expressing mutant p53 protein. Cancer Gene Ther 1996;3(2):121–30.

[28] Tanaka M, et al. MMAC1/PTEN inhibits cell growth and induces chemosensitivity to doxorubicin in human bladder cancer cells. Oncogene 2000;19(47):5406–12.

[29] Minaguchi T, et al. Growth suppression of human ovarian cancer cells by adenovirus-mediated transfer of the PTEN gene. Cancer Res 1999;59(24):6063–7.

[30] Sakurada A, et al. Adenovirus-mediated delivery of the PTEN gene inhibits cell growth by induction of apoptosis in endometrial cancer. Int J Oncol 1999;15(6):1069–74.

[31] Burrows FJ, et al. Purified herpes simplex virus thymidine kinase retroviral particles: III. Characterization of bystander killing mechanisms in transfected tumor cells. Cancer Gene Ther 2002;9(1):87–95.

[32] Moolten FL. Tumor chemosensitivity conferred by inserted herpes thymidine kinase genes: paradigm for a prospective cancer control strategy. Cancer Res 1986;46(10):5276–81.
[33] Huber BE, et al. Metabolism of 5-fluorocytosine to 5-fluorouracil in human colorectal tumor cells transduced with the cytosine deaminase gene: significant antitumor effects when only a small percentage of tumor cells express cytosine deaminase. Proc Natl Acad Sci U S A 1994;91(17):8302–6.
[34] Trinh QT, et al. Enzyme/prodrug gene therapy: comparison of cytosine deaminase/5-fluorocytosine versus thymidine kinase/ganciclovir enzyme/prodrug systems in a human colorectal carcinoma cell line. Cancer Res 1995;55(21):4808–12.
[35] Hanna NN, et al. Virally directed cytosine deaminase/5-fluorocytosine gene therapy enhances radiation response in human cancer xenografts. Cancer Res 1997;57(19):4205–9.
[36] Green NK, et al. Sensitization of colorectal and pancreatic cancer cell lines to the prodrug 5-(aziridin-1-yl)-2,4-dinitrobenzamide (CB1954) by retroviral transduction and expression of the E. coli nitroreductase gene. Cancer Gene Ther 1997;4(4):229–38.
[37] McNeish IA, et al. Virus directed enzyme prodrug therapy for ovarian and pancreatic cancer using retrovirally delivered E. coli nitroreductase and CB1954. Gene Ther 1998;5(8):1061–9.
[38] Chung-Faye G, et al. Virus-directed, enzyme prodrug therapy with nitroimidazole reductase: a phase I and pharmacokinetic study of its prodrug, CB1954. Clin Cancer Res 2001;7(9):2662–8.
[39] Kerr D. Clinical development of gene therapy for colorectal cancer. Nat Rev Cancer 2003;3(8):615–22.
[40] Tsang KY, et al. Generation of human cytotoxic T cells specific for human carcinoembryonic antigen epitopes from patients immunized with recombinant vaccinia-CEA vaccine. J Natl Cancer Inst 1995;87(13):982–90.
[41] Zhu MZ, et al. Specific cytolytic T-cell responses to human CEA from patients immunized with recombinant avipox-CEA vaccine. Clin Cancer Res 2000;6(1):24–33.
[42] Horig H, et al. Phase I clinical trial of a recombinant canarypoxvirus (ALVAC) vaccine expressing human carcinoembryonic antigen and the B7.1 co-stimulatory molecule. Cancer Immunol Immunother 2000;49(9):504–14.
[43] Marshall JL, et al. Phase I study in cancer patients of a replication-defective avipox recombinant vaccine that expresses human carcinoembryonic antigen. J Clin Oncol 1999;17(1):332–7.
[44] Sobol RE, et al. Interleukin 2 gene therapy of colorectal carcinoma with autologous irradiated tumor cells and genetically engineered fibroblasts: a Phase I study. Clin Cancer Res 1999;5(9):2359–65.
[45] Schmidt-Wolf IG, et al. Phase I clinical study applying autologous immunological effector cells transfected with the interleukin-2 gene in patients with metastatic renal cancer, colorectal cancer and lymphoma. Br J Cancer 1999;81(6):1009–16.
[46] Nabel GJ, et al. Direct gene transfer with DNA-liposome complexes in melanoma: expression, biologic activity, and lack of toxicity in humans. Proc Natl Acad Sci U S A 1993;90(23):11307–11.
[47] Rubin J, et al. Phase I study of immunotherapy of hepatic metastases of colorectal carcinoma by direct gene transfer of an allogeneic histocompatibility antigen, HLA-B7. Gene Ther 1997;4(5):419–25.
[48] Sorrentino BP. Gene therapy to protect haematopoietic cells from cytotoxic cancer drugs. Nat Rev Cancer 2002;2(6):431–41.

[49] Mickisch GH, et al. Chemotherapy and chemosensitization of transgenic mice which express the human multidrug resistance gene in bone marrow: efficacy, potency, and toxicity. Cancer Res 1991;51(19):5417–24.
[50] Johnstone RW, Cretney E, Smyth MJ. P-glycoprotein protects leukemia cells against caspase-dependent, but not caspase-independent, cell death. Blood 1999;93(3):1075–85.
[51] Smyth MJ, et al. The drug efflux protein, P-glycoprotein, additionally protects drug-resistant tumor cells from multiple forms of caspase-dependent apoptosis. Proc Natl Acad Sci U S A 1998;95(12):7024–9.
[52] Blakley RL, Sorrentino BP. In vitro mutations in dihydrofolate reductase that confer resistance to methotrexate: potential for clinical application. Hum Mutat 1998;11(4):259–63.
[53] Takebe N, et al. Generation of dual resistance to 4-hydroperoxycyclophosphamide and methotrexate by retroviral transfer of the human aldehyde dehydrogenase class 1 gene and a mutated dihydrofolate reductase gene. Mol Ther 2001;3(1):88–96.
[54] Reese JS, et al. Retroviral transduction of a mutant methylguanine DNA methyltransferase gene into human CD34 cells confers resistance to O6-benzylguanine plus 1,3-bis(2-chloroethyl)-1-nitrosourea. Proc Natl Acad Sci U S A 1996;93(24):14088–93.
[55] Uchida K, et al. Gene therapy strategies for the treatment of spinal cord injury. Ther Deliv 2014;5(5):591–607.
[56] Garg T, et al. Scaffold: a novel carrier for cell and drug delivery. Crit Rev Ther Drug Carrier Syst 2012;29(1):1–63.
[57] Luo J, et al. Gene therapy for bone regeneration. Curr Gene Ther 2005;5(2):167–79.
[58] Wegman F, et al. Non-viral gene therapy for bone tissue engineering. Biotechnol Genet Eng Rev 2013;29:206–20.
[59] Reddi AH. Bone morphogenetic proteins: from basic science to clinical applications. J Bone Joint Surg Am 2001;83-A(Suppl. 1(Pt 1)):S1–6.
[60] Nakajima H, et al. Targeted retrograde gene delivery of brain-derived neurotrophic factor suppresses apoptosis of neurons and oligodendroglia after spinal cord injury in rats. Spine (Phila Pa 1976) 2010;35(5):497–504.
[61] Koelsch A, et al. Transgene-mediated GDNF expression enhances synaptic connectivity and GABA transmission to improve functional outcome after spinal cord contusion. J Neurochem 2010;113(1):143–52.
[62] Zhou Z, et al. IL-10 promotes neuronal survival following spinal cord injury. Exp Neurol 2009;220(1):183–90.
[63] Lv B, et al. Lentivirus-siNgR199 promotes axonal regeneration and functional recovery in rats. Int J Neurosci 2012;122(3):133–9.
[64] Ando T, et al. Photomechanical wave-driven delivery of siRNAs targeting intermediate filament proteins promotes functional recovery after spinal cord injury in rats. PLoS One 2012;7(12):e51744.
[65] Anderson WF. Human gene therapy. Nature 1998;392(6679 Suppl.):25–30.
[66] Naldini L. Ex vivo gene transfer and correction for cell-based therapies. Nat Rev Genet 2011;12(5):301–15.
[67] Copelan EA. Hematopoietic stem-cell transplantation. N Engl J Med 2006; 354(17):1813–26.
[68] Mazurier F, et al. Lentivector-mediated clonal tracking reveals intrinsic heterogeneity in the human hematopoietic stem cell compartment and culture-induced stem cell impairment. Blood 2004;103(2):545–52.
[69] Sauce D, et al. Retrovirus-mediated gene transfer in primary T lymphocytes impairs their anti-Epstein-Barr virus potential through both culture-dependent and selection process-dependent mechanisms. Blood 2002;99(4):1165–73.

[70] Murray LJ, et al. Thrombopoietin, flt3, and kit ligands together suppress apoptosis of human mobilized CD34+ cells and recruit primitive CD34+ Thy-1+ cells into rapid division. Exp Hematol 1999;27(6):1019–28.

[71] Gattinoni L, et al. Removal of homeostatic cytokine sinks by lymphodepletion enhances the efficacy of adoptively transferred tumor-specific CD8+ T cells. J Exp Med 2005;202(7):907–12.

[72] Davidson BL, McCray Jr. PB. Current prospects for RNA interference-based therapies. Nat Rev Genet 2011;12(5):329–40.

[73] Edelstein ML, et al. Gene therapy clinical trials worldwide 1989-2004-an overview. J Gene Med 2004;6(6):597–602.

[74] Burton EA, Fink DJ, Glorioso JC. Replication-defective genomic HSV gene therapy vectors: design, production and CNS applications. Curr Opin Mol Ther 2005;7(4):326–36.

[75] Sadeghi H, Hitt MM. Transcriptionally targeted adenovirus vectors. Curr Gene Ther 2005;5(4):411–27.

[76] Barker SD, et al. Combined transcriptional and transductional targeting improves the specificity and efficacy of adenoviral gene delivery to ovarian carcinoma. Gene Ther 2003;10(14):1198–204.

[77] Waehler R, Russell SJ, Curiel DT. Engineering targeted viral vectors for gene therapy. Nat Rev Genet 2007;8(8):573–87.

[78] Mercier GT, et al. A chimeric adenovirus vector encoding reovirus attachment protein sigma1 targets cells expressing junctional adhesion molecule 1. Proc Natl Acad Sci U S A 2004;101(16):6188–93.

[79] DiPaolo N, et al. Evaluation of adenovirus vectors containing serotype 35 fibers for vaccination. Mol Ther 2006;13(4):756–65.

[80] Croyle MA, et al. PEGylation of a vesicular stomatitis virus G pseudotyped lentivirus vector prevents inactivation in serum. J Virol 2004;78(2):912–21.

[81] Pereboev AV, et al. Enhanced gene transfer to mouse dendritic cells using adenoviral vectors coated with a novel adapter molecule. Mol Ther 2004;9(5):712–20.

[82] Bartlett JS, et al. Targeted adeno-associated virus vector transduction of nonpermissive cells mediated by a bispecific F(ab'gamma)2 antibody. Nat Biotechnol 1999;17(2):181–6.

[83] Wurdinger T, et al. Targeting non-human coronaviruses to human cancer cells using a bispecific single-chain antibody. Gene Ther 2005;12(18):1394–404.

[84] Kreppel F, et al. Combined genetic and chemical capsid modifications enable flexible and efficient de- and retargeting of adenovirus vectors. Mol Ther 2005;12(1):107–17.

[85] Stone D, et al. Development and assessment of human adenovirus type 11 as a gene transfer vector. J Virol 2005;79(8):5090–104.

[86] Stachler MD, Bartlett JS. Mosaic vectors comprised of modified AAV1 capsid proteins for efficient vector purification and targeting to vascular endothelial cells. Gene Ther 2006;13(11):926–31.

[87] Kotterman MA, Schaffer DV. Engineering adeno-associated viruses for clinical gene therapy. Nat Rev Genet 2014;15(7):445–51.

[88] Mingozzi F, High KA. Therapeutic in vivo gene transfer for genetic disease using AAV: progress and challenges. Nat Rev Genet 2011;12(5):341–55.

[89] Assessment of adenoviral vector safety and toxicity: report of the National Institutes of Health Recombinant DNA Advisory Committee. Hum Gene Ther 2002;13(1):3–13.

[90] Schroder AR, et al. HIV-1 integration in the human genome favors active genes and local hotspots. Cell 2002;110(4):521–9.

[91] Woods NB, et al. Lentiviral vector transduction of NOD/SCID repopulating cells results in multiple vector integrations per transduced cell: risk of insertional mutagenesis. Blood 2003;101(4):1284–9.
[92] Wu X, et al. Transcription start regions in the human genome are favored targets for MLV integration. Science 2003;300(5626):1749–51.
[93] Glover DJ, Lipps HJ, Jans DA. Towards safe, non-viral therapeutic gene expression in humans. Nat Rev Genet 2005;6(4):299–310.
[94] Magin-Lachmann C, et al. In vitro and in vivo delivery of intact BAC DNA – comparison of different methods. J Gene Med 2004;6(2):195–209.
[95] Andre F, Mir LM. DNA electrotransfer: its principles and an updated review of its therapeutic applications. Gene Ther 2004;11(Suppl. 1):S33–42.
[96] Udvardi A, et al. Uptake of exogenous DNA via the skin. J Mol Med (Berl) 1999;77(10):744–50.
[97] Raju PA, et al. Assessment of epidermal cell viability by near infrared multi-photon microscopy following ballistic delivery of gold micro-particles. Vaccine 2006;24(21):4644–7.
[98] Olivares EC, et al. Site-specific genomic integration produces therapeutic Factor IX levels in mice. Nat Biotechnol 2002;20(11):1124–8.
[99] Zhang G, et al. Efficient expression of naked dna delivered intraarterially to limb muscles of nonhuman primates. Hum Gene Ther 2001;12(4):427–38.
[100] Telenius H, et al. Stability of a functional murine satellite DNA-based artificial chromosome across mammalian species. Chromosome Res 1999;7(1):3–7.
[101] Fraley R, et al. Introduction of liposome-encapsulated SV40 DNA into cells. J Biol Chem 1980;255(21):10431–5.
[102] Johnson-Saliba M, Jans DA. Gene therapy: optimising DNA delivery to the nucleus. Curr Drug Targets 2001;2(4):371–99.
[103] Kim SW. Polylysine copolymers for gene delivery. Cold Spring Harb Protoc 2012;2012(4):433–8.
[104] Bazile D, et al. Stealth Me.PEG-PLA nanoparticles avoid uptake by the mononuclear phagocytes system. J Pharm Sci 1995;84(4):493–8.
[105] Lungwitz U, et al. Polyethylenimine-based non-viral gene delivery systems. Eur J Pharm Biopharm 2005;60(2):247–66.
[106] Boussif O, et al. A versatile vector for gene and oligonucleotide transfer into cells in culture and in vivo: polyethylenimine. Proc Natl Acad Sci U S A 1995;92(16):7297–301.
[107] Discher DE, Ahmed F. Polymersomes. Annu Rev Biomed Eng 2006;8:323–41.
[108] Martin ME, Rice KG. Peptide-guided gene delivery. AAPS J 2007;9(1):E18–29.
[109] Sokolova V, Epple M. Inorganic nanoparticles as carriers of nucleic acids into cells. Angew Chem Int Ed Engl 2008;47(8):1382–95.
[110] Dang JM, Leong KW. Natural polymers for gene delivery and tissue engineering. Adv Drug Deliv Rev 2006;58(4):487–99.
[111] Haupenthal J, et al. Inhibition of RNAse A family enzymes prevents degradation and loss of silencing activity of siRNAs in serum. Biochem Pharmacol 2006;71(5):702–10.
[112] Prince WS, et al. Pharmacodynamics of recombinant human DNase I in serum. Clin Exp Immunol 1998;113(2):289–96.
[113] Kawabata K, Takakura Y, Hashida M. The fate of plasmid DNA after intravenous injection in mice: involvement of scavenger receptors in its hepatic uptake. Pharm Res 1995;12(6):825–30.
[114] Nishikawa M, Takakura Y, Hashida M. Theoretical considerations involving the pharmacokinetics of plasmid DNA. Adv Drug Deliv Rev 2005;57(5):675–88.

[115] Dykxhoorn DM, Palliser D, Lieberman J. The silent treatment: siRNAs as small molecule drugs. Gene Ther 2006;13(6):541–52.
[116] Bonadio J. Tissue engineering via local gene delivery: update and future prospects for enhancing the technology. Adv Drug Deliv Rev 2000;44(2-3):185–94.
[117] Prabaharan M. Review paper: chitosan derivatives as promising materials for controlled drug delivery. J Biomater Appl 2008;23(1):5–36.
[118] Mansouri S, et al. Chitosan-DNA nanoparticles as non-viral vectors in gene therapy: strategies to improve transfection efficacy. Eur J Pharm Biopharm 2004;57(1):1–8.
[119] MacLaughlin FC, et al. Chitosan and depolymerized chitosan oligomers as condensing carriers for in vivo plasmid delivery. J Control Release 1998;56(1-3):259–72.
[120] Thanou M, et al. Quaternized chitosan oligomers as novel gene delivery vectors in epithelial cell lines. Biomaterials 2002;23(1):153–9.
[121] Saranya N, et al. Chitosan and its derivatives for gene delivery. Int J Biol Macromol 2011;48(2):234–8.
[122] Corsi K, et al. Mesenchymal stem cells, MG63 and HEK293 transfection using chitosan-DNA nanoparticles. Biomaterials 2003;24(7):1255–64.
[123] Lee MK, et al. The use of chitosan as a condensing agent to enhance emulsion-mediated gene transfer. Biomaterials 2005;26(14):2147–56.
[124] Dastan T, Turan K. In vitro characterization and delivery of chitosan-DNA microparticles into mammalian cells. J Pharm Pharm Sci 2004;7(2):205–14.
[125] Zhao X, et al. Transfection of primary chondrocytes using chitosan-pEGFP nanoparticles. J Control Release 2006;112(2):223–8.
[126] Huang M, et al. Transfection efficiency of chitosan vectors: effect of polymer molecular weight and degree of deacetylation. J Control Release 2005;106(3):391–406.
[127] Ishii T, Okahata Y, Sato T. Mechanism of cell transfection with plasmid/chitosan complexes. Biochim Biophys Acta 2001;1514(1):51–64.
[128] Lavertu M, et al. High efficiency gene transfer using chitosan/DNA nanoparticles with specific combinations of molecular weight and degree of deacetylation. Biomaterials 2006;27(27):4815–24.
[129] Strand SP, et al. Molecular design of chitosan gene delivery systems with an optimized balance between polyplex stability and polyplex unpacking. Biomaterials 2010;31(5):975–87.
[130] Koping-Hoggard M, et al. Relationship between the physical shape and the efficiency of oligomeric chitosan as a gene delivery system in vitro and in vivo. J Gene Med 2003;5(2):130–41.
[131] Carrillo C, et al. Chitosan nanoparticles as non-viral gene delivery systems: determination of loading efficiency. Biomed Pharmacother 2014;68(6):775–83.
[132] Centelles MN, et al. Influence of the chitosan nature on the transfection efficacy of DNA-loaded nanoparticles after hydrodynamic administration in mice. J Microencapsul 2010;27(5):460–9.
[133] Nimesh S, et al. Enhanced gene delivery mediated by low molecular weight chitosan/DNA complexes: effect of pH and serum. Mol Biotechnol 2010;46(2):182–96.
[134] Raftery R, O'Brien FJ, Cryan SA. Chitosan for gene delivery and orthopedic tissue engineering applications. Molecules 2013;18(5):5611–47.
[135] Ozbas-Turan S, Akbuga J, Sezer AD. Topical application of antisense oligonucleotide-loaded chitosan nanoparticles to rats. Oligonucleotides 2010;20(3):147–53.
[136] Plapied L, et al. Bioadhesive nanoparticles of fungal chitosan for oral DNA delivery. Int J Pharm 2010;398(1–2):210–8.
[137] Bowman K, et al. Gene transfer to hemophilia A mice via oral delivery of FVIII-chitosan nanoparticles. J Control Release 2008;132(3):252–9.

[138] Kumar M, et al. Chitosan IFN-gamma-pDNA Nanoparticle (CIN) Therapy for Allergic Asthma. Genet Vaccines Ther 2003;1(1):3.

[139] Yang PT, et al. In utero gene delivery using chitosan-DNA nanoparticles in mice. J Surg Res 2011;171(2):691–9.

[140] Ozbas-Turan S, et al. Co-encapsulation of two plasmids in chitosan microspheres as a non-viral gene delivery vehicle. J Pharm Pharm Sci 2003;6(1):27–32.

[141] Yang C, et al. Chitosan/siRNA nanoparticles targeting cyclooxygenase type 2 attenuate unilateral ureteral obstruction-induced kidney injury in mice. Theranostics 2015;5(2):110–23.

[142] Han HD, et al. Chitosan hydrogel for localized gene silencing. Cancer Biol Ther 2011;11(9):839–45.

[143] Oliveira JM, et al. Novel hydroxyapatite/chitosan bilayered scaffold for osteochondral tissue-engineering applications: scaffold design and its performance when seeded with goat bone marrow stromal cells. Biomaterials 2006;27(36):6123–37.

[144] Peng L, et al. Novel gene-activated matrix with embedded chitosan/plasmid DNA nanoparticles encoding PDGF for periodontal tissue engineering. J Biomed Mater Res A 2009;90(2):564–76.

[145] Monaghan M, Pandit A. RNA interference therapy via functionalized scaffolds. Adv Drug Deliv Rev 2011;63(4-5):197–208.

[146] Dehousse V, et al. Comparison of chitosan/siRNA and trimethylchitosan/siRNA complexes behaviour in vitro. Int J Biol Macromol 2010;46(3):342–9.

[147] Zhao X, et al. Thiolated trimethyl chitosan nanocomplexes as gene carriers with high in vitro and in vivo transfection efficiency. J Control Release 2010;144(1):46–54.

[148] Morris VB, Sharma CP. Folate mediated histidine derivative of quaternised chitosan as a gene delivery vector. Int J Pharm 2010;389(1–2):176–85.

[149] Wang J, et al. Reversion of multidrug resistance by tumor targeted delivery of antisense oligodeoxynucleotides in hydroxypropyl-chitosan nanoparticles. Biomaterials 2010;31(15):4426–33.

[150] Kim TH, et al. Mannosylated chitosan nanoparticle-based cytokine gene therapy suppressed cancer growth in BALB/c mice bearing CT-26 carcinoma cells. Mol Cancer Ther 2006;5(7):1723–32.

[151] Kievit FM, et al. PEI-PEG-chitosan copolymer coated iron oxide nanoparticles for safe gene delivery: synthesis, complexation, and transfection. Adv Funct Mater 2009;19(14):2244–51.

[152] Oliveira H, et al. Chitosan-based gene delivery vectors targeted to the peripheral nervous system. J Biomed Mater Res A 2010;95(3):801–10.

[153] Jiang HL, et al. Galactosylated chitosan-g-PEI/DNA complexes-loaded poly(organophosphazene) hydrogel as a hepatocyte targeting gene delivery system. Arch Pharm Res 2010;33(4):551–6.

[154] Zhu AP, et al. Interaction between O-carboxymethylchitosan and dipalmitoyl-sn-glycero-3-phosphocholine bilayer. Biomaterials 2005;26(34):6873–9.

[155] Ravina M, et al. Hyaluronic acid/chitosan-g-poly(ethylene glycol) nanoparticles for gene therapy: an application for pDNA and siRNA delivery. Pharm Res 2010;27(12):2544–55.

[156] Ghosn B, et al. Efficient gene silencing in lungs and liver using imidazole-modified chitosan as a nanocarrier for small interfering RNA. Oligonucleotides 2010;20(3):163–72.

[157] Gao Y, et al. Arginine-chitosan/DNA self-assemble nanoparticles for gene delivery: in vitro characteristics and transfection efficiency. Int J Pharm 2008;359(1–2):241–6.

[158] Rojanarata T, et al. Methylated N-(4-N, N-dimethylaminobenzyl) chitosan for novel effective gene carriers. Eur J Pharm Biopharm 2008;70(1):207–14.

[159] Lu H, et al. Porous chitosan scaffolds with embedded hyaluronic acid/chitosan/plasmid-DNA nanoparticles encoding TGF-beta1 induce DNA controlled release, transfected chondrocytes, and promoted cell proliferation. PLoS One 2013;8(7):e69950.

[160] Eliyahu H, Barenholz Y, Domb AJ. Polymers for DNA delivery. Molecules 2005;10(1):34–64.

[161] Merdan T, et al. Intracellular processing of poly(ethylene imine)/ribozyme complexes can be observed in living cells by using confocal laser scanning microscopy and inhibitor experiments. Pharm Res 2002;19(2):140–6.

[162] Pouton CW, et al. Polycation-DNA complexes for gene delivery: a comparison of the biopharmaceutical properties of cationic polypeptides and cationic lipids. J Control Release 1998;53(1–3):289–99.

[163] Benns JM, et al. pH-sensitive cationic polymer gene delivery vehicle: N-Ac-poly (L-histidine)-graft-poly(L-lysine) comb shaped polymer. Bioconjug Chem 2000 ;11(5):637–45.

[164] Midoux P, Monsigny M. Efficient gene transfer by histidylated polylysine/pDNA complexes. Bioconjug Chem 1999;10(3):406–11.

[165] Kogure K, et al. Development of a non-viral multifunctional envelope-type nano device by a novel lipid film hydration method. J Control Release 2004;98(2):317–23.

[166] Toncheva V, et al. Novel vectors for gene delivery formed by self-assembly of DNA with poly(L-lysine) grafted with hydrophilic polymers. Biochim Biophys Acta 1998;1380(3):354–68.

[167] Wolfert MA, et al. Characterization of vectors for gene therapy formed by self-assembly of DNA with synthetic block co-polymers. Hum Gene Ther 1996;7(17):2123–33.

[168] Tiera MJ, et al. Polycation-based gene therapy: current knowledge and new perspectives. Curr Gene Ther 2011;11(4):288–306.

[169] Lai WF. Cyclodextrins in non-viral gene delivery. Biomaterials 2014;35(1):401–11.

[170] Kettel MJ, et al. Nanogels with high active beta-cyclodextrin content as physical coating system with sustained release properties. ACS Appl Mater Interfaces 2014;6(4):2300–11.

[171] Cryan SA, et al. Cell transfection with polycationic cyclodextrin vectors. Eur J Pharm Sci 2004;21(5):625–33.

[172] Forrest ML, Gabrielson N, Pack DW. Cyclodextrin-polyethylenimine conjugates for targeted in vitro gene delivery. Biotechnol Bioeng 2005;89(4):416–23.

[173] Pun SH, et al. Cyclodextrin-modified polyethylenimine polymers for gene delivery. Bioconjug Chem 2004;15(4):831–40.

[174] Ren LL, et al. Math1 gene transfer based on the delivery system of quaternized chitosan/Na-carboxymethyl-beta-cyclodextrin nanoparticles. J Nanosci Nanotechnol 2010;10(11):7262–5.

[175] Azzam T, et al. Polysaccharide-oligoamine based conjugates for gene delivery. J Med Chem 2002;45(9):1817–24.

[176] Hosseinkhani H, et al. Dextran-spermine polycation: an efficient nonviral vector for in vitro and in vivo gene transfection. Gene Ther 2004;11(2):194–203.

[177] Mack KD, et al. A novel method for DEAE-dextran mediated transfection of adherent primary cultured human macrophages. J Immunol Methods 1998;211(1–2):79–86.

[178] Park IK, et al. Galactosylated chitosan-graft-dextran as hepatocyte-targeting DNA carrier. J Control Release 2000;69(1):97–108.

[179] Sun YX, et al. Synthesis of (Dex-HMDI)-g-PEIs as effective and low cytotoxic nonviral gene vectors. J Control Release 2008;128(2):171–8.

[180] Chuang VT, Kragh-Hansen U, Otagiri M. Pharmaceutical strategies utilizing recombinant human serum albumin. Pharm Res 2002;19(5):569–77.

[181] Fischer D, et al. Cationized human serum albumin as a non-viral vector system for gene delivery? Characterization of complex formation with plasmid DNA and transfection efficiency. Int J Pharm 2001;225(1–2):97–111.

[182] Orson FM, et al. Genetic immunization with lung-targeting macroaggregated polyethyleneimine-albumin conjugates elicits combined systemic and mucosal immune responses. J Immunol 2000;164(12):6313–21.

[183] Tsutsui JM, Xie F, Porter RT. The use of microbubbles to target drug delivery. Cardiovasc Ultrasound 2004;2:23.

[184] Taniyama Y, et al. Local delivery of plasmid DNA into rat carotid artery using ultrasound. Circulation 2002;105(10):1233–9.

[185] Elzoghby AO. Gelatin-based nanoparticles as drug and gene delivery systems: reviewing three decades of research. J Control Release 2013;172(3):1075–91.

[186] Magadala P, Amiji M. Epidermal growth factor receptor-targeted gelatin-based engineered nanocarriers for DNA delivery and transfection in human pancreatic cancer cells. AAPS J 2008;10(4):565–76.

[187] Zwiorek K, et al. Delivery by cationic gelatin nanoparticles strongly increases the immunostimulatory effects of CpG oligonucleotides. Pharm Res 2008;25(3):551–62.

[188] Leong KW, et al. DNA-polycation nanospheres as non-viral gene delivery vehicles. J Control Release 1998;53(1–3):183–93.

[189] Bhavsar MD, Amiji MM. Gastrointestinal distribution and in vivo gene transfection studies with nanoparticles-in-microsphere oral system (NiMOS). J Control Release 2007;119(3):339–48.

[190] Bhavsar MD, Amiji MM. Development of novel biodegradable polymeric nanoparticles-in-microsphere formulation for local plasmid DNA delivery in the gastrointestinal tract. AAPS PharmSciTech 2008;9(1):288–94.

[191] Bhavsar MD, Tiwari SB, Amiji MM. Formulation optimization for the nanoparticles-in-microsphere hybrid oral delivery system using factorial design. J Control Release 2006;110(2):422–30.

Synthetic cationic glycopolymers for gene delivery

4

M. Ahmed[1], *R. Narain*[2]
[1]California Institute of Technology, Pasadena, CA, USA; [2]University of Alberta, Edmonton, AB, Canada

4.1 Introduction

The development of non-toxic, efficient gene delivery vector holds great promise to cure a number of genetic and malignant disorders [1–3]. The cationic polymers are well known for their DNA condensation properties and are widely synthesized and explored for their gene delivery efficacies. The limited success of these earlier non-viral agents in research and clinics led to exploration of less toxic, targeted gene delivery vectors. The use of carbohydrates as gene delivery vector dates back to 1965, but the use of synthetic glycopolymers as gene delivery therapeutics is a relatively new approach [1]. The concept of glycopolymer synthesis and their usage for a variety of applications ranging from raw materials in industry to biomedicines was first introduced in the 1960s. The synthesis of glycopolymer-macromolecules with carbon backbone and pendant carbohydrate moieties was reported in 1961. Initially, vinyl-bearing carbohydrates were polymerized by free radical polymerization to yield glycopolymers. However, the earlier macromolecules produced were less well defined, involving tedious reaction procedures, complex protecting group chemistries, and low yields [2]. On the other hand, synthesis of amine-based polymers was extremely challenging, requiring complex chemistries and stringent reaction conditions [3–5]. Hence, it was not until the late 2000s that cationic cyclodextrin polymers carrying small interfering ribonucleic acid (siRNA) as delivery vectors were commercialized under the brand name of CALAA-01 and were first used in clinical trials for cancer treatment in humans [6]. The progress in the field of chemistry and polymer chemistry offered facile synthetic approaches with desirable yields to produce cationic glycopolymers of well-defined and desired morphologies, hence allowing their applications in medicines. The evolutionary steps in the development of cationic glycopolymers of desired molecular weights and architectures are described in detail elsewhere [1–3]. The focus of this chapter is to emphasize the breakthroughs made in the field of nucleic acid delivery using synthetic cationic glycopolymers.

4.2 Evaluation of glycosylated cationic polymers for gene delivery

The role of glycosylation in gene transfer efficacies has been a topic of interest during the last two decades. Glycosylation of cationic polymers was found to alter

Polymers and Nanomaterials for Gene Therapy. http://dx.doi.org/10.1016/B978-0-08-100520-0.00004-7
© 2016 Elsevier Ltd. All rights reserved.

its physiochemical and biological properties for their application in gene expression. Ring opening polymerization (ROP) is one of the widely used methods for the synthesis of cationic polymers and glycopolymers [3–5]. Most commonly used cationic polymers such as poly(ethyleneimine) (PEI) and poly L-lysine (PLL) are produced by ROP. Due to high gene expression of PEI, this vector now serves as gold standard to evaluate gene expression efficacies *in vitro*. However, poor control over molecular weights and resulting high polydispersities are a major drawback of this vector. Glycosylation of cationic polymers (PEI, PLL) was extensively studied to evaluate the gene delivery efficacies and targeting capabilities of cationic glycopolymers [1–3]. The structures of typical glycated cationic polymers used to study gene delivery profiles are shown. (Scheme 4.1)

4.2.1 Extent of glycosylation

In general, high glycosylation of cationic vectors was associated with their low gene expression, possibly due to a decrease in amine content and hence buffering capacity. An experimentally determined critical glycosylation (30–50% substitution of primary amines) of cationic vectors was shown to improve their gene expression in certain cell lines bearing lectins as surface receptors [7]. Non-enzymatic glycation of dendrimers and PEI by Maillard reaction is a straightforward approach to producing glycated transfection reagents. The degree of glycation of dendrimers was calculated by the consumption of glucose in the solution. The increase in glycation density of polymers was not related to the increase in their biological function (gene expression), possibly due to the change in their physiochemical properties (size and net charge, DNA binding, and release efficacy), or the change in ligand-receptor binding efficacies due to steric crowding. The experimentally determined critical amount of sugar on the surface of polymers was required to obtain optimum increases in gene expression profile. The glycated cationic polymers were stable under physiological conditions, and showed no change in gene expression [8]. Mannosylated-PAMAM G2 dendrimers were prepared and degree of mannosylation on gene expression efficacies was studied. As expected, for all synthetic polymers a critical experimentally determined cationic to sugar ratio is required for optimum *in vitro* targeting and gene expression. The cationic glycopolymers of known molecular weights, sugar content, and degree of polymerization were prepared by reversible addition fragmentation chain-transfer polymerization (RAFT). The copolymers of similar cationic content but high sugar residues also showed decreased gene expression [9]. Further increase in sugar content led to a decrease in gene expression due to several reasons such as, decreased DNA binding and protection efficacies, and aggregation of polyplexes on the target receptor, hence leading to slow or poor cellular uptake [7–9].

4.2.2 Glycosylation and intracellular routing of polyplexes

Others supported the role of glycosylation in intracellular trafficking and nuclear translocation [10,11]. For example, a detailed comparative study on lactosylated and mannosylated PLL for intracellular trafficking of polyplexes was performed in

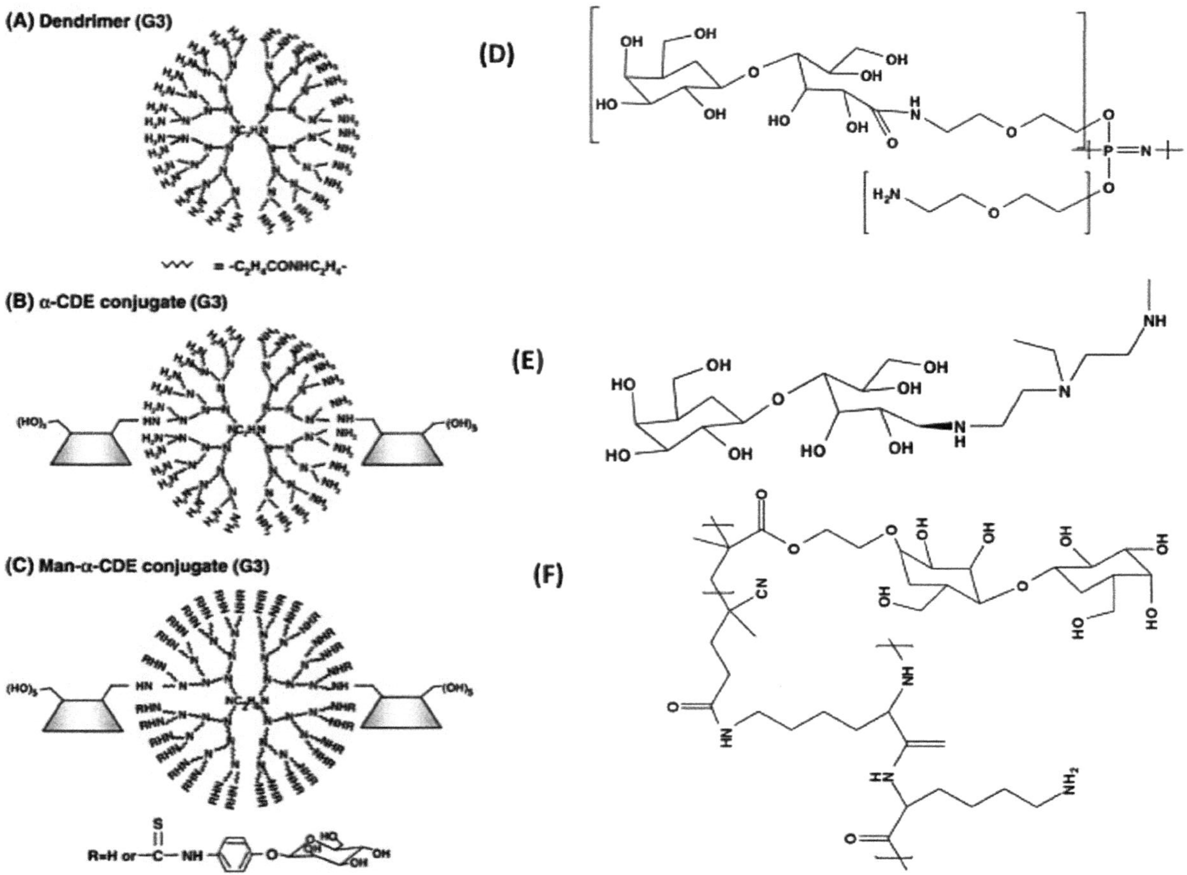

Scheme 4.1 Chemical structures of glycosylated cationic polymers. (a-c) glycated dendrimers [7], (d) lactosylated poly (2-(2-aminoethoxy)ethoxy) phosphazene, (e) lactosylated PEI, and (f) lactosylated PDMAEMA-g-PLL.

pulmonary epithelial cells. In comparison to lactosylated vectors, mannosylated vectors showed poor nuclear entry, high lysosomal accumulation, and poor DNA transcription efficacies [7]. The nuclear targeting capabilities of glycoplexes were analyzed. It was found that the poor entry of mannosylated-PLL in nucleus of human airway epithelial cells, as compared to lactosylated PLL, was due to their poor nuclear localization, despite the high cellular uptake of the former. This nuclear uptake of lactosylated PLL was inhibited by wheat germ agglutinin (WGA), which binds to *N*-acetyl glucosamine (NAG) residues of nuclear pore complex binding [12].

4.2.3 Ternary complexes formulations

PEI-DNA polyplexes coated with anionic glycopolymers were also formulated and were studied for their cellular uptake, gene expression, and toxicity profiles. The idea was to study the biological effect of degree of glycosylation, without compromising the amine content of PEI. The pH-responsive anionic glycopolymers were prepared by the RAFT polymerization approach and were studied for their endosomal escape capabilities. The ternary complexes of PEI prepared using linear and hyperbranched glycosylated anionic polymers showed specific cell-targeting efficacies *in vitro*. The linear anionic polymer-based ternary complexes showed efficient targeting in human embryonic kidney cells, while hyperbranched anionic glycopolymers based analogs showed high gene expression in HepG2 cells [13].

4.2.4 Limitations of glycoslation of cationic systems in vivo

The intracellular localization and DNA delivery efficiencies of lactosylated-PEI in mouse lung post-nasal instillation were studied. Despite the polyplexes localization in lung alveolar cells, the gene expression was significantly low *in vivo*. The poor gene expression was associated with poor DNA protection ability and intracellular trafficking of polyplexes *in vivo* [14,15]. However, the reports limited to one type of polymer construct need to be evaluated with caution and should not be over-generalized. Moreover, the physical and physiochemical properties of cationic polymer conjugates may play a significant role in their biological behavior. It was not feasible to investigate these properties due to the poor control on the polymer molecular weights.

4.3 Evaluation of cationic glycopolymers for gene delivery

Reineke and Narain groups performed significant work on the synthesis of cationic glycopolymers and evaluation of their gene delivery efficacies. Reineke and colleagues prepared series of cationic glycopolymers of relatively narrow polydispersities using poly D-glucaric acid and primary amines of varying lengths in an effort to study structure activity relationship of well-tailored cationic glycopolymers [16]. Narain and co-workers prepared well-defined cationic glycopolymers of known molecular weights, narrow polydispersities, desirable shapes, and architectures by RAFT polymerization and studied their structure-activity relationship in terms of gene expression abilities [9,17,18]. The idea was to control the synthetic parameters of polymers such as molecular weights, cationic content, type and degree of glycosylation, and degree of polymerization and explore their affect on gene expression profiles in different cell lines.

4.3.1 Effect of cationic content in gene delivery

Poly(D-glucaramidoamines) (PGAs) were prepared by poly-condensation of esterified D-glucaric acids and amine containing monomers of varying amine content (3-6 amine per monomer) at fixed degree of polymerization (D*p*). The increase in amine content of monomer at fixed D*p* leads to decrease in polyplexes size and increase in gene expression [16]. It was further confirmed by Narain and colleagues that an increase in primary amine content in cationic glycopolymers of fixed D*p* increases the gene expression *in vitro* [9].

4.3.2 Effect of type of amine

Tertiary and quaternary amines bearing comb-shaped cationic glycopolymer were prepared by atom transfer radical polymerization approach (ATRP). Initiator-modified dextran was polymerized by ATRP in the presence of DMAEMA (tertiary amine), followed by its partial quaternization. The presence of quaternary amines in cationic glycopolymers was related to low gene expression and relatively high toxicities as compared to tertiary amine bearing cationic copolymers [19].

4.3.3 Effect of degree of polymerization (Dp)

The effect of degree of polymerization of trehalose-based cationic polymer on DNA condensation ability was studied, and it was determined that D*p* is not directly related to DNA complexation efficacies, protection, and release as determined by isothermal calorimetry (ITC), pico green assays, and agarose gel electrophoresis. The stability of polyplexes in physiological solution was also investigated. In general, all polymers showed greater aggregation in serum-free media, as a function of time. However, the addition of serum led to well-defined polyplexes with copolymers of all sizes possibly due to shielding affects of trehalose present on polyplexes surface. These copolymers showed high gene expression in Hela cells, while no correlation between gene expression and D*p* was observed in H9c2(2-1) cell line [20]. In contrast, others reported a direct correlation between D*p* of cationic glycopolymers of linear architecture at fixed carbohydrate to cationic content (1:1) and gene expression. The increase in D*p* was associated with increase in gene expression at low N/P ratios [9]. However, the trend did not hold true for hyperbranched cationic glycopolymers. The increase in D*p* of hyperbranched cationic glycopolymers was not directly related to the increase in gene expression and was dependent on carbohydrate type and amine content of the copolymer [17].

4.3.4 *Effect of hydroxyl chemistry*

The effect of hydroxyl stereochemistry on gene expression of cationic glycopolymers was investigated. Open-chain, esterified carbohydrate monomers D-glucarate (D4), D-manarate (M4), and meso-galactarate (G4) were used to prepare cationic glycopolymers by the polycondensation approach, as discussed above. At fixed cationic content, the degree of biding of polymers with DNA was inversely associated with the release of DNA from these copolymers, showing manarate-based copolymers fastest release of DNA at low heparin concentrations. However, the transfection efficacy was G4>D4>M4 in tested cell lines [21]. NF-κB ODN decoys were used as payload for the proof of PGAs as potent delivery vector to treat physiological disorders. The pericardial injection of polyplexes in mice showed localization of payload in heart, without significant uptake in liver or induction of any immune reaction over a 48-hour time period of study (Figure 4.1) [22].

Glucarate-based polymers were later commercialized under the brand name of *Glycofect* [23]. Uekema and colleagues focused on the development of **α,β,γ,** CD-based PAMAM G2 dendrimers and studied their gene expression profiles [22]. The results showed comparatively high gene expression of α-CD based PAMAM dendrimers, G2 and G3, as compared to other CD conjugated counterparts [24].

4.4 Role of hydrogen bonding between glycopolymers and DNA

The role of hydrogen bonding of carbohydrate-based vectors, in DNA packaging, polyplex internalization, and cell surface binding was studied. Apart from electrostatic interactions between DNA and cationic polymer, H-bonding was found to facilitate DNA compaction and the internalization of polyplexes due to their interactions with cell surface glycoproteins [25].

4.4.1 *Intracellular trafficking of cationic glycopolymers*

Polyglycoamidoamines (PGAs) showed preferential cellular uptake by their association with sulfated GAGs [25]. The cell line (pgsA-745) lacking sulfated GAGs showed decreased cellular uptake of PGAs. In contrast, little effect of presence of heparin sulfate on PGAs internalization was reported. Furthermore, interactions of GAGs with polyplexes was found to induce conformational changes, by causing "loosening of polyplexes," which was confirmed by quartz crystal microbalance (QCM) and by *in vitro* assays. Moreover, the interactions between GAGs and PGAs were dependent on the hydroxyl group stereochemistry; glucarate-based cationic glycopolymers showed high GAGs dependent cellular internalization as compared to tartarate based polyplexes [25, 26].

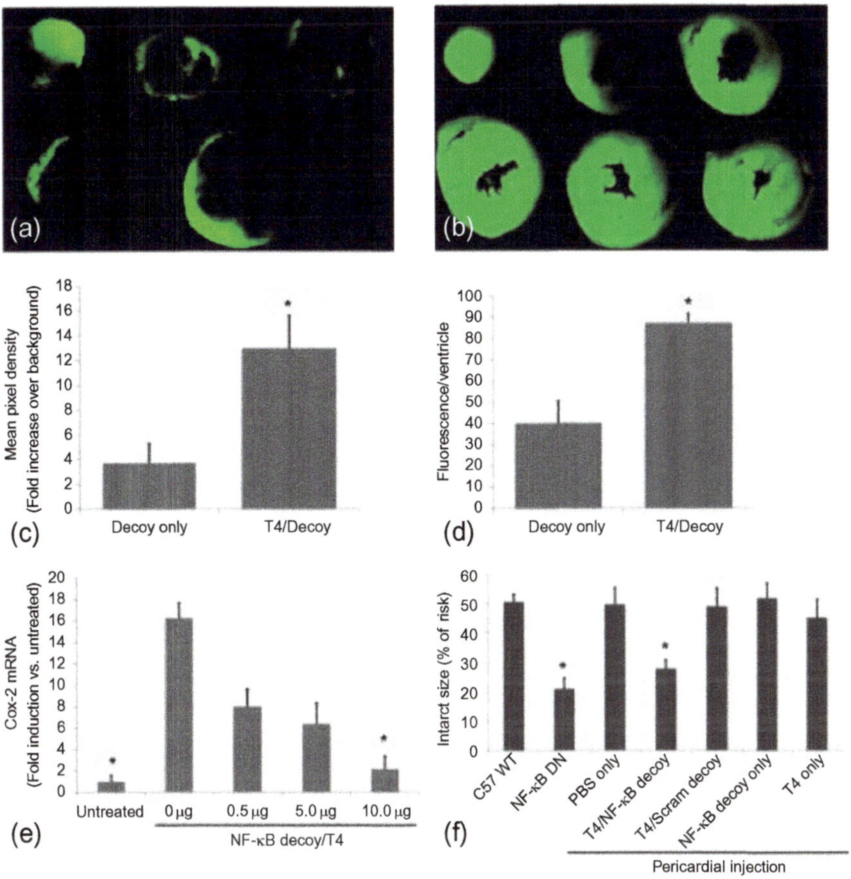

Figure 4.1 Pericardial injection of NF-κB decoy using cationic glycopolymer vector in mice. (a,b) Cardiac sections from mice, 24-hour post-injection showing Alexa-fluor 488 labeled NF-κB decoy. (c,d) quatitation of uptake in myocardium, (e) analysis of Cox-2 mRNA level post-injection, and (f) assesment of infract size post 10 μg delivery of glycopolymer-ODN complexes in percardium [22].

4.4.2 End-group modification

Poly-trehalose oligoethylamine copolymers were prepared by combination of click chemistry and ATRP. Diazido-trehalose was clicked with oligo ethylamine polymer prepared by ATRP to produce cationic glycopolymers of varying molecular weights and end-functionalities. PEGylated and triphenylacetamide bearing cationic glycopolymers thus produced were compared with their parent analog to study the affect of end-group functionalities on gene expression. No significant impact on DNA condensation efficacies and gene expression was observed as a function of end group modification on the studied cell lines [27].

4.4.3 Types of carbohydrates

The DNA and siRNA delivery efficacies of commercially available cationic glycopolymers, namely CALAA-01 (a cyclodextrin based system), and *Glycofect* were evaluated. CALAA-01 and Glycofect were also compared with trehalose-based cationic polymers of similar charge content to determine a correlation between carbohydrate size and type of NA. The chemical structures of these cationic glycopolymers are depicted in Scheme 4.2. Glycofect and other cationic glycopolymers bearing small carbohydrate content showed high DNA delivery efficacies, while CALAA-01 showed high siRNA delivery efficacies *in vitro*. This trend was attributed to relatively weak binding capability of siRNA with short carbohydrate chain-length containing cationic copolymers, indicating the role of carbohydrates' size in polyplexes formulation. The molecular weight of cationic glycopolymers and presence of free polymer was found to play a critical role in siRNA knock-down efficacies *in vitro*. Especially in the case of CD based polymers, endosomal disruption in the presence of free polymers, hence causing enhanced siRNA delivery and gene knock-down, efficacies were attributed to the ability of CD to remove cell-membrane cholesterol. As discussed above, an increase in the number of secondary amines in cationic glycopolymers from 1-4 linearly increases the gene expression, further increase in the number of secondary amine however, plateau gene expression [16]. The trend was strictly dependent upon the type of carbohydrate used and did not hold true for trehalose-based cationic polymers [28].

Narain and colleagues studied the effect of small sugars (glucose and galactose residues) on gene expression efficacies. The hybranched cationic glycopolymers bearing pendant glucose and galatose residues were prepared by RAFT. Interestingly, the gene expression was found to be less dependent on sugar residues and cationic content, and D*p* of copolymers dominated their biological effect [17].

4.4.5 Type of stealth layer

Adamantane conjugated PEG and poly(2-methacrylamide-2-deoxy-glycopyranose) (MAG) were used to compare the role of two potent and widely used stealth layers on a CDP-based nanoparticle system. PEGylated and glycosylated gene delivery systems are probably the most studied types of nanocarriers to reduce the toxicities and provide colloidal stability to cationic polymers. The comparison between the two systems in terms of cellular uptake, gene expression, and toxicities *in vitro* provides interesting insights about the future of gene delivery therapeutics. The cellular uptake of these two systems was cell-line dependent. These two nanocarriers showed similar gene delivery efficacies, but low toxicities were observed for glycosylated nanocarriers [29].

4.4.6 Effect of polymer shape

Narain and colleagues prepared cationic glycopolymers of well-defined shape and molecular weights by RAFT. The cationic glycopolymers bearing glucose and galactose-pendant moieties of linear and branched architectures were studied for their

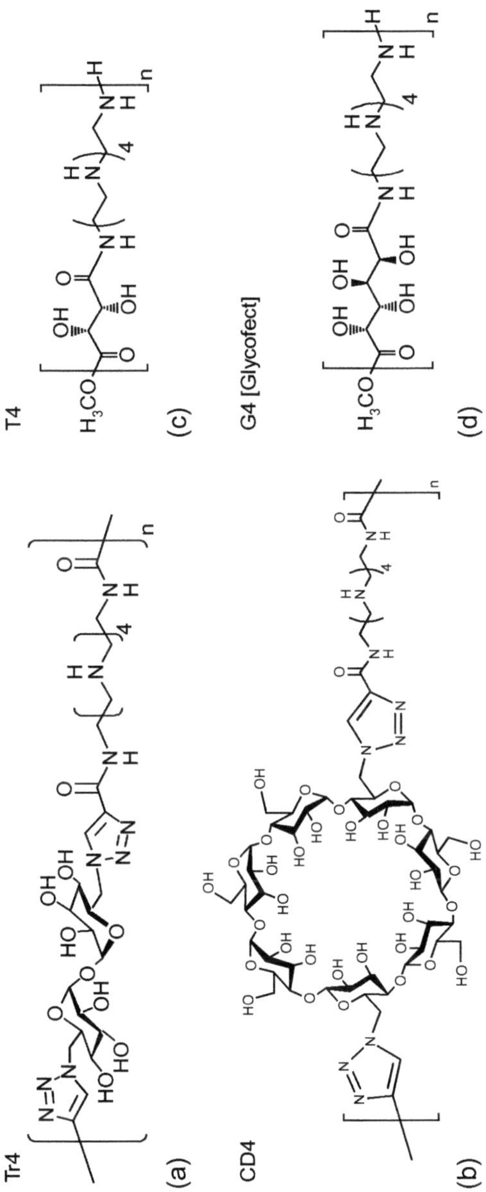

Scheme 4.2 Structure of cationic glycopolymers bearing different types of sugar moieties [28].

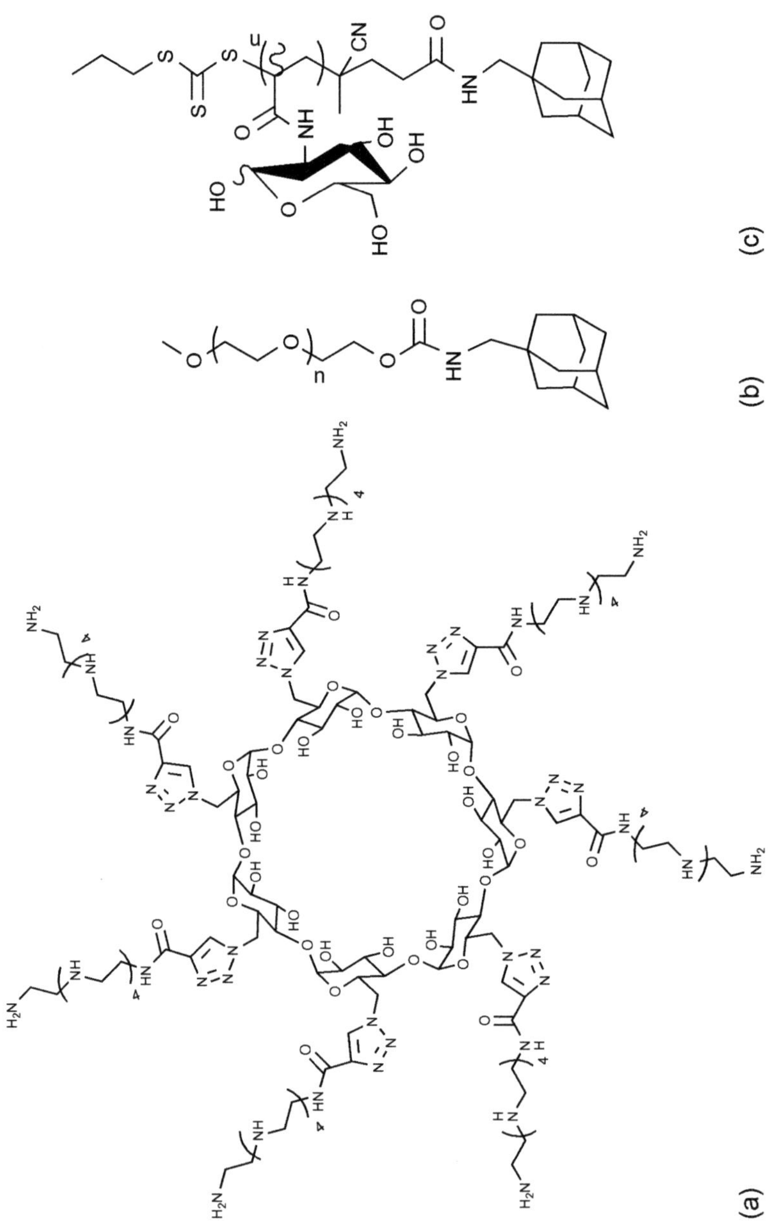

Scheme 4.3 Chemical structures of CDP nanocarrier (a) PEGylated, (b) glycated, and (c) stealth layers [29].

gene expression profiles in a variety of cell lines. In general, hyperbranched cationic glycopolymers showed high gene expression and low toxicities, as compared with the linear cationic glycopolymers [7,19]. Star-shaped cationic glycopolymers bearing CD core and PDMAEMA arms were prepared by ATRP. CD initiator was polymerized by DMAEMA and PEGMA to produce copolymer of varying molecular weights. The incorporation of CD and PEGMA in DMAEMA polymer reduced their toxicity along with the improvement in gene expression [30] (Scheme 4.3).

4.4.7 Effect of polymer architecture

Tailor-made cationic glycopolymers of known molecular weight and architectures were prepared and their gene expression profiles were studied using β-galactosidase activity based reporter assay. The cationic glycopolymers of block, statistical, and *block-statistical* architectures were compared. The statistical cationic glycopolymers showed high gene expression accompanied by low toxicities in the presence of serum proteins, as compared to their block counterparts. The difference in gene expression of the two types of copolymers was dependent on their polyplexes biophysical properties. The polyplexes produced by statistical copolymers were larger in size, as compared to their block copolymer-based polyplexes. Moreover, the statistical copolymer-based polyplexes showed slight aggregation in serum-containing media, possibly due to the exposure of amine content on polyplex surface. The high gene expression of statistical copolymers in the presence of serum proteins is thought to be aided by the interactions between serum proteins and cell surface glycans [9,17]. Carefully designed *block-statistical* copolymers were synthesized to maintain high gene expression accompanied with no aggregation of copolymers. The short *block-statistical* cationic glycopolymer-based polyplexes maintained their physiological stability in serum containing media, without compromising their gene expression [18] (Scheme 4.4).

Cationic glycopolymers of random architecture bearing pendant tertiary amine *N,N*-dimethyl amioethylmethacrylate (DMAEMA) and carbohydrate moieties (glucose or galactose) were prepared by radical polymerization approach. The polymers ranging from 30–86 kDa and varying sugar content were studied for their Apo B siRNA delivery efficacies *in vivo*. The cationic and carbohydrate content in the copolymers were tuned to obtain a group with minimum toxicity and high siRNA delivery efficacies. Interestingly, gene delivery profiles revealed high toxicities of glucose-bearing polymers as compared to their galactose analogs in mouse epithelial cells [31]. It should be noted that no such trend of toxicity of hyperbranched glycosylated and galactosylated cationic glycopolymers was observed in human liver cancer cell line [17]. The high molecular weight galactose-bearing cationic glycopolymer showed some ApoB knock-down efficacies in mouse liver. However, the effect was similar to homopolymer of PDMAEMA and no targeting was observed, possibly due to aggregation of polyplexes in mouse liver. APoB is a glycoprotein high concentrations of which are involved in atherosclerosis, hyperchlostremia, and myocardial infractions. It is thought that non-specific accumulation of vectors in liver and lungs is the major issue with low gene knock-down efficacies *in vivo* [31].

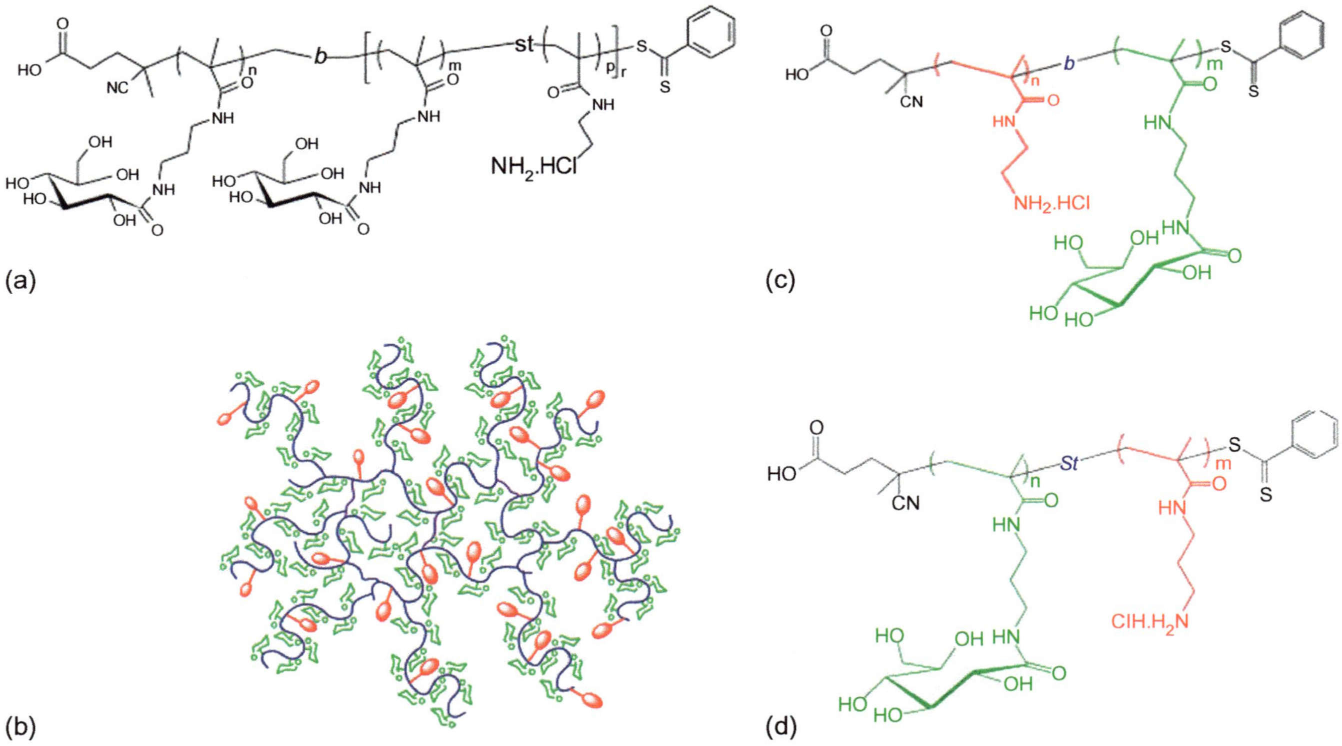

Scheme 4.4 Chemical Structure of "linear" and hyperbranched cationic glycopolymers of varying architectures. (a) linear block-statistical cationic glycopolymer, (b) hyperbranched cationic glycopolymer, (c) block copolymer, and (d) statistical cationic glycopolymer [9, 17, 18].

4.5 Degradable cationic glycopolymers

Reduction sensitive comb-shaped polymers bearing non-ionic dextran polymer backbone and disulfide modified DMAEMA and PEGMA were prepared. These cationic glycopolymers showed efficient cleaving of side chains under reducing conditions and are thought to be potent degradable vectors for physiological applications [32]. Degradable cationic glycopolymers bearing polyester-based linkers were also prepared to facilitate nucleic acid release at acidic pH [33]. Thermoresponsive cationic glycopolymers were prepared by RAFT and were cross-linked with acid labile cross-linker to devise a dual responsive (pH and thermoresponsive) system for efficient gene delivery. These degradable cationic glyconanogels showed excellent DNA condensation efficacies at physiological temperature (37°C), and degradation of vectors at acidic pH (4–5). The cationic glyconanogels showed high DNA and siRNA delivery efficacies in liver cancer cells [34,35].

4.5.1 Cationic glycopolymer formulations in vial

Formulation of polyplexes for *in vivo* applications requires high concentration of polyplexes, which usually leads to aggregation in isotonic solution. The conjugation of cationic vector with hydrophilic residues such as glycopolymers or polyethyleneglycol (PEG) aids in the formation of discrete polyplexes at high concentrations of nucleic acids, by masking the surface charge of polyplexes. In addition, storage of polyplexes and shelf-life are two basic requirements for their successful implication in clinical settings. Polytrehalose-based polymers offer great opportunity to make lyophilized polyplexes for long-term storage without affecting the biological properties of cargo by decreasing the phase transition energy of the solution. Trehalose-based cationic polymers maintained their gene knock-down efficacies before and after lyophilization [36].

4.5.2 Lectin-mediated targeting of cationic glycopolymers

Glycotargeting is a technique for the delivery of nucleic acids (NAs) by receptor-mediated endocytosis (RME) due to the interactions between cell-surface lectins and carbohydrate vectors. This magic bullet approach has been reported to account for enhanced cellular delivery of polyplexes. The complexation of cargo with glycosylated polymers can be achieved either directly *via* a covalent linkage between cargo and the polymer or by non-covalent immobilization (H-bonding, electrostatic interactions) of NAs with polymer. The latter approach has been discussed above in detail. Glycosylated NAs occur naturally in simple organisms and are shown to offer protective and immunogenic properties to the NAs. Lectins are proteins that specifically interact with cell-surface glycans. Two possible approaches for lectin-mediated targeting are: 1) conjugation of drugs with lectins, which can be identified by cell-surface glycan, or 2) attachment of drug with sugar molecules for their preferential uptake by cell-surface lectins. The second approach is mostly utilized to avoid the cost and complexity of preparing lectin-conjugated carriers. However, the glycans required for efficient binding to lectins are complicated in structure and synthesis. Asialo-orosomucoid targeted pLL- polyplexes were used for targeting of asialoglycoprotein receptor (ASGPR) in hepatoma cells *in vitro.* In addition, lactosylated polyplexes have also been extensively used for hepatocytes targeting. Non-covalent conjugation of glycans with NAs is usually achieved by the electrostatic interactions between DNA and cationic counterpart of the polymer. In another study, glycoplexes were prepared by combining biotinylated oligonucleotides with streptavidin-modified mannose. These complexes showed slight success in cellular localization *in vitro.* The glycan-DNA complexes were also prepared by conjugation of lactose derivatives with DNA intercalating agent "spermidine-bisacridine." These trilactosylated bisacridine based complexes showed binding affinity toward ASGPR and ricin lectin, but poor gene expression was observed possibly due to poor membrane disrupting ability of the complexes. Covalent linking of glycans with nucleic acids was successfully achieved using a number of chemical approaches such as oxime bond formation, phosphoramidite chemistry, diazo coupling, to name a few. The covalent conjugates maintained their ligand binding affinity but no gene delivery efficacies were reported [37]. Carbohydrate recognition domain (CRD) containing lectins are located intracellularly, which are responsible for intracellular trafficking and are secreted on plasma membrane such as C-type lectins, Gal-receptors, Mannose-receptors, and I-type lectins. Examples of membrane lectins, which mediate endocytosis, are ASGPR and Mannose-6-phosphate receptor/insulin like growth factor-II-receptor (M6P/IGFIIR). DC-SIGN receptor is another type of mannose-specific lectin present on dendritic cells, which is extensively used for antigen presentation to immune cells. In addition, nuclear lectins are being explored as a toll to target nuclear entry by glycofection, but no reports so far show successful glycofection and nuclear localization of complexes. Galactosylated and mannosylted PLL was used to selectively target parenchymal and non-parenchymal cells of liver, respectively, after intravenous injection of nucleotides. Mannosylated PLL was also used to target macrophages. Glycosylation of PEI was used as a tool to study lectin-specific gene delivery. A number of *in vitro* experiments ensure galactose and mannose-specific gene delivery in hepatocytes and macrophages, respectively, but the fate of these vectors for *in vivo* applications is still unclear. Cystic fibrosis transmembrane receptor (CFTR) gene delivery was investigated for clinical applications of glycofection. Although relatively high gene expression in CF cells was reported a decade ago using galactosylated PLL there are still no reports in clinics [7].

The crucial role of lectins in targeted drug delivery was well elaborated by Gabbius and colleagues. The diversity of glycated molecules is much higher as compared to the amino acid or nucleic acid base analogs in nature. This is simply due to the possibility of a large number of chemical bonding and OH group substitutions, which provide larger heterogeneity to glycated molecules. Hence, the term "sugar code" was introduced to obtain successful targeted delivery of cargo molecules in living systems. However, it is clear that a number of problems and pitfalls can complicate the successful translation of these sugar codes as targeted therapeutics, which includes identification of new ligands, study of their structures, and topological changes in their structure in the presence of cargo material or chemical bonds. C-type lectins such as (DC-SIGN receptor, MBL, and ASGPR) of hepatocytes, macrophage, and dendritic cells are one example of lectins, which are successfully used for targeted gene delivery [38].

Receptors for advanced glycation end products (RAGE) were targeted with glycated dendrimers for efficient gene transfer *in vitro*. RAGE are members of IgG superfamily of cell-surface molecules, which are able to interact with AGE ligands involved in diabetes and other pathophysiological processes including Alzheimer's, tumorogenesis, and inflammatory diseases. AGE are generated non-enzymatically by the reaction between high concentration of reducing sugars and amino groups present in lipids, proteins, and nucleic acids and forming Schiff bases and Amadori compounds. The binding of AGE to RAGE results in activation of signal transduction pathways, including Rac/Cdc24, Ras, and Jak/Stat pathway, and ubiquitous activation of this pathway contributes to pathological conditions. CHO-K1 cells stably expressing RAGE and unmodified CHO-K1 cells were used to study RAGE-dependent signal transduction pathways, cellular uptake mechanism, and gene expression. Glycated dendrimers showed enhanced gene expression in RAGE containing cells, and showed RAGE-dependent uptake of glycated polyplexes. The glycated polymers that were unable to stimulate RAGE were also unable to mediate high transfection. However, no conclusive data was obtained from *in vivo* studies [8]. The modification of G2 PAMAM-CD conjugates with galactose moieties provided a useful tool to target ASGPR expressing cell lines *in vitro* [39]. Mannosylated PAMAM-CD and PEI conjugates showed preferential gene expression in mice kidney [40], However, both galactosylated and mannosylated G2 PAMAM dendrimers showed no receptor-dependent gene expression *in vivo*. To investigate this issue the author extended the spacer length between sugar moieties and PAMAM-CD conjugates and lactose-targeted PAMAM-G2 dendrimers were formulated. Slightly improved gene expression in liver hepatocytes post-lactosylated-dendrimers injection was observed [41]. In order to improve hepatocytes targeting, lactosylated and mannosylated targeted PAMAM-G3 dendrimers were prepared and subjected to *in vitro* gene expression analysis [37,41].

Methacrylate modified closed-ring lactose was polymerized by RAFT, followed by the grafting of PLL by amidation reaction. The cationic glycopolymer of low molecular weight showed preferential gene delivery in HepG2, while their high molecular weight analog showed preferential gene expression in NIH3T3 cells [42].

Galactosylated cationic polymers were also prepared by the combination of polycondensation and click chemistry. Length of linker in galactose targeting was explored but effect was found to be mainly dominated by cationic charge of polyplexes [43].

4.6 Future perspective

The synthesis of cationic glycopolymers and the study of their gene delivery profiles have been the focus of intensive research from more than a decade. The significant progress made in the field so far has highlighted the role of glycopolymers as biocompatible non-toxic stealth layer, which can outcompete PEGylated nanoparticles. In addition, it is possible to produce a variety of well-defined cationic glycopolymers with known shapes, molecular weights, and architectures, which further influence their gene expression. However, the engineering of glycated polymers for gene delivery efficacies is not a straightforward approach for several reasons. The excess of glycation of cationic polymers decreases their gene expression due to several reasons such as the aggregation of polyplexes and their interactions with nucleic acids, hence hindering facile DNA release in cytosol. In addition, glycotargeting is a complex process due to the presence of variety of lectins intracellularly and extracellularly. Mannosylated and galactosylated cationic polymers have shown some targeting properties thus far *in vitro*, but *in vivo* no conclusive data has been obtained. One of the major challenges in the field of gene delivery is poor gene expression of cationic vectors *in vivo*. The colloidal stability and targeting properties are two main extracellular barriers of polyplexes after *in vivo* administration. Although cationic glycopolymer-based complexes seem to maintain their stability under physiological conditions, targeting efficacies are poor and possibly non-specific interactions of these glycans with cell-surface lectins is the main reason. The well-designed cationic glycopolymers are shown to work well *in vitro,* but their fate under *in vivo* conditions is unclear. The detailed study exploring the role of novel glycopolymer architectures in gene delivery efficacies *in vitro* is required. The field of gene delivery holds great promise to cure many lethal diseases, but the progress made in the field thus far has been slow. A significant amount of work is required to evaluate the status of these engineered molecules as clinical therapeutics.

References

[1] Ahmed M, Narain R. Progress of RAFT based polymers in gene delivery. Progr Polymer Sci 2013;38:767–90.
[2] Pearson S, Chen G, Stenzel MH. Synthesis of glycopolymers. CH1.
[3] Ahmed M, Narain R. Cationic glycopolymers. CH3.
[4] Ahn C-H, Chae SY, Bae YH, Kim SW. Synthesis of biodegradable multi-block copolymers of poly(L-lysine) and poly(ethylene glycol) as a non-viral gene carrier. J Control Release 2004;97:567–74.
[5] Brissault B, Kichler A, Guis C, Leborgne C, Danos O, Cheradame H. Synthesis of linear polyethylenimine derivatives for DNA transfection. Bioconjugate Chem 2003;14:581–7.
[6] Davis ME, Zuckerman JE, Chung HJC, David S, Anthony T, Christopher AA, et al. Evidence of RNAi in humans from systemically administered siRNA via targeted nanoparticles. Nature 2010;464:1067–70.
[7] Roche AC, Fajac I, Grosse S, Frison N, Rondanino C, Mayer R, et al. Glycofection: facilitated gene transfer by cationic Glycopolymers. Cell Mol Life Sci 2003;60:288–97.
[8] Giron-Gonzalez MD, Morales-Portillo A, Salinas-Castillo A, Lopez-Jaramillo FJ, Hernandez-Mateo M, Santoyo-Gonzalez F, et al. Engineered glycated amino dendritic polymers as specific nonviral gene delivery vectors targeting the receptor for advanced glycation end products. Bioconjugate Chem ASAP 2014;.
[9] Ahmed M, Narain R. The effect of polymer architecture, composition, and molecular weight on the properties of glycopolymer-based non-viral gene delivery systems. Biomaterials 2011;32:5279–90.

[10] Duverger E, Pellerin-Mendes C, Mayer R, Roche AC, Monsigny M. Nuclear import of glycoconjugates is distinct from the classical NLS pathway. J Cell Sci 1995;108(Pt 4):1325–32.
[11] Duverger E, Carpentier V, Roche AC, Monsigny M. Sugar dependent nuclear import of glycoconjugates from the cytosol. Exp Cell Res 1993;207:197–201.
[12] Klink DT, Chao S, Glick MC, Scanlin TF. Nuclear translocation of lactosylated poly-L-lysine/cDNA complex in cystic fibrosis airway epithelial cells. Mol Ther 2001;3:831–41.
[13] Ahmed M, Narain R. Cell line dependent uptake and transfection efficiencies of PEI-anionic glycopolymer systems. Biomaterials 2013;34:4368–76.
[14] Fajac I, Thevenot G, Bedouet L, Danel C, Riquet M, Figarella C, et al. Uptake of plasmid/glycosylated polymer complexes and gene transfer efficiency in differentiated airway epithelial cells. J Gene Med 2003;5:38–48.
[15] Grosse S, Thevenot G, Aron Y, Duverger E, Abdelkarim M, Roche A-C, et al. *In vivo* gene delivery in the mouse lung with lactosylated polyethylenimine, questioning the relevance of *in vitro* experiments. J Control Release 2008;132:105–12.
[16] Liu Y, Wenning L, Lynch M, Reineke TM. New poly(D-glucaramidoamine)s induce DNA nanoparticle formation and efficient gene delivery into mammalian cells. J Am Chem Soc 2004;126:7422–3.
[17] Ahmed M, Narain R. The effect of molecular weight, compositions and lectin type on the properties of hyperbranched glycopolymers as non-viral gene delivery systems. Biomaterials 2012;33:3990–4001.
[18] Ahmed M, Jawanda M, Ishihara K, Narain R. Impact of the nature, size and chain topologies of carbohydrate-phosphorylcholine polymeric gene delivery systems. Biomaterials 2012;33:7858–70.
[19] Wang ZH, Li WB, Ma J, Tang WT, Xu FJ. Functionalized nonionic dextran backbones by atom transfer radical polymerization for efficient gene delivery. Macromolecules 2011;44:230–9.
[20] Srinivasachari S, Liu Y, Prevette LE, Reineke TM. Effects of trehalose click polymer length on pDNA complex stability and delivery efficacy. Biomaterials 2007;28:2885–98.
[21] Liu Y, Reineke TM. Hydroxyl stereochemistry and amine number within poly(glycoamidoamine)s affect intracellular DNA delivery. J Am Chem Soc 2005;127:3004–15.
[22] Tranter M, Liu Y, He S, Gulick J, Ren X, Robbins J, et al. *In vivo* delivery of nucleic acids via glycopolymer vehicles affords therapeutic infarct size reduction *in vivo*. Mol Ther 2012;20:601–8. http://www.techulon.com/products.
[23] Wada K, Arima H, Tsutsumi T, Hirayama F, Uekama K. Enhancing effects of galactosylated dendrimer/a-cyclodextrin conjugates on gene transfer efficiency. Biol Pharm Bull 2005;28:500–5.
[24] Prevette LE, Kodger TE, Reineke TM, Lynch ML. Deciphering the role of hydrogen bonding in enhancing pDNApolycation interactions. Langmuir 2007;23:9773–84.
[25] Mclendon PM, Buckwalter DJ, Davis EM, Reineke TM. Interaction of poly(glycoamidoamine) DNA delivery vehicles with cell-surface glycosaminoglycans leads to polyplex internalization in a manner not solely dependent on charge. Mol Pharm 2010;7:1757–68.
[26] Kizjakina K, Bryson JM, Grandinetti G, Reineke TM. Cationic glycopolymers for the delivery of pDNA to human dermal fibroblasts and rat mesenchymal stem cells. Biomaterials 2012;33:1851–62.
[27] Xue L, Ingle NP, Reineke TM. Highlighting the role of polymer length, carbohydrate size, and nucleic acid type in potency of glycopolycation agents for pDNA and siRNA delivery. Biomacromolecules 2013;14:3903–15.

[28] Buckwalter DJ, Sizovs A, Ingle NP, Reineke TM. MAG versus PEG: incorporating a poly(MAG) layer to promote colloidal stability of nucleic acid/"click cluster" complexes. ACS Macro Lett 2012;1:609–13.

[29] Xu FJ, Zhang ZX, Ping Y, Li J, Kang ET, Neoh KG. Star-shaped cationic polymers by atom transfer radical polymerization from β-cyclodextrin cores for nonviral gene delivery. Biomacromolecules 2009;10:285–93.

[30] Tachibana Y, Munisso MC, Kamata W, Kitagawa M, Harada-Shiba M, Yamaoka T. Quick nuclear transportation of siRNA and *in vivo* hepatic ApoB genesilencing with galactose-bearing polymeric carrier. J Biotech 2014;175:15–21.

[31] Wang Z-H, Zhu Y, Chai M-Y, Yang W-T, Xu F-J. Biocleavable comb-shaped gene carriers from dextran backbones with bioreducible ATRP initiation sites. Biomaterials 2012;33:1873–83.

[32] Lee C-C, Liu Y, Reineke TM. Glucose-based poly(ester amines): synthesis, degradation, and biological delivery. ACS Macro lett 2012;1:1388–92.

[33] Sunasee R, Wattanaarsakit P, Ahmed M, Lollmahommed FB, Narain R. Biodegradable and nontoxic nanogels as nonviral gene delivery systems. Bioconjugate Chem 2012;23:1925–33.

[34] Ahmed M, Wattanaarsakit P, Narain R. Cationic glyco-nanogels for epidermal growth factor receptor (EGFR) specific siRNA delivery in ovarian cancer cells. Polym Chem 2013;4:3829–36.

[35] Sizovs A, Xue L, Tolstyka ZP, Ingle NP, Wu Y, Cortez M, et al. Poly(trehalose): sugar-coated nanocomplexes promote stabilization and effective polyplex-mediated siRNA delivery. J Am Chem Soc 2013;135:15417–24.

[36] Yan H, Tram K. Glycotargeting to improve cellular delivery efficiency of nucleic acids. Glycoconj J 2007;24:107–23.

[37] Yamazaki N, Kojima S, Bovin NV, Andre S, Gabius S, Gabius H-J. Endogenous lectins as targets for drug delivery. Adv Drug Deliv Rev 2000;43:225–44.

[38] Wada K, Arima H, Tsutsumi T, Hirayama F, Uekama K. Enhancing effects of galactosylated dendrimer/a-cyclodextrin conjugates on gene transfer efficiency. Biol Pharm Bull 2005;28:500–5.

[39] Wada K, Arima H, Tsutsumi T, Chihara Y, Hattori K, Hirayama F, et al. Improvement of gene delivery mediated by mannosylated dendrimer/a-cyclodextrin conjugates. J Control Release 2005;104:397–413.

[40] Arima H, Yamashita S, Mori Y, Hayashi Y, Motoyama K, Hattori K, et al. *In vitro* and *In vivo* gene delivery mediated by Lactosylated Dendrimer/α-Cyclodextrin Conjugates (G2) into Hepatocytes. J Control Release 2010;146:106–17.

[41] Arima H, Chihara Y, Arizono M, Yamashita S, Wada K, Hirayama F, et al. Enhancement of gene transfer activity mediated by mannosylated dendrimer/α-cyclodextrin conjugate (generation 3, G3). J Control Release 2006;116:64–74.

[42] Zhou D, Li C, Hu Y, Zhou H, Chen J, Zhang Z, et al. Glycopolymer modification on physicochemical and biological properties of poly(l-lysine) for gene delivery. Int J Bio Macromol 2012;50:965–73.

[43] Lee C-C, Gradinetti G, McLendon PM, Reineke TM. A polycation scaffold presenting tunable "click" sites: conjugation to carbohydrate ligands and examination of hepatocyte-targeted pDNA delivery. Macromol Biosci 2010;10:585–98.

Engineering peptide-conjugated non-viral delivery systems for enhanced SiRNA and DNA therapeutics

5

S. Quan, P. Kumar, R. Narain
University of Alberta, Edmonton, AB, Canada

5.1 Introduction

The control of cellular uptake, intracellular trafficking, and nuclear targeting of plasmid-based therapies is crucial for enhancing the efficacies of DNA therapeutics. While gene therapy is limited by several factors such as insufficient nuclear penetration and non-specific cytoplasmic transport, a number of viral and non-viral gene delivery systems have been engineered to overcome the potential intracellular problems [1]. The most effective gene therapy strategies rely on the modification of viral vectors, such as adenoviruses, retroviruses, and herpes virus [2–4]. Although currently less efficient than their viral counterparts, non-viral gene delivery systems have been under intense evaluation as a safer alternative for gene therapy due to their inherent properties associated with less systemic problems such as immunogenicity, cytotoxicity, and insertional mutagenesis [5]. Some of the most effective non-viral gene therapy vectors used in research include polyethyleneimine, dendrimers, chitosan, and various polypeptide and oligopeptide sequences [6–8]. In particular, peptide-based delivery vectors offer many advantageous features over other non-viral strategies that include the ability to efficiently condense plasmid DNA and RNA into tightly packed vesicles by electrostatic interactions, target specific cell-surface receptors in protein-protein interactions, escape the endosomal membrane, and penetrate the nucleus to deliver plasmid cargo [5]. Peptide sequences can be engineered to contain basic residues to enhance the cationic nature, which can bestow more efficient DNA condensation capabilities and stability in serum, or can be optimized to replicate viral protein and membrane transduction domain sequences for specific cellular targeting [9–11]. In recent developments, peptide-based gene delivery systems are showing increased potential as commercially viable strategies due to their incredibly tunable versatility and functionalization.

5.2 Lysine-based peptides

Poly-L-lysine (PLL) has been widely used as a non-viral gene delivery agent due to its inherent biodegradable properties and primary ε-amine groups, which have

Polymers and Nanomaterials for Gene Therapy. http://dx.doi.org/10.1016/B978-0-08-100520-0.00005-9
© 2016 Elsevier Ltd. All rights reserved.

electrostatic interactions with negatively charged phosphate groups of DNA, first described in 1975 by Laemmli et al. [12]. Synthesis of PLL occurs by the conversion of the ε-amine protected group to N-carboxy-(N-benzyloxycarbonyl)-L lysine anhydride, which undergoes ring-opening polymerization using a primary amine initiator [13]. Interestingly, at physiological pH, all primary amine groups of PLL are protonated that aid in the escape from the endosomal compartment once internalized. However, only PLL structures with molecular weights greater than 2500 Da can effectively condense with DNA to form stable complexes, which implicates the importance of the presence of primary amine groups needed for DNA complex formation [5,13]. The length of PLL greatly influences the particle size and stability of DNA complexes, with higher molecular weight having a greater cationic charge and the ability to form more stable and tightly bound DNA complexes compared to low molecular weight PLL [5]. Despite the high efficiency of DNA complexation, studies have shown that increasing molecular weight also increases cytotoxicity and aggregation *in vitro* and *in vivo* [14]. However, this cytotoxicity has been reduced with the incorporation of imidazole groups and the use of dendrimer structural conformations [15].

In recent findings, Haratono and co-workers reported the synthesis of large pore cubic mesoporous silica nanoparticles functionalized with poly-l-lysine designed for carrier gene delivery (Figure 5.1) [16]. Interestingly, the mesoporous silica nanoparticles (MSN-P) contained multiple cavities that were 28 nm in size, which allowed for covalent immobilization of low molecular weight PLL (2.9 kDa) on the outer surface and inner cavities of the nanoparticles. First, epoxysilane was reacted with the silanol groups present on the silica surface under reflux conditions with toluene and to immobilize PLL on silica surface by ring-opening nucleophilic addition to the epoxide site. Notably, the nanoparticle modification with PLL significantly increased the binding capacity of oligo DNA-Cy3 as compared to terminal amino-functionalized and native unmodified silica particles. Ultimately, the PLL functionalized silica nanoparticles demonstrated superior transfection of HeLa cell as compared to its amine-functionalized equivalent. They also exhibited strong binding capacity to negatively charged nucleic acid and relatively low cytotoxicity at concentrations up to 100 μg/mL.

In other works, Liu and co-workers described the preparation of star-shaped cyclodextrin-PLL derivatives (CD-PLLD) using 'click' coupling of per-6-azido-β-cyclodextrin with propargyl focal point PLL for the co-delivery of chemotherapy drug Docetaxel (DOC) and MMP-9 siRNA (pMR3) for cancer therapy [17]. Interestingly, CD-PLLD/DOC/pMR3 complex induced more significant apoptosis and decreased invasiveness in nasopharyngeal epithelial carcinoma in comparison to DOC or pMR3 used alone. Indeed, no complicated micellization and encapsulation processes were observed. In addition, these star-shaped copolymers demonstrated greater blood biocompatibility and lower cytotoxicity as compared to hyperbranched PEI (25 kDa) delivery vector standards that are well established. Ma et al. on a similar platform developed a multifunctional polymers for gene and drug co-delivery using cyclodextrin- derived PLL dendrons prepared by click chemistry [18]. It was found that the PLL conjugate

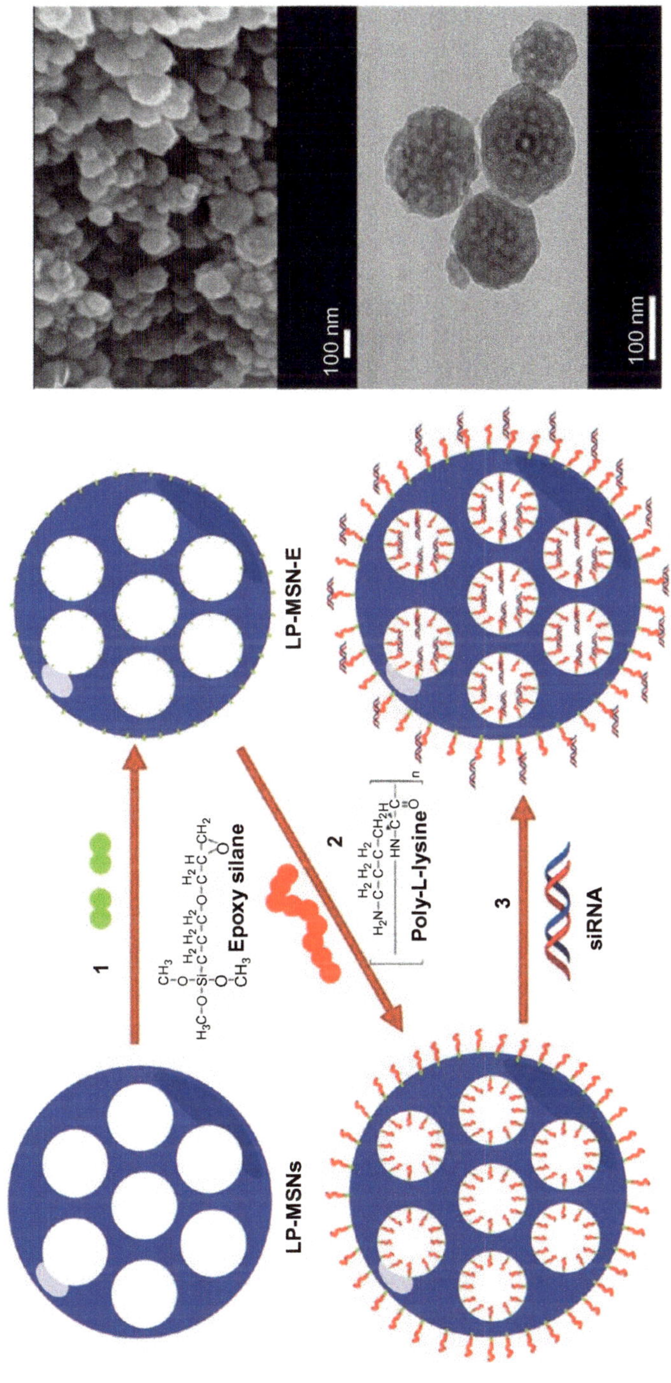

Figure 5.1 Schematic of the steps for the poly-l-lysine grafting and gene loading into LP-MSN-P and SEM images of LP-MSN-P. Reprinted with permission from [16]. Copyright 2012 American Chemical Society.

could form colloidally stable nanoplexes with pEGFP-C2 plasmid DNA, and demonstrated high transfection efficacy in MCF-7 human breast cancer cells. Moreover, the multifunctional PLL conjugate showed loading of water-soluble methotrexate (MTX) anticancer drug and sustained release behavior over 25 hours, demonstrating the versatile ability to co-deliver gene therapeutics with water-insoluble antimetabolite chemotherapeutics. Similar studies by Zheng and colleagues described the self-assembly of polypeptide micellar nanoparticles using a triblock copolymer poly(ethylene glycol)-b-poly(l-lysine)-b-poly-(l-leucine) for the systematic co-delivery of docetaxel (DOX) and siRNA-Bcl-2 toward MCF7 breast cancer xenografts in murine models (Figure 5.2) [19]. This innovative study showed that upon self-assembly, the hydrophobic poly(l-leucine) core entrapped DOC while the PLL polypeptide cationic backbone allowed for electrostatic binding with siRNA-Bcl-2. These conjugate systems functionalized with PLL demonstrates the incredible synergistic applications in combined delivery of plasmid DNA and anticancer chemotherapeutics using non-viral gene and drug delivery systems.

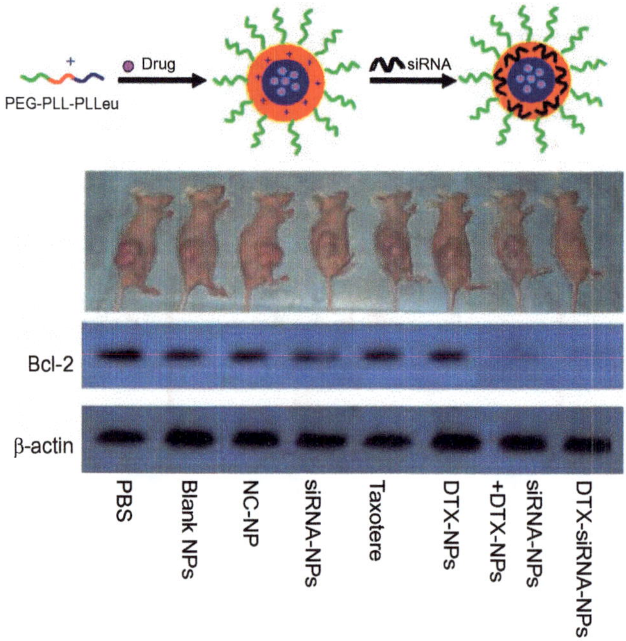

Figure 5.2 Schematic of the self-assembled cationic micelle and loading of siRNA and drug and a representative mouse from each group and Bcl-2 expression of tumor tissue from each group determined by Western Blot Analysis. Dose per Injection: DTX 0.5 mg/kg and siRNA 0.2 mg/kg.
Reprinted with permission from [19]. Copyright 2013 Elsevier Ltd.

5.3 Histidine-based peptides

Endosomal entrapment remains one of the greatest barriers for efficient gene transfection and nuclear localization [20]. Recent findings have shown an innovative strategy to attain endosomal escape from the endosomal membrane, which is to buffer against the proton pump gradient with the addition of poly-L-histidine (PLH) [21]. As protons are pumped in by the ATP-dependent proton pump, the imidazole group of histidine, which has a pKa of ~6.0, becomes protonated, swells, and increases the osmotic pressure within the membrane causing it to rupture [22]. However, at physiological pH, the histidine charge remains neutral, thereby imparting selective membrane disruption in the acidic late endosome compartments [23].

In recent studies, Brevet et al. functionalized mesoporous silica nanoparticles (MSN) with L-histidine for the delivery of luciferase gene plasmid DNA (pDNA) *in vitro* and *in vivo* [24]. Interestingly, pDNA complexes with MSN-His much more efficiently in comparison to MSN-NH2 or MSN-imidazole derivatives and further protects it from DNase I degradation. Their study found that MSN-His also showed increased cellular uptake and escape compared to MSN-NH2 or MSN-imidazole indicating the significance associated with conjugating the histidine molecule. In contrast, a novel comprehensive study by Hwang and colleagues demonstrated the importance of grafting monomeric l-histidine compared to poly(l-histidine) with the same number of constituent groups in the gene transfection of plasmid DNA (Figure 5.3) [25]. Experiments showed the grafted polymeric histidine arrangement demonstrated greater ability to escape the endosomal compartment and trigger the release of YOYO1-pDNA bound plasmid, leading to nuclear localization in comparison to the monomeric grafter counterpart. Thus, while the PLL chain was grafted with equivalent components of l-histidine, Hwang and co-workers discovered that the poly-(L-histidine) graft contributed to higher local charge density at endosomal pH creating stronger membrane buffering capacity in comparison to grafting monomeric L-histidine [25].

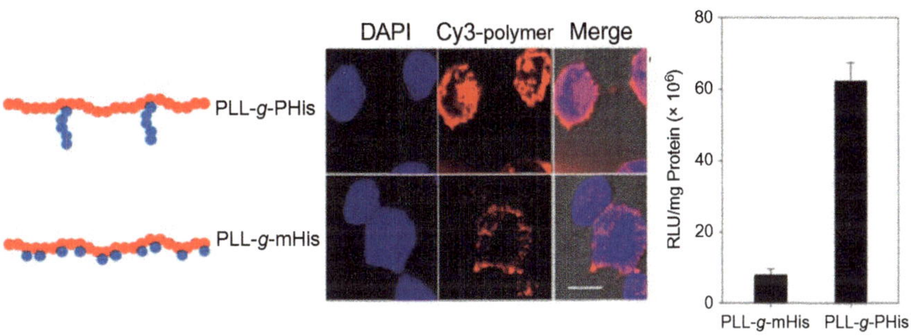

Figure 5.3 Schematic of monomeric and polymeric L-Histidine grafted on PLL, confocal images of PLL-g-PHis and PLL-g-mHis at 50 μg/mL treated HeLa Cells after 1.5 h incubation, luciferase expression of polyplexes mixed with pLuc plasmid at N/P ratios of 6 transfection in HeLa cells.

Reprinted with permission from [25]. Copyright 2014 American Chemical Society.

In general, the individual properties of amino acid conjugates can greatly improve the efficiency of gene delivery vectors. However, the synergistic effects of dual- and triple-functionalized delivery scaffolds have not been adequately explored. A novel study by Wang and co-workers reported the synthesis of single-, dual-, and triple-functionalized G5 PAMAM dendrimers with histidine, arginine, and phenylalanine using a one-pot approach [26]. By exploiting the properties of each individual amino acid, the dendrimer benefitted from the synergistic activity of arginine in forming stable tertiary polyplexes with EGFP plasmid, phenylalanine improving cellular uptake efficiency, histidine increasing pH-buffering capacity, and endosomal escape. Interestingly, while these dual- and triple-functionalized dendrimers showed minimal toxicity in transfected NIH3T3 cells, the combination strategy also demonstrated comparable transfection efficacy to several commercially available reagents such as Lipofectamine 2000, JetPEI, and Polyfect (Figure 5.4). In another study, Goncalves and co-workers used a commercially available liposome (LIP100) and modified the surface with histidinylated polyethyleneimine (His-PEI) by standard mixing and ethanol evaporation method for the delivery of LucGL3 siRNA knockdown of luciferase [27]. Similarly, the LIP100/His-PEI complex with LucGL3 siRNA demonstrated increased inhibition efficiency compared to unmodified LIP100, commercially available Lipofectamine 2000 and INTERFERin. These studies indicate the superior advantages of incorporating the pH-buffering properties of histidine in non-viral and commercially available gene delivery vector as an effective strategy to improve transfection efficiency.

5.4 Arginine-based peptides

The intracellular delivery of exogenous macromolecules using membrane-permeable carriers offers us the opportunity to establish novel methods for controlling cellular uptake of bioactive materials. Biomedical research has largely used arginine to enhance the therapeutic efficacy of gene and drug delivery. It addresses the disadvantages underlying the poor penetration of the cell membrane of viral and non-viral vectors. One of the most widely used peptide vectors in penetrating the cell membrane is the arginine-rich peptide segment from the transactivation (TAT protein) derived from HIV-1 virus, first described by Harrington and colleagues [28]. The sequence of TAT (48-60), which is crucial for the translocation, GRKKRRQRRRPPQ, across cellular membrane is highly basic and contains six arginine and two lysine residues in short peptide sequence [29]. The studies revealed that deletion of one arginine residue from the TAT sequence decreases translocation efficiency by approximately 50%, while deletion of two arginine residues results in 75% loss in activity [30]. Empirically, researchers have found that oligopeptides with less than five arginine residues significantly impair internalization rates and oligopeptides composed of less than three arginine residues show complete loss of translocation capabilities, thus implicating the dependency of chain-length on internalization rates [31]. In addition, studies by Futaki and co-workers have shown that branched arginine-rich peptides translocate as effectively as linear chained arginine sequences, which indicates that the number of arginine's present is a more critical determinant in nuclear translocation than tertiary

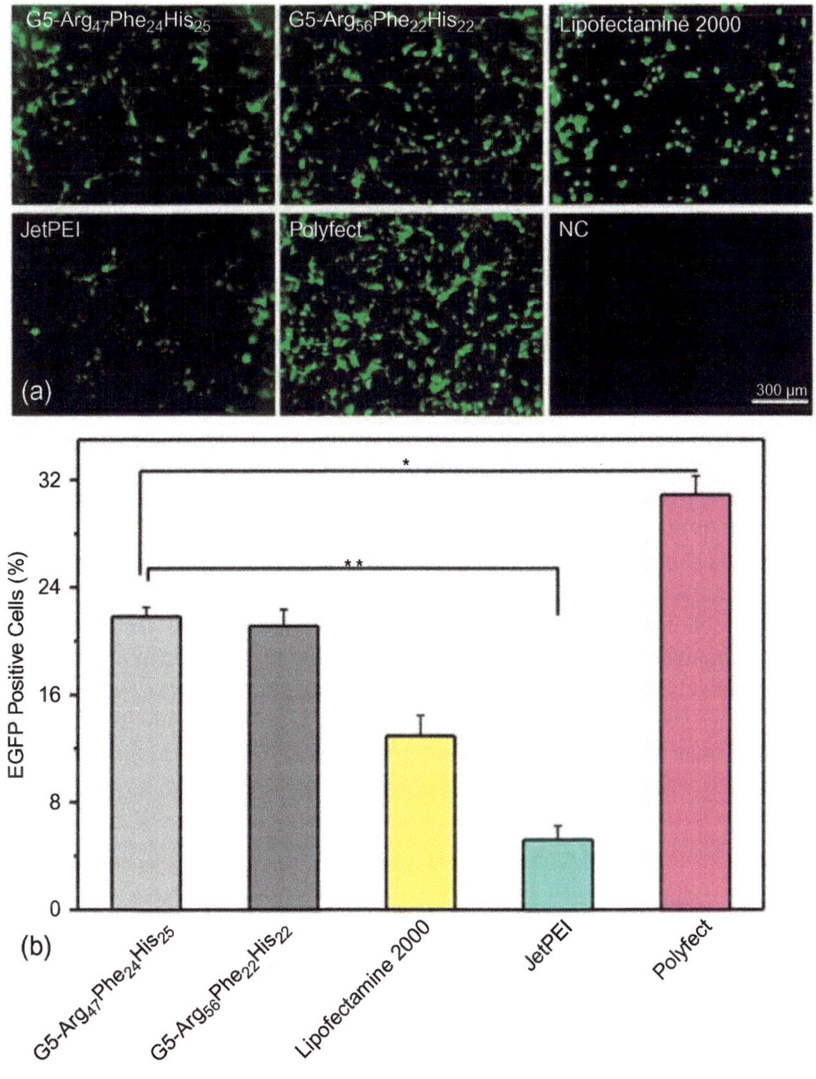

Figure 5.4 Comparisons of triple-functionalized dendrimers with three commercial transfection reagents (JetPEI, PolyFect and Lipofectamine 2000) on transfection efficacy. N/P ratios for G5-Arg47Phe24His25 and G5-Arg56Phe22His22 are 30:1 and 32:1, respectively. Gene transfection experiments for the commercial reagents were optimized according to the manufacturer's protocols. (a) Fluorescence microscopy images of the transfected cells and (b) EGFP expressions in NIH 3T3 cells measured by flow cytometry. Reprinted with permission from [26]. Copyright 2014 Elsevier Ltd.

peptide conformation and spatial cationic charge disposition [32]. However, while the cellular entry and translocation properties of arginine-rich peptides have been globally documented, the exact molecular mechanism of translocation remains elusive [33].

The advantages of functionalizing polymers with arginine residues to optimize the cell-penetrating properties have been explored by multiple groups in the past decade. In recent reports, Liu and co-workers demonstrated the capacity for triethanolamine (TEA) core poly(amidoamine) (PAMAM) dendrimers of high generations as highly effective nanodelivery systems for siRNA [34,35]. They combined the unique TEA-core PAMAM dendrimers with terminal arginine functionalization to incorporate the cell penetrating properties (Figure 5.5) [36]. Most notably, the generation 4 dendrimer bearing terminal arginine functionalization formed 130 nm stable polyplexes and led to improved uptake of siRNA compared to the non-arginine functionalized counterparts. The addition of the arginine residues on the dendrimer surface drastically improved the cytotoxicity profile of the nanoparticle and demonstrated excellent gene silencing of HSP27 and anticancer effects by caspase-3 cleavage of prostate cancer PC-3 models *in vivo* and *in vitro*. In a similar study, Liu and colleagues reported the facile synthesis of an arginine-decorated amphiphilic dendrimer composed of a hydrophobic alkyl chain and a hydrophilic PAMAM dendron with pendant arginine groups for siRNA delivery [37]. The arginine-functionalized PAMAM dendron was able to form stable colloidal polyplexes with HSP27 siRNA and anti-CD4 DsiRNA for gene silencing in human prostate cancer PC-3 cells and human hematopoietic CD34+ stem cells, respectively, in comparison to the corresponding non-arginine bearing controls. These findings show that the decoration of dendrimer surfaces with arginine residues can be used as an effective strategy to improve delivery efficacy and bind with various different siRNA plasmids as a versatile gene delivery system.

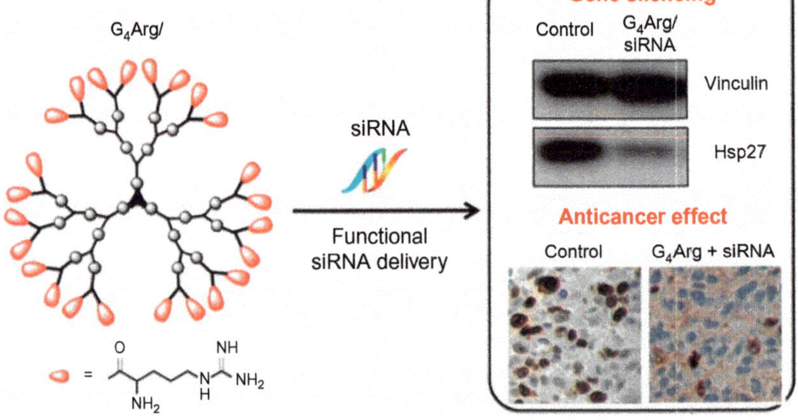

Figure 5.5 Schematic for chemical structure of arginine terminated TEA-core PAMAM generation 4 dendrimer and Gene silencing of Hsp27 of G4Arg/SiRNA complex by Western Blot Analysis and immunohistochemistry staining using K*i*-67.

Reprinted with permission from [36]. Copyright 2014 American Chemical Society.

In a novel study utilizing the cell-penetrating properties of arginine, Florinas and co-workers designed a non-viral VEGFR siRNA carrier using a combination of arginine-grafted bioreducible polymer, microbubbles, and ultrasound for targeting ovarian cancer cell lines [38]. The microbubbles with a perfluoro crown ether gas core loaded with siRNA demonstrated significant knock-down of EGFR in A2780 ovarian cancer cell lines only after a 10-minute transfection and ultrasonication treatment. They reported a faster transfection and siRNA uptake with the combined efficacy of the cell-penetrating properties of arginine and burst release of siRNA loaded in microbubbles after sonication *in vitro* in comparison to modern transfection modalities. Other recent findings from Wang and colleagues reported the development of a strategy for the co-delivery of apoptotic AVPI peptide and p53 pDNA as apoptosis inducing adjuvant treatment for therapy-resistant breast cancer [39]. The AVPI tetrapeptide has poor cell-permeability, therefore limiting its value in therapeutic applications. However, an AVPI derivative synthesized by Wang and co-workers that incorporated a terminal octa-arginine (R_8) sequence further allowed nanocomplex formation with p53 pDNA. In this study, the co-administration of AVPIR_8/p53 pDNA nanoparticles enabled full arrest of tumor growth *in vivo* and *in vitro* when combined with reducing DOX dosages, thereby resulting in an effective and safe chemotherapeutic treatment in drug-resistant breast cancer.

5.5 Cell-penetrating and fusogenic peptides

Bypassing the cell membrane is a critical step, but the efficient transport of peptide-DNA complexes across the nuclear membrane remains one of the most challenging hurdles in DNA delivery. Recently, the discovery of cell-penetrating peptides (CPPs) with the inherent property to translocate across the nuclear membrane has gained high attention due to their ability to significantly enhance gene knock-down from siRNA. CPPs are peptides that promote and facilitate the uptake of a wide range of macromolecules, including large proteins, multi-functionalized vectors, and non-viral/viral plasmids for gene therapy (Table 5.1). Although efficient delivery of DNA and siRNA can be achieved by CPPs, the transfection efficacy rarely exceeds the commercially available standard of Lipofectamine 2000 [40]. A common characteristic of CPPs is the presence of highly basic amino acids, such as arginine and lysine, which suggests the importance of these cationic motifs in delivering exogenous macromolecules. High cellular translocation proficiency with CPPs have been documented in the past decade, but the mechanism by which CPPs permeate the cellular membrane remains controversial. One advantageous feature of CPPs that is widely utilized for DNA and gene delivery vectors is the incorporation of peptide sequences that are known to function as nuclear localization signals (NLS). The transport of macromolecules between the cytoplasm and nucleus is generally facilitated by the interaction between specific NLS and transport receptors that translocate cargo across the nuclear membrane through nuclear pores.

Table 5.1 **Overview of cell penetrating peptides with their sequence, origin, and function in biomedical applications**

Identification	Sequence	Origin	Function	Ref.
Cell-Penetrating Peptides (CPP)				
TAT	GRKKRRQRRRPPQ	HIV-TAT domain	Translocation across cell membrane	41
Penetratin	RQIKIWFQNRRMKWKK	Antennapedia homeodomain	Translocation across cell membrane	42
Transportan	CLIKKALAALAKLNIKLLYGASNLTWG	Peptide fusion between galanin and mastoparan	Translocation across cell membrane	43
CADY	GLWRALWRLLRSLWRLLWKA	Synthetic	Translocation across cell membrane	44
NLS	CGYGPKKKRKVGG	SV40 NLS peptide	Nuclear localization	45
VP22	NAATATRGRSAASRPTQRPRAPARSASRPRRPVQ	Herpes simplex virus (HSV)	Nuclear localization	46
MPG	GALFLGWLGAAGSTMGAPKKKRKV	HIV glycoprotein 41 protein	Nuclear localization	47
R6H4	RRRRRRHHHH	Synthetic	Cell Penetration and Endosomal Escape	48
Fusogenic Peptides				
GALA	WEAALAEALAEALAEHLAEALAEALEALAA	Synthetic	pH-Responsive Endosomal Escape	49
KALA	WEAKLAKALAKALAKHL AKALAKALKACEA	Synthetic	Membrane transduction	50
HA-2	GLFGAIAGFIENGWEGMIDGWYG	Influenza Virus	Endosomal Membrane Fusion	51

Lee and co-workers described the gene delivery efficacy of PAMAM dendrimers conjugated with a nuclear localization signal, RRRK peptide, originating from mouse fibroblast growth factor 3 (FGF3) [52]. PAMAM is one of the most widely used non-viral gene delivery vectors, but it results in relatively low transfection efficiency as compared to that of PEI [6,53]. Interestingly, the PAMAM-FGF3 conjugated vector showed a transfection efficiency of luciferase expression plasmid (pLuc) similar to that of PEI in HEK293 and HeLa, but most notably demonstrated significant increase in luciferase gene expression in NIH3T3 mouse fibroblast cells, from which FGF3 peptide is originated, with relatively decreased cytotoxicity. These results imply the introduction of FGF3 NLS peptide potentially improves gene delivery efficacy in various cell lines, but also demonstrate a great potential for enhancing gene expression using targeting peptides on cell lines from which they originated [52]. In other works, Chu and colleagues reported the co-delivery of Bcl-2 siRNA and anticancer drug Ellipticine (EPT) with nanoparticles modified with cell penetrating peptide stearyl-(histidine)16-(arginine)8 (STR-H16-R8), which imparted cationic character and pH-buffering properties [54]. The synthesized EPT nanoparticles were prepared by solvent evaporation method, followed by the incubation with Bcl-2 siRNA to form complexes with STR-H16-R8. In this study, the knock-down of Bcl-2 siRNA efficiencies were comparable to commercially available Lipofectamine 2000, while also decreasing the toxicity usually associated with common transfection reagents. Interestingly, the knock-down of Bcl2 gene, a regulator protein affecting cellular apoptosis and necrosis, further sensitized the A549 lung carcinoma cells to the chemotherapeutic toxicity of EPT. This study demonstrated a novel dual-capacity delivery system for increasing cytotoxic effects of CPP-modified nanoparticles by combining the synergistic effects with siRNA gene silencing and chemotherapeutic drugs for enhanced efficacy.

Another study by Yang and colleagues reported the dual-modification of cationic liposomes with photolabile-caged cell penetrating peptide (pcCPP) and asparagine-glycine-arginine (NGR) targeting peptide for the delivery of c-myc siRNA in CD13 positive tumors [55]. The pcCPP was synthesized by functionalizing the positive charges of the lysine residues on the CPP (CGRRMKWKK) with 4,5-dimethoxy-2-nitrobenzyl groups. This is a near-infrared (NIR) two-photon excitation responsive group that can be cleaved by NIR adsorption and most recognizably used in applications for controlling functionality of materials in living cells [56]. In conjunction with the pcCPP, the NGR peptide motif has been found to be very efficient in targeting highly expressed CD13-positive tumors. Astonishingly, when the liposomes were loaded with c-myc siRNA, the NGR moiety was selectively bound to CD13-positive tumors on the cell surface and accumulated in tumor tissue, but could not penetrate the tumor membrane. However, upon illumination with NIR at the tumor site, the photo-labile group was cleaved and the uncaged CPP interaction with the tumor cell membrane, along with the synergistic effect of NGR targeting, caused rapid endocytosis and uptake of the liposome carrying siRNA. In systemic administration in mice, this novel gene delivery system was able to effectively silence c-myc expression and delayed tumor progression in CD13-positive tumors.

5.6 Conclusion and future prospects

Recent developments in amino acid and peptide conjugated delivery systems have shown promising potential in improving modern gene and drug delivery systems. The versatile and unique properties of individual amino acids along with the synergistic behavior of conjugated peptides demonstrate strong applications in therapy and diagnosis tools. However, while peptide conjugates have ultimately been shown to increase biocompatibility, reduce systemic toxicity, and contribute to compartmentalization escape and intracellular localization processes, the commercialization of peptide conjugates in the biomedical field remains a challenge. By exploiting novel technical strategies for pharmaceutical development, the engineering of peptide-conjugated delivery systems may provide a viable option for non-viral gene and drug therapies in the near future.

References

[1] Mahat R, Monera O, Smith L, Rolland A. Curr Opin Mol Ther 1999;1(2):226–43.
[2] Kozarsky KF, Wilson JM. Curr Opin Genet Dev 1993;3(3):499–503.
[3] Mulligan RC. Science 1993;260(5110):926–32.
[4] Thomas CE, Ehrhardt A, Kay MA. Nat Rev Genet 2003;4(5):346–58.
[5] Martin ME, Rice KG. AAPS J 2007;9(1):E18–29.
[6] Pack DW, Hoffman AS, Pun S, Stayton PS. Nat Rev Drug Discov 2005;4(7):581–93.
[7] Dufès C, Uchegbu IF, Schätzlein AG. Adv Drug Deliv Rev 2005;57(15):2177–202.
[8] Erbacher P, Zou S, Bettinger T, Steffan A-M, Remy J-S. Pharm Res 1998;15(9):1332–9.
[9] Luo D, Saltzman WM. Nat Biotechnol 2000;18(1):33–7.
[10] Zanta MA, Belguise-Valladier P, Behr J-P. Proc Natl Acad Sci 1999;96(1):91–6.
[11] Gupta B, Levchenko TS, Torchilin VP. Adv Drug Deliv Rev 2005;57(4):637–51.
[12] Laemmli U. Proc Natl Acad Sci 1975;72(11):4288–92.
[13] Mintzer MA, Simanek EE. Chem Rev 2008;109(2):259–302.
[14] Richardson SW, Kolbe HJ, Duncan R. Int J Pharm 1999;178(2):231–43.
[15] Ohsaki M, Okuda T, Wada A, Hirayama T, Niidome T, Aoyagi H. Bioconjug Chem 2002;13(3):510–7.
[16] Hartono SB, Gu W, Kleitz F, Liu J, He L, Middelberg AP, et al. ACS Nano 2012;6(3):2104–17.
[17] Liu T, Xue W, Ke B, Xie M-Q, Ma D. Biomaterials 2014;35(12):3865–72.
[18] Ma D, Zhang H-B, Chen Y-Y, Lin J-T, Zhang L-M. 305-311. J Colloid Interface Sci 2013;405.
[19] Zheng C, Zheng M, Gong P, Deng J, Yi H, Zhang P, et al. Biomaterials 2013;34(13):3431–8.
[20] El-Sayed A, Futaki S, Harashima H. AAPS J 2009;11(1):13–22.
[21] Benns JM, Choi J-S, Mahato RI, Park J-S, Kim SW. Bioconjug Chem 2000;11(5):637–45.
[22] Langlet-Bertin B, Leborgne C, Scherman D, Bechinger B, Mason AJ, Kichler A. Pharm Res 2010;27(7):1426–36.
[23] Varkouhi AK, Scholte M, Storm G, Haisma HJ. J Control Release 2011;151(3):220–8.
[24] Brevet D, Hocine O, Delalande A, Raehm L, Charnay C, Midoux P, et al. Int J Pharm 2014;471(1):197–205.
[25] Hwang HS, Hu J, Na K, Bae YH. Biomacromolecules 2014;15(10):3577–86.

[26] Wang F, Wang Y, Wang H, Shao N, Chen Y, Cheng Y. Biomaterials 2014;35(33):9187–98.
[27] Gonçalves C, Berchel M, Gosselin M-P, Malard V, Cheradame H, Jaffrès P-A, et al. Int J Pharm 2014;460(1):264–72.
[28] Harrington W, Rodgers M. Annu Rev Cell Biol 1985;1:41.
[29] Schmidt N, Mishra A, Lai GH, Wong GC. FEBS Lett 2010;584(9):1806–13.
[30] Futaki S, Suzuki T, Ohashi W, Yagami T, Tanaka S, Ueda K, et al. J Biol Chem 2001;276(8):5836–40.
[31] Tung C-H, Weissleder R. Adv Drug Deliv Rev 2003;55(2):281–94.
[32] Futaki S, Nakase I, Suzuki T, Youjun Z, Sugiura Y. Biochemistry 2002;41(25):7925–30.
[33] Futaki S, Nakase I, Tadokoro A, Takeuchi T, Jones A. Biochem Soc Trans 2007;35(4):784.
[34] Liu Xx, Rocchi P, Qu Fq, Zheng Sq, Liang Zc, Gleave M, et al. ChemMedChem 2009;4(8):1302–10.
[35] Liu X, Liu C, Laurini E, Posocco P, Pricl S, Qu F, et al. Mol Pharm 2012;9(3):470–81.
[36] Liu C, Liu X, Rocchi P, Qu F, Iovanna JL, Peng L. Bioconjug Chem 2014;25(3):521–32.
[37] Liu X, Liu C, Zhou J, Chen C, Qu F, Rossi JJ, et al. Nanoscale. 2015.
[38] Florinas S, Nam HY, Kim SW. Mol Pharm 2013;10(5):2021–30.
[39] Wang H, Wang H, Liang J, Jiang Y, Guo Q, Peng H, et al. Mol Pharm 2014;11(10):3352–60.
[40] Lundberg P, El-Andaloussi S, Sütlü T, Johansson H, Langel Ü. FASEB J 2007;21(11):2664–71.
[41] Sugita T, Yoshikawa T, Mukai Y, Yamanada N, Imai S, Nagano K, et al. Biochem Biophys Res Commun 2007;363(4):1027–32.
[42] Zorko M, Langel Ü. Adv Drug Deliv Rev 2005;57(4):529–45.
[43] Meade BR, Dowdy SF. Adv Drug Deliv Rev 2007;59(2):134–40.
[44] Rydström A, Deshayes S, Konate K, Crombez L, Padari K, Boukhaddaoui H, et al. PLoS One 2011;6(10):e25924.
[45] Wang F, Wang Y, Zhang X, Zhang W, Guo S, Jin F. J Control Release 2014;174:126–36.
[46] Chen H-CG, Chiou S-T, Zheng J-Y, Yang S-H, Lai S-S, Kuo T-Y. Arch Virol 2011;156(5):803–15.
[47] Morris MC, Deshayes S, Heitz F, Divita G. Biol Cell 2008;100(4):201–17.
[48] Jiang T, Zhang Z, Zhang Y, Lv H, Zhou J, Li C, et al. Biomaterials 2012;33(36):9246–58.
[49] Baio JE, Schach D, Fuchs AV, Schmüser L, Billecke N, Bubeck C, et al. Chem Commun 2015;51(2):273–5.
[50] Yoo H, Mok H. Arch Pharm Res 2015;38(1):129–36.
[51] Chen P, Jafari M, Xu W, Chen B, Pan R, Karunaratne N. Peptide sequence design and use thereof for peptide-mediated sirna delivery; 2014, US Patent 20,140,350,082.
[52] Lee J, Jung J, Kim Y-J, Lee E, Choi JS. Int J Pharm 2014;459(1):10–8.
[53] Eichman JD, Bielinska AU, Kukowska-Latallo JF, Baker JR. Pharmaceutical science & technology today 2000;3(7):232–45.
[54] Chu D, Xu W, Pan R, Chen P. RSC Adv 2015;5(26):20554–6.
[55] Yang Y, Yang Y, Xie X, Wang Z, Gong W, Zhang H, et al. Biomaterials 2015;48:84–96.
[56] Aujard I, Benbrahim C, Gouget M, Ruel O, Baudin JB, Neveu P, et al. Chemistry-a European Journal 2006;12(26):6865–79.

Dendrimers for gene therapy

6

M.J. Santander-Ortega, M.V. Lozano, I.F. Uchegbu, A.G. Schätzlein
University College of London, London, UK

6.1 Introduction

Since their first application as transfection agents, dendrimers have emerged as a promising synthetic vectors for gene therapy, often demonstrating improved efficiency with limited toxicity ratio compared to many systems based on linear and branched polymers [1–3].

As for other polyelectrolytes, the nucleic acid condensation capacity of dendrimers depends on the attractive electrostatic interactions between the anionic phosphorous groups of the nucleic acid backbone and the cationic groups of the dendrimer (typically amine groups). This interaction leads to the condensation of the nucleic acids and polymer into nanoparticles, also known as dendriplexes [4]. Dendrimers efficiently complex different types of nucleic acids (NAs) such as DNA [1,5–7], RNA [7–9], or ODN [10–12], often with improved transfection efficiency and lower toxicity than linear or branched polymers [1–3].

Dendrimers differ from related cationic polymers in a number of ways, as they are hyperbranched polymers with a well-defined 3D structure and molecular weight; see Figure 6.1 [13,14]. Their spherical shape, high density of superficial functional groups, and controlled molecular structure (Mw/Mn 1.01–1.05) are due to the synthetic route used. In contrast to stochastic polymers, dendrimers are synthesized systematically in a step-wise fashion [15–18].

The complexation for a given dendrimer depends on the ratio of dendrimer and nucleic acid as well as on the bulk conditions (e.g., pH or ionic strength) that modulate the dendrimer-gene interactions. The molecular structure of the nucleic acid can also play an important role, i.e., properties of complexes will differ between, e.g., siRNA and plasmid DNA [7,19].

The ability of dendriplexes to facilitate the uptake and intracellular and nuclear trafficking of NA has been widely demonstrated [20–23]. Similar to many other vector systems dendriplexes, when used *in vitro*, dendrimers show excellent transfection efficiency, achieving even the levels obtained with lentivirus [24]. However, the main barrier for the wider clinical application of this technology is their much more limited efficacy when used *in vivo*. In comparison with cells in a petri dish, under *in vivo* conditions dendriplexes are exposed to a much more complex environment where interactions with components of the vascular compartment modulate complex properties and the majority of cells encountered are off-target.

The chemical versatility of dendrimers has been the source of much interest in this context as it facilitates grafting of various functional groups in high density and allows the fine-tuning of the molecule to improve *in vivo* performance. In fact, this chemical

Polymers and Nanomaterials for Gene Therapy. http://dx.doi.org/10.1016/B978-0-08-100520-0.00006-0
© 2016 Elsevier Ltd. All rights reserved.

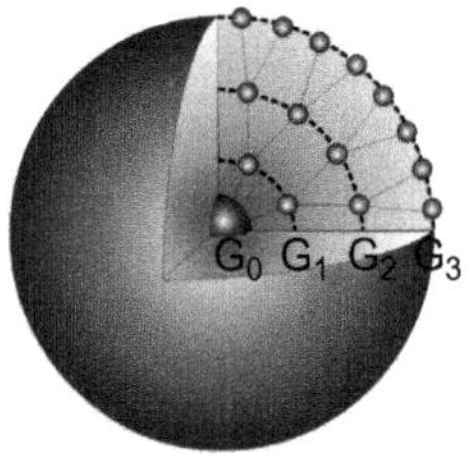

Figure 6.1 Model dendrimer structure representing the number of functional groups as a function of the dendrimer generation.

modification of dendrimers has been used with some success to improve condensation capacity, promote colloidal stability, and target the nanostructure towards the desired cells *in vivo* [25–31].

However, while dendrimer chemistry modulates the physical chemistry of the complex and thus its biological properties, these correlations are complex and interdependent. This makes rational development of such carriers non-trivial. For example, chemical modifications designed to modulate a specific property, such as increased condensation capacity or colloidal stability, may result in unwanted modification of the *in vivo* biodistribution of the dendriplexes and loss of efficiency [22,32,33].

Interestingly, complex modifications may not always be required. For example, un-modified low-molecular weight poly propylenimine (PPI) dendrimers show highly selective tumor gene expression after systemic intravenous administration [34,35]. These observations are unlikely attributable to the positively charged complexes alone but also are at least partly due to the specific interaction of these nanoparticles with serum proteins after their administration. The rapid coating of the nanostructure by serum proteins will lead to the formation of a protein corona around the dendriplexes that will affect both their pharmacokinetic and pharmacodynamics, and which complicates the systematic development of such carriers further [36–39].

In the present chapter we will focus on the correlation of dendrimers chemistry and physical chemistry to *in vitro/in vivo* performance of dendriplexes highlighting properties that make the dendrimers such successful synthetic vectors.

6.2 Dendrimer structure and chemistry

Initially we will describe dendrimer structure and chemistry using some of the more frequently used dendrimers as examples (Figure 6.2 and 6.3).

6.2.1 PAMAM

Since Haensler and co-workers initially used them as transfection agents in 1993, poly(amido amine) ("PAMAM") dendrimers have been the most widely studied dendrimers for gene therapy [2,14]. This cationic polyelectrolyte displayed enhanced transfection efficacy and lower toxicity than other polycations used for synthetic

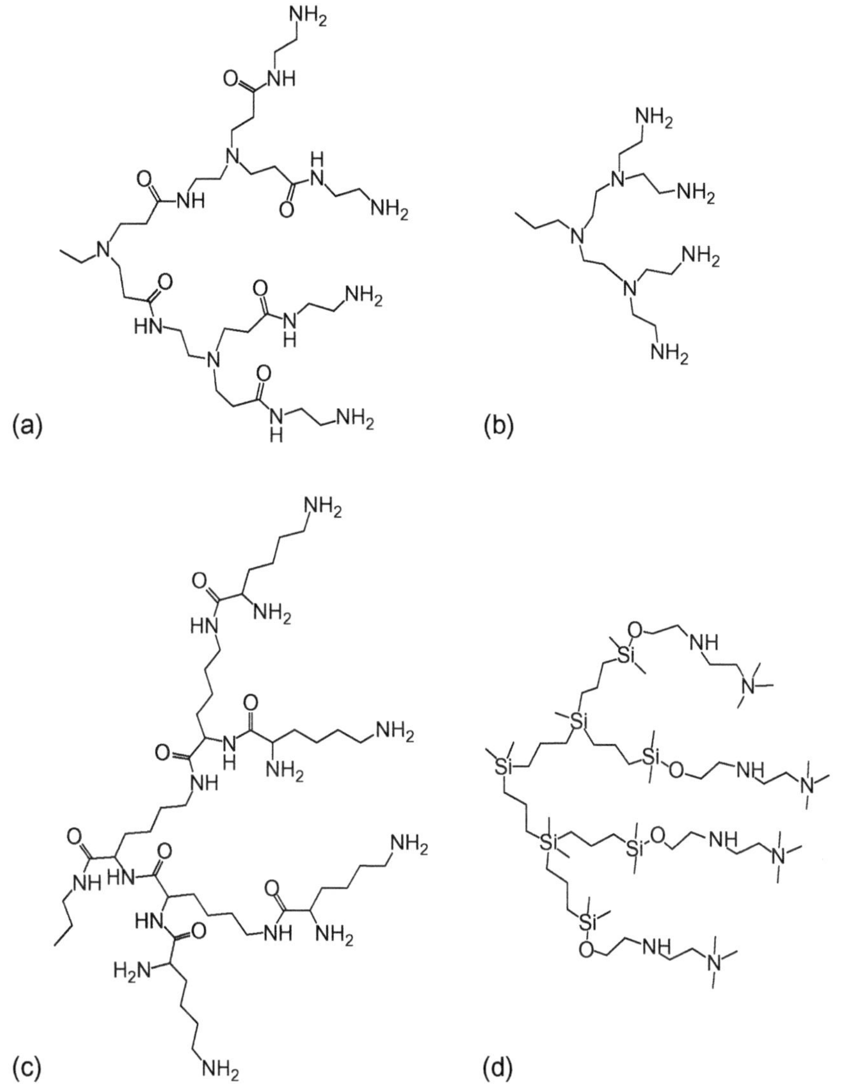

Figure 6.2 Examples of 2D molecular structure of G2 dendrons (half-molecule) of PAMAM (a), PPI (b), PLL (c), and CBS (d) dendrimers.

vector formulations at that time, such as linear poly lysine (PLL), polyethylenimine (PEI), or DEAE-dextran [2,40]. Interestingly, studies carried out by Bielinska and co-workers [41] demonstrated that only a small fraction of the whole formulation (10–20%), composed of soluble low-density dendriplexes, was responsible for 90% of the transfection efficiency of the formulation.

Key dendrimer properties that have been linked to transfection efficiency of these molecules are molecular weight, inner architecture, dendrimer-gene ratio, etc., and

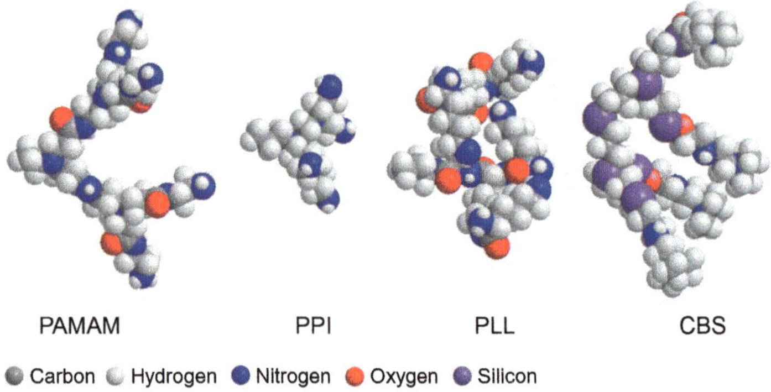

Figure 6.3 Examples of 3D molecular structure of G2 dendrons (half molecule) as a function of the building blocks, amido amine (PAMAM), propylenimine (PPI), L-lysine (PLL) and propylensilane (CBS). Minimal energy conformation was reached by MM2 calculations (Chembiodraw 12). Illustration clearly shows the different molecular architecture of the dendrimers as a function of their buildings blocks.

some general rules have been suggested. For example, as a general rule, increase in the dendrimer molecular weight will lead to more effective transfection agents but also to more toxic synthetic vectors [1,42,43].

6.2.2 PPI

This type of dendrimer in gene therapy is based on a polypropylenimine (PPI) architecture with a butylenediamine inner core. Originally these dendrimers were thought to be too toxic to be used in synthetic vector formulations [44–46] but work carried out in our group demonstrated that for the lower generation PPI dendrimers (G1-G4) a balance of efficiency versus toxicity exists that allows their exploitation as vectors [47]. Interestingly, G3 PPI dendrimers have shown great potential for the formulation of different NAs for cancer gene therapy [14,48], and have shown selective tumor accumulation [34,49].

6.2.3 Peptide

Peptide dendrimers are based on peptidyl branching units. Poly-l-lysine (PLL) dendrimers, described by Denkewalter and co-workers in 1981 [50], are the most frequently used dendrimers of this category. Dendritic PLL (dPLL) has shown improved efficiency compared to linear PLL, the first polyelectrolyte used in gene therapy [3]. The change from linear to hyperbranched architecture has resulted in a weaker dendrimer-gene interaction and an improved endosomal escape due to the lower pKa of the amine groups in the dendrimer [3]. Despite these improvements, in comparison with others dendrimers, PLL dendrimers have shown lower endosomal escape capacity. This handicap has been overcome by the insertion of fusogenic and other peptides, such as arginine and histidine, which promote both cellular uptake and endosomal escape [51,52].

The most relevant clinical results obtained with PLL dendrimers in gene therapy have been published by Marano and co-workers for the long-term treatment of choroidal neovascularization (CNV), an age-related macular degeneration process [53], using anti-VEGF oligodeoxynucleotide (ODN-1) as a therapeutic gene [12,53].

6.2.4 Other dendrimers

Carbosilane dendritic macromolecules modified with terminal amino groups have been recently proposed as suitable dendrimers for gene therapy [54,55]. Despite their initial promising potential, the fast hydrolysis of Si-O bonds under physiological conditions has prompted design of a second-generation prototype with the insertion of phenyl groups. The lower reactivity of Si-C bonds reduced the degradation of the compound and led to a controlled release of the NA [55]. Complexes of G2 Carbosilane dendrimers (CBS) and siRNA have been used to silence the expression of glyceraldehyde 3-phosphate dehydrogenase (GAPDH) in human leukemia T lymphocytes and peripheral blood mononuclear cells (PBMC) with low toxicity [54]. CBS G2-NN16 dendrimers condensing specific siRNA sequences have also been used to modulate gene expression in macrophages [56]. CBS dendrimers have also shown a great potential to transfect neurons, classically considered as difficult to transfect cell type, reaching efficiencies similar to those obtained with lentivirus and much higher than those obtained with Lipofectamine [24]. For more details about the use of CBS dendrimers in gene therapy see [8].

Another interesting category is glycodendrimers, a family of dendritic macromolecules that include carbohydrates in their structure [57]. Glycodendrimers can present *i)* surface associated saccharides, *ii)* inner core saccharides, and *iii)* saccharide building units. Surface associated saccharides are of special interest due to their characteristics as cell targeting and active uptake promoters. Mannosylated PAMAM-α-CD have shown gene expression in kidneys after tail vein injection in mice, whereas mannosylated PLL dendrimers presented a main activity in liver [58,59]. More detailed information about glycodendrimers can be found elsewhere [60].

6.3 Dendrimer architecture

Dendrimers present a hyperbranched structure created by a step-wise synthesis based on the addition of the monomeric units layer by layer (see Figure 6.1). There are two different strategies for the synthesis of these molecules, the divergent (from core to shell) and the convergent strategy (from shell to core) [14], where each new additional cycle is known as generation [61]. The divergent strategy tends to be less efficient because incomplete and side reactions eventually lead to the accumulation of structural defects. In contrast, the convergent approach limits the number of side reactions, enables easier purification of intermediate products, and hence minimizes structural defects. Despite the potential defects in their structure, both approaches allow the production of molecules with a well-defined architecture with low polydispersity ($Mw/Mn < 1.01$-1.05) [15]. This type of architecture leads to a gradient of functional groups from the inner core (low density) to the shell of the molecule (high density) [15,16,18,62,63].

Interestingly, starting from intact PAMAM dedrimers, Szoka and Tang have shown that the thermal degradation of the dendrimer led to a 50-fold increase in the transfection capacity of the dendrimers compared to intact PAMAM [64]. Hydrolysis of the dendrimer amide groups leads to a reduction of branch density and intramolecular repulsions. This increase in the molecule's flexibility has been proposed as the main reason behind the improved performance of the "activated" dendrimer [65,66].

To confirm the role of the polyelectrolyte architecture in the performance of the corresponding synthetic vector, Wagner and co-workers studied the efficiency of vectors formulated with linear PEI (LPEI), branched PEI (BPEI), PAMAM, and activated PAMAM [65,67,68]. After tail vein injection in mice all the systems were mainly accumulated in the lungs. Despite the similar biodistribution, the transfection level was clearly dependent on the polyelectrolyte architecture. While intact PAMAM presented negligible gene expression, LPEI showed predominant lung gene expression [67]. In line with the results obtained by Szoka and co-workers, activated PAMAM presented a significant increase in lung gene expression [65]. Considering the similar biodistribution of the three dendriplexes, differences in their gene expression could potentially be attributed to their release capacity [64,66,69–71]. The fractured PAMAM dendrimer has since been used widely as it is commercially available as Superfect® from Qiagen [66,72].

6.4 Dendrimer generation

As shown in Figure 6.1, each new layer of monomeric units added to the dendrimer core represents a new generation. Bearing this in mind, the number of functional groups in the dendrimer surface at a specific generation (F_G) depends on N_b^G, where N_b is the monomer multiplicity and G the dendrimer generation [73,74]. Exponential increase of the number of functional groups with respect to the dendrimer generation will affect the molecule shape and its interaction with both, the nucleic acids and the biological membranes after administration [75,76]. Interestingly, for PAMAM dendrimers, each new generation leads to cationic $-NH_2$ bearing dendrimers superficial groups while each intermediate half generation (G2.5, G3.5,…) present terminal anionic $-COO^-$ carboxylate groups.

While low-generation dendrimers tend to adopt a planar-elliptical shape, increase in the generation results in an arrangement of the dendrimer architecture towards a spherical shape characterized by a higher branch density in the outer shell than in the inner core [14,76,77]. Molecular structure of dendrimers can be affected by their interaction with the solvent. Factors such as pH, ionic strength (even specific type of ions), or solubility can lead to the folding of the shell branches toward the inner core [78–80], resulting in a uniform or even inverse functional groups distribution [14,81].

The affinity of the dendrimer to NA is also affected by the dendrimer generation [47,82,83]. The use of low-generation dendrimers leads to more fluid structures. Due to the relatively lower binding affinity the dendrimer can be re-accommodated depending on the nucleic acid conformation [83,84]. However, for high-generation dendrimers the higher number of functional groups per dendrimer molecule favors a tight dendrimer-nucleic acid interaction. This higher binding affinity promotes the dendrimer-gene binding by wrapping the nucleic acid around the dendrimer [47]; see Figure 6.4. Differences in the dendrimer-gene affinity of low- and high-generation

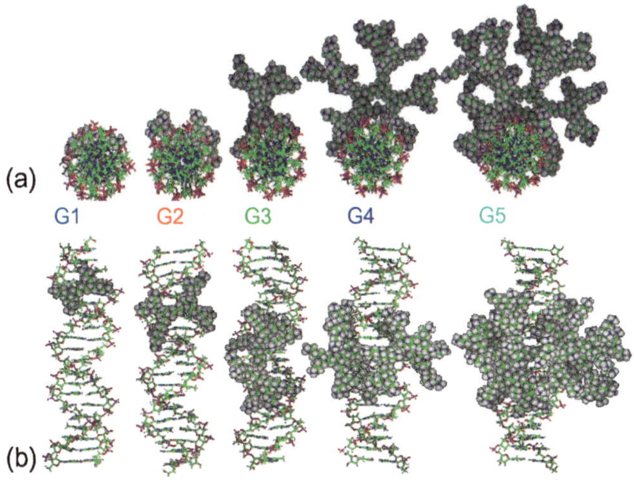

Figure 6.4 (a) Molecular models of PPI-DNA binding (top view). From left to right: generation 1–5. (b) Molecular models of the dendrimer-DNA binding (side view). From left to right: generation 1–5. Molecular modeling using the Insight II software (Accelrys, California) showed a clear increment in simulated intermolecular energies (Kcal mol^{-1}) as a function of the dendrimer generation PPI G1 ~ -3.9, PPI G2 ~ -9.6, PPI G3 ~ -17.6, PPI G4 ~ -31.3 and PPI G5 ~ -61.7 [47].

dendrimers have been confirmed by the analysis of PAMAM dendriplexes formation/gene release in media with different salinity [82,83] with low-generation dendriplexes being much more prone to be destabilized by electrolytes compared to higher-generation dendrimers.

Another drawback of lower-generation dendrimers stems from their tendency to form electro-neutral dendriplexes with nucleic acids. Due to the lack of superficial charge (ζ-potential ≈ 0) these complexes will present a high tendency to aggregate even in low salinity media [84]. By increasing the dendrimer generation, the exponential growth in the number of charged functional groups per dendrimer facilitates the overcharging of the genes during the complexation. This creates cationic dendriplexes that are colloidally stable due to the inter-particles electrostatic repulsions [85].

Despite the advantages of higher-generation dendrimers to promote the formation of strong dendrimer-NA interactions and colloidally stable dendriplexes, the efficiency of the dendriplexes also depends on *i)* their interaction with biological fluid components and living cells, *ii)* intracellular trafficking, and *iii)* their ability to release the cargo in the target site.

In dendriplex formulations the dendrimers are typically present in excess to ensure formation of fully charged colloidally stable particles. However, in this environment, the presence of the highly cationic free dendrimer molecules can lead to strong interaction with the mainly anionic biological fluids and biomembranes, which may result in some toxicity. As well as leading to a dramatic increase in the formulations cytotoxicity, the increase in free cationic groups will also affect the biodistribution of the complexes [10,14,44,47,67,86]. In addition, once in the target, the tight polyelectrolyte-gene interaction can reduce the chances of efficient release of the NA [70,71].

Therefore, design of efficient dendriplexes requires a balance to be struck between complex stability in biological fluids and NA release at the target site. At the same time it is mandatory to minimize the undesirable interactions with the anionic components of biological fluids and membranes.

Bearing this in mind we recently designed a novel core-shell strategy based on nanostructures with a dendriplex core based on low-generation dendrimers (PPI G3) optimized for DNA condensation/release, surrounded by a shell of hyaluronic acid that *i)* provides colloidal stability under physiological conditions, *ii)* reduces the cytotoxicity of the nanostructure, and *iii)* promotes its uptake by $CD44^+$ tumorigenic cells [87].

6.5 Chemical modifications

Despite the excellent properties of dendrimers as nucleic acid condensation agents, their applicability is hampered for two main reasons *i)* non-specific, and potentially toxic, interaction with biological membranes due to their cationic character and *ii)* lack of tissue specificity.

Dendrimer molecular architecture allows for facile covalent modification of terminal groups resulting in modulation and control of interactions with biological membranes and blood serum components, tissue targeting [25–31], and tuning of the NA condensation and release [26,88], and tuning of the NA condensation and release [26,88], see Figures 6.5 and 6.6.

Superficial cationic functional groups are thought to play an important role in the potential toxicity displayed by dendrimers. Different alternatives have been explored to transform these reactive superficial groups into others with a more inert character. Chemical modifications to obtain hydroxyl or acetylated end-capped dendrimers have shown a clear reduction in the cytotoxicity of PAMAM and PPI dendrimers [89–91].

Polyethylene glycol (PEG) derivatives are probably the most frequently used to reduce the interaction of nanoparticles with membranes and macromolecules [92–95]. The effect of molecular weight and degree of substitution of covalently PEG modified terminal dendrimer groups has been studied in detail [96]. Low and intermediate molecular weight PEG derivatives have shown to be more efficient than acetylation reducing the cytotoxicity of the dendrimer [25,97]. However, high degrees of modification with longer PEG derivatives lead to dendrimers aggregation, increased cytotoxicity, and reduced uptake [33].

Although these covalent modifications have shown to be an efficient tool to avoid the toxicity associated with the cationic superficial groups of dendrimers, their substitution by more inert groups also affects the condensation capacity of the dendrimer, and thus the efficiency of the dendriplex. To compensate for the loss of condensation capacity some authors have also quaternized the inner amino groups resulting in a permanent positive charge to help maintain the molecule's NA condensation capacity with a view to retain high *in vitro* transfection efficiency even in serum and to allow tumor accumulation *in vivo* after intravenous administration [89,98].

Quaternization of superficial amino groups with methyl moieties has also been used to improve the condensation capacity of lower generation (G1-G4) PPI dendrimers [32,49]. Interestingly, the degree of amine substitution is also linked to reduced toxicity ($R\text{-}NH_3^+ > R_2\text{-}NH_2^+ > R_3\text{-}NH^+ > R_4\text{-}N$ [99].

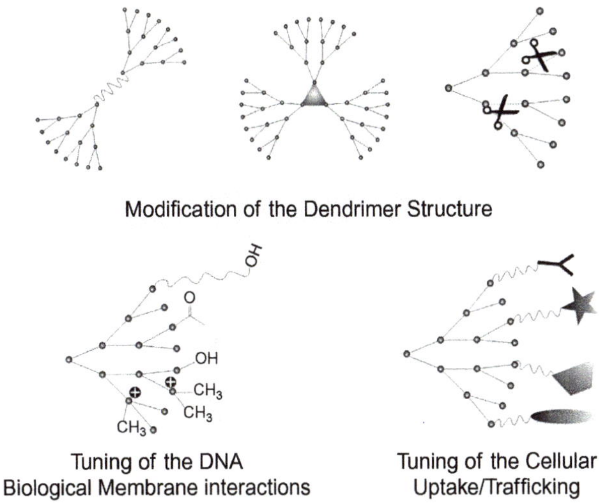

Figure 6.5 Schematic representation of some chemical modifications of dendrimers. 1) Modification of the Dendrimer Structure: Among others, such modifications include the insertion of spacers (either rigid or flexible) in the inner core as well as the thermal degradation of the branches [114,140,203]. 2) Tuning of the DNA and Biological Membrane Interactions: Screening of the strong interaction of dendrimers with biological membranes has been modulated by the grafting of PEG derivatives, acetylation of terminal amino groups, or insertion of hydroxyl groups, whereas improvement of the genes' condensation capacity has been done by the quaternization of inner or superficial amino groups [25,32,89–91,97,98]. 3) Tuning of the Cellular Uptake and Trafficking: Cellular internalization has been promoted by the insertion of targeting motifs (top), fusigenic peptides, cyclodextrins, or hydrophobic molecules (bottom) [20,51,52,100,105,106,224].

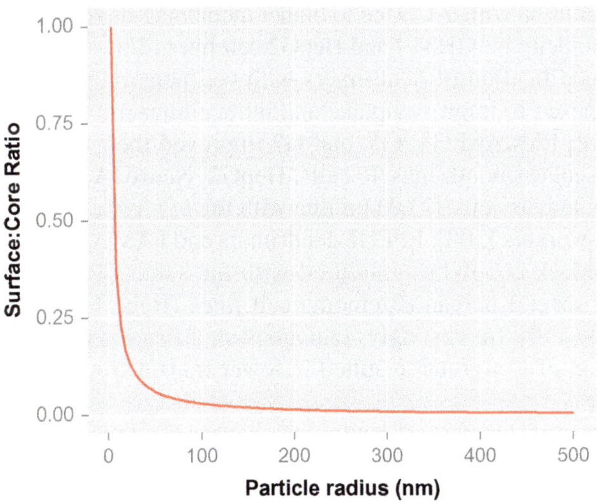

Figure 6.6 The relationship of the relative number of molecules of a nanoparticle "hidden" in the core of a particle vs. those exposed on the surface depends on particle size and can be estimated: the surface area of a sphere is given by $4\pi r^2$. The volume of a sphere by $\frac{4}{3}\pi r^3$. The surface to volume ratio of a sphere therefore is given by $\frac{4\pi r^2}{\frac{4}{3}\pi r^3} = 3/r$

Hydrophobic modification of dendrimers can also considerably modify the interaction of dendriplexes with cells [100,101]. For example, Yoo and co-workers report that PAMAM dendrimers labeled with the hydrophobic dye Oregon Green 488 improved transfection efficiency over the unmodified dendrimer [20].

Similarly, low degrees of covalent modification of PAMAM G5 dendrimers with cholesterol molecules also improve the efficacy of the dendriplexes [102]. However, a higher degree of modification resulted in the loss of the solubility and the condensation capacity of the dendrimers with a corresponding reduction in their transfection capacity [102]. A similar effect was observed by the modification of PAMAM G4 dendrimers with the hydrophobic phenylalanine amino acid [103].

More recently, PAMAM G4 dendrimers modified with PEG_{2000}-DOPE have been synthetized for optimized siRNA condensation and release [104]. These compounds are thought to form micellar structures with the phospholipid at the core improving the compatibility and the cell penetration of the dendriplex. PAMAM-PEG_{2000}-DOPE dendriplexes mixed with PEG_{2000}-DOPE micelles led to an assembled system stable in serum with significant higher uptake of siRNA and then a better gene-silencing level in comparison with un-modified PAMAM G4 dendriplexes [104]. In parallel to the condensation of siRNA in the dendriplex units, this assembled system allowed the incorporation of the anticancer drug doxorubicin in the counterpart micellar units.

The improved transfection from α-cyclodextrines (α-CD) modified PAMAM G3 has also been attributed to the membrane disruptive properties of α-cyclodextrines (α-CD), which could lead to a more efficient endosomal escape [105,106]. The optimal effect was achieved for intermediate degrees of modification as the increase in the degree of substitution with α-CD led to higher membrane-disruption capacity but also to higher cytotoxicity in NIH3T3 and HepG2 cell lines [22].

Covalent modification of dendrimers with cell-penetrating peptides (CPP) has also been proposed to improve uptake and intracellular trafficking [107]. Grafting of L-arginine to PAMAM G4, G5, and G7 improved their transfection efficiency in primary vascular smooth muscle cells, HepG2, Neuro2A, HUVECs, and SMCs with relatively low toxicity [27,31]in line with the *in vivo* results recently obtained by Liu and co-workers [108]. PPI G2 dendrimers and PAMAM-PEG3400-PAMAM G5 dendritic block copolymer modified with the same CPP confirmed this positive effect in several human carcinoma cell lines (Hela, HEK 293, HepG2, and HUVECs) [109,110]. Interestingly, analogous modification of the dendrimers with L-lysine instead of L-arginine resulted in lower transfection efficiency of the formulation [27].

Another structural element of dendrimers that has been modified is the inner core. Zhou and coworkers used PAMAM G2 with a triethanolamine inner core instead of the classical ethylenediamine for siRNA delivery [43,111–113]. This modification is thought to create a more flexible molecule that could favorably interact with siRNA allowing gene silencing of hsp2 in prostate cancer cells [111]. A PAMAM hybrid dendrimer with a rigid polyphenylenevinylene (PPV) core surrounded by three PAMAM G1 dendron branches also demonstrated improvements in rat cerebellar granular neurons (CGNs) [114]. These results, together with the enhanced efficiency displayed by the fractured PAMAM Superfect®, suggest that the flexibility of the inner core could be a key element for rational design of new vectors, aiming to reduce the number of surface amine groups while improving the structural mobility by either, the insertion of inner core spacers or by the reduction of the branches density by thermal degradation.

Chemical modifications to promote the triggered endosomal disassembly and release of the complex have also been explored in some detail in a number of *in vitro* studies [115–117]. For example, Baigude and co-workers have shown the specific design of dendriplexes (iNOP-7DS) for facilitating siRNA intracellular release. In this molecule, polylysine dendrimers G4 surface groups have been modified with lipid chains and spacers with reducible disulfide bonds containing terminal $-NH_2$ groups [118,119]. The incubation of the dendriplexes with reducing agents (glutathione or dithiothreitol) cleaved the disulfide bond leading to the loss of the terminal amino groups and then the release of the associated gene [118]. Efficiency of the formulation was confirmed *in vitro* by the silencing of ApoB transcription in FL83B mouse hepatocytes (90% of expression inhibition). Injection of ApoB siRNA- iNOP-7DS to mice led to 40% inhibition of ApoB expression in liver and a 45% reduction of the protein content in blood [118]. Additionally, intravenous administration of the same nanostructure condensing anti-miR-122 ODN, instead of ApoB siRNA, specifically reduced the liver-expressed miR-122 in mice by ~80% and reduced the plasma cholesterol concentration [120].

Although all these modifications are designed to modify specific aspects of dendriplexes, e.g., condensation capacity, cytotoxicity, etc., such modifications usually affect more than one aspect of the final formulation. Addition of methyl groups to superficial amino groups of PPI G2 dendrimers improved their condensation capacity and reduced their cytotoxicity *in vitro*, but also modified their biodistribution after tail vein injection in mice, promoting their accumulation in the liver instead that in the lungs [32]. Analogously grafting of α-cyclodextrin to PAMAM G3 dendrimers favored the gene expression in the spleen *in vivo* [22]. These results suggest that although some modifications are designed to modulate specific aspects of dendriplexes, their potential effect on the modification of the final fate of the formulation after administration should be carefully considered. These mixed effect often make it very challenging to clearly demonstrate the mechanism behind the observed structure activity relationships.

The versatile architecture of dendrimers also allows the anchoring of moieties for targeting. Poly-l-lysine dendrigrafts, a class of dendrimers that combine features of dendrimers and hyperbranched polymers, (DGL) covalently modified with Dermorphin, a ligand of the μ-opioid receptors present in brain capillary endothelial cells, have been used to complex anti-Ask1 shRNA to target the Ask gene, an important gene in cerebral ischemia reperfusion injury [121]. Intravenous administration in a rat model led to a significant reduction of apoptotic cells death and reduced the infarct area in brain in comparison with non-targeted dendriplexes [121].

Another example of ligand-targeted dendriplexes is the PPI G5 targeted to ovarian cancer using luteinizing hormone–releasing hormone (LHRH) peptide to deliver siRNA against CD44 mRNA enhancing the effect of paclitaxel (Tax) chemotherapy [35]. Intravenous co-administration of the PPI-LHRH-siRNA and PPI-LHRH-Tax dendriplexes to nude mice bearing metastatic human ovarian cancer xenografts (from cells isolated from malignant ascites from patients with advanced ovarian carcinoma) resulted in a reduction of the xenografts tumor volumes [35]. Individually, either PPI-LHRH-siRNA or PPI-LHRH-Tax dendriplexes were only able to slow down the tumor growth.

Chemical modifications of dendrimers have also been used for theranostic applications, i.e., to create nano-structures for simultaneous therapeutic and diagnostic use [122]. PEGylated PPI G5 dendrimers targeted with LHRH motifs have been used to decorate super paramagnetic iron oxide (SPIO) nanoparticles [123] with the dendritic shell carrying siRNA and the SPIO core acting as a MRI contrast agent. Such multifunctional system would allow the tracking of the particles *in vivo* for the diagnose imaging of tumor nodules by the selective targeting of the particles and the associated therapeutic treatment of the tumor. Once diagnosed it is quite likely though that the prolonged therapy would be limited to the therapeutic agent alone to minimize unnecessary exposure to the SPIOs.

6.6 Dendriplexes formulation

6.6.1 Dendrimer-nucleic acid interactions

Complexation of nucleic acid by dendrimers relies on the electrostatic interaction between the dendrimer and the nucleic acid, as is the case for all polymer-based complexes. This interaction depends on the bulk conditions such as pH, ionic strength, and the inherent physicochemical properties of both polyelectrolytes [85,124].

The bulk pH determines the degree of protonation (α) of the cationic groups of the dendrimer, then, pH values close or higher than the pKa of the cationic groups of the dendrimers will compromise the dendrimer-nucleic acid interaction resulting in incomplete condensation and potentially rapid dendriplexes aggregation by bridging [74,85,125]. Quantification of the enthalpy of the interaction under different conditions provides useful guidance for optimization of the process. For example, for PPI G3 dendrimers the effect of α was quantified at three conditions pH 11 (α=0.05), pH 7.4 (α=0.6), and pH 3 (α=1) [85]. The apparent binding enthalpy (ΔH_{app}) increased with decreasing pH in a linear manner from 0 $Kcalmol^{-1}$, with no binding occurring at pH 11, to 12.8 $kcalmol^{-1}$ at pH 3, at which all amines within the dendrimer structure were protonated and available for binding. AFM analysis of the dendriplexes confirms that dendriplexes formulated at α=0.6 (pH 7.4) are compact while those formulated at pH 10.5 (α=0.1; $\Delta Happ \approx 0$) have a flower petal shape with uncondensed nucleic acid loops in the dendriplexes surface [85].

These types of interactions also affect the internal complex structure. PAMAM G4 formed at low degrees of protonation ($\alpha \approx 0.1$; pH 9.1) present a squared columnar phase arrangement of DNA while the slight increase in the dendrimer charge ($\alpha \approx 0.3$; pH 8) leads to a slight wrapping of the DNA around the dendrimer under a hexagonal packaging fashion [97]. Finally, for high α values, from 0.6 (pH 5.3) up to 0.9 (pH 3.4), the strong interaction between the highly charged groups of the dendrimer and the gene results in a nucleosome-like structures with the DNA strands tightly wrapped around the dendrimer molecules [126].

Understanding of the influence of complexation conditions on dendriplexes morphology and structure are fundamentally important. Dendriplex surface charge and morphology has direct bearing on the control of colloidal stability while the internal structure is likely to be important for stability and timely nucleic acid release. For example, PAMAM G4 complexes with the strongest interaction show a reduced NA release and lower transfection activity [126].

Therefore, the protonation dependent interactions should be optimized considering three main aspects *i)* NA condensation capacity, *ii)* colloidal stability of dendriplexes, and *iii)* gene release capacity [85,87,126].

Another factor that can modulate these electrostatic interactions is the ionic strength of the medium, as higher concentrations of ions will lead to screening of electrostatic charges by counter ions. Dendriplexes formulated with low-generation dendrimers (G1-G4) are more prone to suffer the effect of the high bulk ionic strength leading to increase in their hydrodynamic mean size [82,83]. Despite this effect, our own experience suggests that a minimum salt concentration is recommended to formulate dendriplexes with a narrow size distribution [85]. This observation can be due to the acceleration of the polyelectrolyte interchange reaction in the presence of low concentrations of electrolytes in the condensation medium [127].

Together with the bulk properties, dendrimer-gene interaction is affected by the inherent properties of the nucleic acid. PPI G3 dendriplexes used to complex either siRNA (21 bp) or shRNA (3176 bp) presented similar down regulation capacity of ITCH ligase after incubation with MiaPaca-2 cancer cells [7]. However, those dendriplexes formulated with siRNA presented lower long-term stability than shRNA dendriplexes. The results suggest that RNAi molecular weight is a key driver of dendriplexes dissembling as has been shown for interpolyelectrolyte complexes [7,19,124,128].

6.6.2 Dendrimer-nucleic acid mass ratio

Another key parameter in the formation of the complexes is the ratio of nucleic acid to the polymer. As electrostatic interactions are the main driver of the complexation the ratio is typically not expressed as mass ratio (w/w) but based on the relative numbers of theoretical positive (typically amine groups, N) and negative charges (NA phosphorous groups, P), i.e., the N/P ratio.

Depending on the ratio distinct types of complexes are formed [84]. Low N/P ratios lead to the formation of highly soluble anionic aggregates. Under this scenario, anionic NA aggregates form around the low abundance dendrimer inducing a minor modification of the NA conformation. With increasing N/P ratio, colloidally unstable electroneutral complexes are formed near N/P ratios that lead to charge neutralization. Finally, at high N/P ratios, the overcharging of the anionic NA by the excess of dendrimer leads to the formation of "soluble" cationic complexes. It is these cationic dendriplexes that are thought to be the most effective for NA delivery [41,129].

Small-angle X-ray scattering (SAXS) has shown that the N/P ratio also affects the internal complex structure [130,131]. For low and intermediate degrees of protonation ($\alpha < 0.65$) an increase in the N/P ratio induces a change in the DNA packaging from a square to a hexagonal rod lattice (columnar mesophase) [130,131].

Similar to other polymer-based complexes [132–134], colloidally stable complexes exist when there is an excess of dendrimer that exists in its non-complexed form [85,135–137]. This free cationic dendrimer in the medium significantly influences colloidal stability, transfection efficiency, and the toxicity of the formulation [85,136]. In optimizing a formulation it is thus necessary to strike a balance between the positive role of the free dendrimer in the transfection efficiency of the formulation vs. its negative effect on colloidal stability and cytotoxicity.

Free excess dendrimer in the formulation also modulates the interactions of the formulation with the surrounding bulk medium after administration, e.g., formation of the protein corona, which in turn modify the pharmacokinetic and pharmacodynamics of the dendriplexes [36,37,85].

6.7 *In vitro* interaction with cells

In the controlled environment of a petri dish the interaction of cationic dendriplexes with the negatively charged cell membranes and the effect of any modifications on uptake and intracellular trafficking of dendriplexes and associated NA can be studied selectively. These studies have shown that modifications such as ligand targeting, the use of permeation enhancers, or fusogenic motifs have can improve the efficiency of the formulation, while others such as grafting of PEG moieties tend to create more inert particles with better colloidal stability but reduced internalization capacity [25,27,33,51,52,138].

These interactions with cells will further be affected by the presence of the excess free dendrimer molecules in the medium; cell culture studies have shown that the free dendrimer plays a role in internalization and intracellular trafficking of the complexes [135]. In addition, the presence of a high concentration of high generation dendrimers in the medium can lead to the transient pore formation in the cell membrane. Free dendrimer thus can increase the cellular membrane permeability but, at the same time, the disruption of the cellular membrane can lead to cytotoxic side effects [25,85,87,139].

Thus, dendriplexes formulations, in contrast to classical colloidal systems (composed of the colloid phase and the aqueous phase), should actually be considered as triphasic systems, composed of colloidal particles, aqueous phase, and the free dendrimer [85,135]. One of the challenges in developing these systems stems from the fact that concentrations of these free molecules are likely to change significantly upon intravenous injection where rapid dilution in the venous blood occurs.

6.7.1 Cellular uptake

Adsorptive endocytosis seems to be the main pathway for dendriplexes uptake, but the specific internalization mechanism is not fully understood [140]. Nanoparticles are internalized in cells by three main endocytosis pathways: clathrin-mediated endocytosis, macropinocytosis, and caveolae [141]. Recent studies using different endocytosis inhibitors have shown that clathrin-mediated endocytosis and macropinocytosis present a minimal role on dendriplexes internalization [126,142]. Regarding the caveolae endocytosis pathway, its role on dendriplexes internalization is not totally clear. Caveolae invaginations are enriched in cholesterol and sphingolipids [143], where cholesterol is involved in the membrane-trafficking process, signaling, protein and lipids sorting as well virus binding and internalization [144–148]. Manunta and co-workers have shown that the interaction of dendriplexes with cholesterol present in caveolae invaginations has a clear effect on their internalization [149,150]. Even though these results suggest that cholesterol of caveolae invaginations represent an important role in dendriplexes internalization, other studies carried out by Perumal and co-workers using filipin, an inhibitor of caveolae pathway that acts through the chelation of cholesterol [151], suggest the role of this pathway to be cell-line dependent [142]. Specifically, filipin had opposite effects when used in human lung carcinoma A549 cells (uptake inhibition) and HT1080 human fibrosarcoma cells (uptake increase). Furthermore, this situation is complicated by the fact that in some cell lines, inhibition of one endocytosis pathway can be counterbalanced by an increase in activity of other internalization pathways [152]. Taken together it would seem that uptake pathways for dendriplexes should be considered as a cell-line specific process [142].

Interestingly, independent of the cell-type specific uptake mechanism, the presence of an excess of free polyelectrolyte in the medium leads to faster and more efficient internalization of the complexes and also affects the intracellular trafficking of the complexes [135]. Cellular uptake and intracellular traffiking of dendriplexes is graphically described in Figure 6.7.

6.7.2 Endosome escape

Once the dendriplexes have been internalized the endosomal uptake mechanisms discussed above it is thought that they need to escape from the endosomal vesicles to avoid degradation and become active [153,154]. Escape from the endosome and release into the cytosol (i.e., RNAi) or transport to the cell nucleus (i.e., expression plasmids) are thus crucial steps of the transfection process.

In this context, it has been proposed that polymers act as a "proton sponge" that facilitates swelling and bursting of the endosome [155]. Briefly, during the acidification of the endosome the buffer capacity of dendrimers forces the v-ATPase to pump a large amount of H^+ to reach the acidic endosomal pH. This process leads to the influx of one Cl^- ion and a molecule of water per H^+ pumped inside the endosome [156]. Subsequent swelling end up in the collapse of the vesicle and finally in the disruption of the endosome and the release of the dendriplexes into the cytosol [155,157]. Thus, the buffer capacity of dendrimers plays an important role in their ability to escape the endosome through the so-called proton sponge effect [158].

The important role of this step for the overall efficiency of the transfection process is exemplified by a study in which PEGylation of PLL dendrimers was used successfully increase steric stabilization of these dendriplexes [28]. However, *in vitro* experiments carried out with retinal pigment epithelial cells, RPE-D407, also showed it resulted in low transfection levels, partially due to the inability of the modified dendrimers to induce the proton sponge effect and escape from the endosome [28].

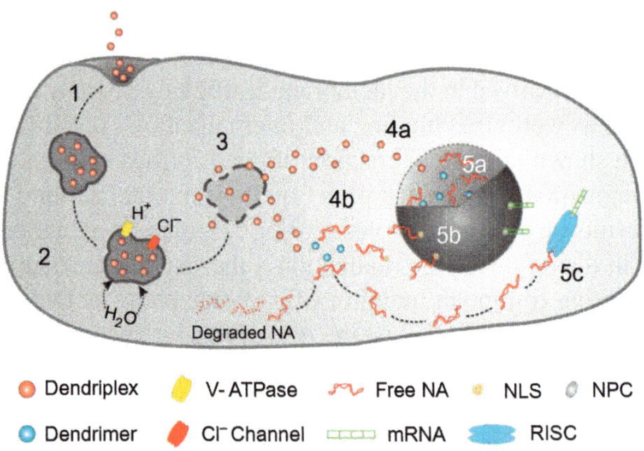

Figure 6.7 Cellular uptake and intracellular trafficking of dendriplexes. 1) While clathrin-mediated endocytosis and macropinocytosis present a minimal role on dendriplexes internalization [126] caveolae endocytosis could play an important role on dendriplexes uptake [149,150]. Nevertheless, recent studies suggest that the dendriplexes uptake pathway should be considered as a cell-specific process [142]. 2) During endosomal transport the buffer capacity of dendrimers forces V-ATPase to pump a large amount of protons to acidify the endosomal vesicle [226]. 3) Inclusion of protons leads to the influx of one chloride and one molecule of water per proton, through the subsequent swelling and disruption process the endosome releases its content into the cytosol [155,157]. 4) Once in the cytosol NA can either remain condensed as dendriplexes (a), or they can be released in the cytosol where they can be rapidly degraded by cytosolic enzymes [165] (b). 5) Condensed dendriplexes can access the nucleus machinery during the cells division when the nucleus membrane disappears [20,21,23,173] (a), or if they have been already released in the cytosol, their binding to NLS can promote their nucleocytoplasmic active transport through the NPC [167,168] or (b) alternatively for RNAi they can be incorporated to the RISC complex to silence the target mRNA [9,159,164] (c).

6.7.3 Cytosolic release

Nucleic acids such as siRNA and miRNA are active in the cytosol by cleaving or inhibiting the translation of the target mRNA [159]. For this reason complexes with RNAi should release their cargo into the cytosol. This process is affected by polyelectrolyte molecular weight and charge as these are key factors in the condensation-release balance of the nanostructure [160].

Interaction with, and displacement of the NA from the dendriplex by other negative species present in the cytosol may be important in promoting NA release [161–163]. Once free in the cytosol, RNAi will be incorporated into the cell's RNA interference machinery, more specifically into the RNA-induced silencing complex (RISC). While siRNA will exclusively bind to its complementary mRNA target, miRNA can inhibit translation of a wider range mRNA sequences as its pairing is imperfect; for a detailed description see [9,159,164].

6.7.4 Nucleic acid release in nucleus

NAs such as expression plasmids are not active in the cytosol but first require transcription into mRNA that occurs by DNA processing enzymes in the nucleus. Unless protected during trafficking through the cytosol the NAs are likely to undergo enzymatic degradation, which has been linked to rapid coating of the NA with cytosolic proteins [70,165].

The cell's nucleic acid processing machinery is separated from the cytosol by the nuclear envelope, a selective membrane that only allows the passive diffusion of certain low molecular weight molecules (smaller than 40 KDa) through the nucleus pore complexes (NPC). All larger molecules require tightly regulated active transport typically mediated by protein tags that act as nuclear localization sequences (NLS) [166,167].

The NPCs regulate the transport of small molecules through the nucleus membrane, but the binding of specific proteins containing NLS to nucleic acids present in the cytosol allows their recognition by importin proteins that mediate the active transport into the nucleus through the NPC [167,168]. Active nucleo-cytoplasmic transport can be inhibited by anti-nucleoporin antibodies and wheat germ agglutinin (WGA) [169]. Interestingly, these also inhibit transport of pDNA into the nucleus suggesting the NPC to be a key part of the trafficking of DNA between the cytosol and the nucleus [167,170]. The discovery of new NLS opens new possibilities for active nucleocytoplasmic transport of DNA [171]. Some of these alternatives to improve the nucleus import of DNA of non-viral vectors have been summarized by Wagstaff and Jans [172].

Studies with lipoplex-mediated transfection suggest a strong correlation of this transport to plasmid size with expression levels dropping dramatically for plasmids above ~ 1kB [166]. In fact, only around 1–10% of pDNA present in the cytosol is estimated to reach the nucleus [70].

These results suggest that despite to the potential active nucleocytoplasmic transport of DNA across the NPC, DNA is more efficiently transported into the nucleus in the form of a complex [20,21,23,173].

The only other way for molecules to gain access to the nuclear enzymes is during mitosis and meiosis when the nuclear membrane is temporarily disassembled [174,175]. This alternative mechanism would be independent of plasmid size and its involvement in the transfection process would help to explain the higher transfection efficiency frequently observed for rapidly dividing cells compared to non-mitotic cells.

Once the complex reaches the nucleus, the release capacity of the nanostructure plays an important role in the final formulation efficiency [70]. Studies carried out with Lipofectamine® lipoplexes and PEI polyplexes have shown that even when PEI polyplexes were more efficient at reaching the cell nucleus, the tighter condensation of the DNA in the polyplexes hampered its release and resulted in lower gene expression levels in comparison with the lipoplexes [70,71].

This highlights the fact that besides the design of colloidally stable prototypes able to protect the nucleic acids from degradation and facilitate its cellular trafficking, it is mandatory to be able to modulate the capacity of the nanostructure to deliver its cargo at the proper time and place.

6.8 *In vivo* administration

6.8.1 *Interaction with complex biological fluids*

Biological fluids, and in particular the blood, contain various cell types and a varied range of proteins, macromolecules, salts, and solutes leading to a much more complex environment.

The high surface/volume ratio is one of the main physico-chemical characteristics of colloidal systems including dendriplexes. Diving up the mass of a particle into smaller particles leads to dramatic increases of surface area,, e.g., a cube with 1 cm sides divided into particles with side a 1/10th of the original size yields 1000 smaller particles and increases the surface are from 6 cm^2 to 600 m^2. This sort of dramatic increase of surface area is associated with a much higher potential for interaction and confers such systems a high level of reactivity [176,177]. Such interactions are relevant to the *in vitro* environment where it is likely to promote particle interaction with the cells. However, in the much more complex environment typically encountered *in vivo* this high level of reactivity can become a challenge, leading to non-specific interactions with non-target cells and with serum components, which dramatically affect dendriplexes biodistribution and furthermore their biological effect [36,178].

The interaction with serum proteins occurs very rapidly and extensively once dendriplexes enter the blood stream [37–39,178]. The adsorption of proteins to the surface of dendriplexes will modify their surface properties [37,179,180], with consequences on their physicochemical stability and biological interactions, as this coating will rule out the fate of the dendriplexes. For instance, *partial* coating of positively charged dendriplexes with negative serum proteins transforms creates heterogeneously charged particles with positive (non-coated) and negative (coated) domains leading to bridging flocculation mechanisms which promote particle aggregation [181] and entrapment of the aggregates in the narrow lung capillaries, the first microvascular bed encountered after the intravenous administration of the dendriplexes [62,65,182].

On the other hand, *complete* coating of particles with serum protein can actually improve the colloidal stability of the particles under physiological conditions [183]. This protein coat, known as the protein corona, not only affects colloidal properties but depending on the specific protein composition can also affect its further biological properties such as biodistribution [37,39,176,177,181]. The composition of the protein corona is determined by the initial physicochemical properties of the particles, such as size, ζ-potential, and hydrophobicity, as well as the concentration and affinity toward the particle surface of the specific proteins of the blood at the specific time and place of injection. Therefore, the differences of the particle dilution in plasma proteins experienced upon administration as a bolus or by infusion explain the differences in the protein corona composition observed [176,178]. Because of the effect of the protein corona on pharmacokinetics and pharmacodynamics of nanoparticles it is important to find means by which to carefully control this process [36,178].

Indeed, the non-specific interactions with blood components and the formation of a corona are likely to interfere with the function of ligand-targeted dendriplexes; the adsorption of serum proteins could potentially mask the ligands and binding of opsonins could facilitate rapid clearance from the blood via the reticuloendothelial system (RES) [94,95]. However, serum protein interactions are not inherently linked to negative effects; they may also produce targeting of the nanoparticles themselves [184]. This is the case of nanoparticles formulated with the surfactant Tween 80, which present an apolipoprotein E (ApoE) enriched protein corona and favored transport across the blood brain barrier due to receptor mediated binding [185–187]. The understanding gained by the analysis of the protein corona composition led to the identification of these targeting moieties for covalent modification of the particles toward their active targeting [188].

The modification of the particle surface with hydrophilic polymers, such as polyethylene glycol (PEG) or polyethylene oxide (PEO), creates a steric barrier that helps to minimize protein adsorption [94,189]. First applied to liposomes [92,93], and nowadays broadly applied to different colloidal platforms, including dendriplexes [97], hydrophilic polymer surface modification has a clear benefit on their biological performance. However, optimization of systems for one purpose frequently leads to detrimental effects in other areas as has been seen with the loss of condensation capacity for some stabilized dendriplexes [75,76,89,98].

The interaction of the complexes with serum components may not only affect stability and biodistribution of dendriplexes may also affect the toxicity of such formulations. For example, dendriplexes may interact with erythrocytes [190,191] leading to clumping or disruption of the membranes of erythrocytes and hemoglobin release [192,193]. Coating of dendriplexes can alleviate some of these effects leading to a three orders of magnitude reduction of haemolytic activity [85].

6.8.2 Biodistribution

The interactions with blood components discussed in the previous section may have implications in the biodistribution of the dendriplexes, like the lung accumulation of the dendriplexes, due to their partial coating with serum proteins followed by the formation of aggregates [181], which are physically entrapped in the lung capillaries [62,65,67,68,182].

Pharmacokinetics and biodistribution are critical factors affecting therapeutic efficacy of these vector systems, as is the case of PAMAM dendrimers, which are quickly cleared from the blood stream and tend to accumulate in liver, kidneys, and pancreas [194–197]. It is important to be aware that the biodistribution profiles of the dendrimers do not necessarily have any bearing on the distribution of the dendriplexes, as the nucleic acid complexation creates dramatically larger aggregates with modified surface characteristics.

Furthermore, the biodistribution of the vector does not necessarily predict the distribution of the pharmacodynamic effect, i.e., expression or down-regulation of genes due to pDNA or siRNA, respectively. Cell- and tissue-specific differences in regulation can be exploited to increase specificity of a vector system and even to achieve tumor targeting [34,48,49,67]. But even for general strong promoters such as CMV differences in response exist [67]. In addition, vector specific factors may affect the distribution of gene expression. For example, linear and branched PEI complex and

PAMAM dendriplexes encoding pCMV-Luciferase after their tail vein injection in mice bearing xenograft neuroblastoma tumors show similar biodistribution, but while the PEI polyplexes showed primarily gene expression in the lung the dendriplexes showed higher levels of expression in the tumor [67]. This study also gave some hints about the importance of the optimization of each one of the formulation components in the final performance of the system, as long-term gene expression in the tumor was achieved thanks to the use of modified plasmid promoters to avoid the *in vivo* inactivation of CMV through the methylation of its CpG sequence [198–200].

Tumor preferential biodistribution is the goal to achieve for an optimized cancer gene therapy, and some dendriplexes may be of help by their tendency to accumulate in tumors [7,34,35,47,49]. We have previously reported that systemic administration of dendriplexes made from unmodified PPI G3 dendrimers and a CMV plasmid carrying the gene for the sodium iodide symporter (hNIS) led to a highly selective gene expression in xenograft tumors. Based on molecular imaging of the hNIS reporter using ^{99}Te and qPCR quantitative analysis 90% of the gene expression was localized in the xenograft tumor [34]. This apparent targeting was applicable to a number of different tumor types and also worked selectively when animals carried another tumor xenograft in the contralateral flank. Others groups have subsequently also found a preferential distribution from dendriplexes [67]. It is plausible that the inherent physicochemical properties of unmodified PPI dendriplexes induce the formation of a protein corona, which modulates the biodistribution of the dendriplexes preferentially toward the tumor. Further suggestions of an intrinsic tumor targeting by dendriplexes comes from Shah and co-workers, who compared the efficiency of targeted and non-targeted PPI G5 dendriplexes in mice bearing human ovarian carcinoma xenografts and found a tumor-favored biodistribution and tumor growth suppression effect by the unmodified dendriplexes, although a targeted derivative proved more efficient [35].

These findings strongly suggest that the protein corona plays an important role in the biodistribution profile and therapeutic performance of dendriplexes. These observations raise the additional question of species-specific corona formation and its effects on pharmacokinetics and biodistribution may make clinical translation of such system challenging.

6.9 Efficaciousness after *in vivo* administration

6.9.1 Local administration

One approach that minimizes these factors would be based on the choice of administration route. The local administration of the dendriplexes, when available, potentially minimizes the challenges and variability introduced buy the interaction with complex biological media. For example, *in situ* injection into the quadriceps of neonatal mice has been the method used for G5 and G6 PAMAM dendrimers modified with different molar ratios of PEG_{5000} (4%, 8%, and 15%) for intramuscular gene therapy [201]. The results showed that 5 μg/g of body weight of 8% of PEG_{5000} modified dendriplexes achieved the most efficient muscular gene expression after 48 h, which showed good correlation to the *in vitro* experiments in 293A cells.

Ocular gene delivery can also offer the benefits of local administration, as achieving an efficient and significant gene expression in the eye by the systemic route is highly challenging. The contribution of dendrimers to the ocular delivery of nucleic acid is supported by the enhancement of the corneal permeation and sustained release of the cargo that these structures provide [202]. Studies performed with activated PAMAM dendrimers (Superfect®) condensing either placZ or pTNFR-Ig in rabbit and human corneas *ex vivo* showed a gene expression of around 10% of the corneal cells population [203]. Nevertheless, the most relevant study of dendrimers for ocular gene delivery was carried out by Marano and co-workers using modified PLL dendrimers and anti-VEGF oligodeoxynucleotide (ODN-1) as therapeutic gene [12,53]. The initial *in vitro* studies in human RPE cell line (D407) pointed to a promising candidate for ocular gene delivery, as it transfected almost 90% of the cells in comparison to the commercial transfection agent (Cytofectin) as well as affecting high reduction of the VEGF mRNA levels. These results were smartly supported by a short-term *in vivo* proof of concept performed by fluorescein angiography in a rat model of choroidal neovascularization (CNV). Significant severity reduction of laser-induced CNV was achieved by ODN-1 dendriplexes for up to 2 months post-injection [53]. This evidence was later on reinforced by the long-term (4–6 months) inhibition of CNV, with ODN-1 penetration in all of the retinal cell layers down to the RPE and no adverse reaction nor toxicity of the formulations confirmed by immunohistochemistry and regular ophthalmological examinations [12].

Local administration has also been applied in cancer therapy in order to avoid the biological barriers associated to a systemic gene therapy. In this regard, dendriplexes have contributed significantly to the advances achieved by synthetic vectors. For example, PAMAM dendrimer complexes of pGEG-FasL (an Epstein-Barr virus (EBV)-based plasmid with the FasL gene) administered by intratumoral injection (3x 10 μg of pDNA) into PC-3-derived xenografts produced a significant growth suppression of the tumors [204], an effect that was enhanced further in combination therapy with cisplatin. Similarly, folate-targeted PAMAM G5 dendrimers efficiently transfected established rat C6 intracraneal gliomas with therapeutic antisense oligonucleotides of EGFR (EGFR-ASODN) after intracranial injection with superior efficacy compared to non-targeted PAMAM or oligofectamine [205]. Further examples highlight the potential advantages of this approach [11,206,207]. However, one has to be clear that for the treatment of advanced and metastatic cancer, responsible for the majority of cancer related deaths, the administration route needs to enable treatment of remote and metastatic tumors. These conditions are likely to require systemic administration, ideally combined with efficient and selective targeting of the tumor [75,76].

6.9.2 Systemic administration

In the context of systemic administration it is essential that the vector systems reach the tumor in sufficient quantity. Ideally, the biodistribution should also be selected to minimize the risk of side effects resulting from off target toxicity. We have shown that unmodified dendriplexes of PPI G3-pDNA can deliver therapeutic genes such as the potentially highly toxic TNFα to mice bearing xenograft tumors by tail vein injection [49]. This approach resulted in a 100% response without re-growth over 17 weeks and a remarkable 80% complete regression of the treated tumors after a single course of treatment [49]. The system also allowed therapeutic interventions based on the down-regulation of repressors of tumor suppressor genes, i.e., expression of a p53 peptide fragment leads to de-repression of the pro-apoptotic p73 and thus tumor growth repression [208]. Thus, there is clear evidence that non-targeted dendrimer systems have the capability to deliver even toxic genes quite safely and provide efficacious anti-tumor therapies. The explanation for this may be an intrinsic affinity of dendrimer-based complexes for tumors as discussed in section 5.2. The underlying mechanism for this selectivity remains unclear but the formation of a dendrimer-specific protein corona that mediates tumor targeting could offer a plausible explanation.

The use of specific targeting ligands aims to provide additional selectivity for tumor targeting, which may be more easily controlled than any serum protein mediated effects. Good examples of this approach are provided by the use LHRH used to target PPI G5 complexes with siRNA to A549 xenograft tumors [209]. The LHRH moiety is covalently linked to the dendrimer using a targeting PEG_{5000} spacer and allows accumulation of siRNA and dendrimer in the tumor. Non-targeted dendriplexes produced accumulation of the dendrimer in liver and kidney, whereas siRNA concentrated in the tumor but at a lower extent than actively targeted dendriplexes, suggesting that the dendriplexes disassemble in the bloodstream.

Koppu and co-workers also applied the active targeting approach by using transferrin (Tf), a glycoprotein involved in iron metabolism, which is over-expressed in some types of tumors [210], for the modification of PPI G3 dendriplexes [48]. The intravenous administration of these surface-modified dendriplexes condensing therapeutic TNFα plasmid facilitated the gene expression mainly in the tumor. Despite the different biodistribution pattern and the slightly higher accumulation of the Tf-targeted dendriplexes compared to the non-targeted system both formulations displayed quite similar tumor regression over one month, with long-term survival of 100% of the animals (90% complete response, 10% partial response).

It is not always clear how much selectivity active targeting adds to a specific vector system; the challenges related to unraveling the mechanisms and their relative contributions to the overall biodistribution makes rational development of such systems difficult. The observed biodistribution is the result of a complex multifactorial system. The active targeting of the ligand works together with other factors including physicochemical modifications due to chemical polymer modifications, corona composition potentially modulated by free excess dendrimer, haemodynamic effects due to aggregation, etc., In addition, ligand/receptor specific considerations come into play. For example, the targeting ligands used may be in competition with endogenous ligand,, e.g., the endogenous levels of Tf under physiological conditions almost completely saturates the Tf receptors giving little chances to Tf-active modified dendriplexes to interact with the receptors and specifically deliver their cargo in the tumor tissue [211,212]. Furthermore, most of the targeted receptors are not exclusive to the targeted site, e.g., Tf-receptors are also over-expressed in other tissues like brain capillary endothelia cells, where dendriplexes may non-specifically accumulate instead of doing so in the tumor tissue.

On the other hand, it may be possible to exploit the physiological distribution of endogenous receptors to target other sites. For example, the expression of Tf-receptors and similar receptors in brain capillary endothelia cells offers an opportunity to overcome the blood brain barrier, a physical, metabolic, and physiological barrier that naturally protects this tissue against potentially risky exogenous agents [213–217].

Huang and co-workers have modified PEG_{3400} spacer PAMAM G5 dendrimers with terminal Tf motifs for the complexation of the pGL2 reporter gene, and achieved enhanced gene expression in the brain after tail vein injection in mice when compared to the un-modified dendriplexes [218]. The Tf-targeted carriers also had a changed biodistribution and gene expression profile, showing those Tf-modified dendriplexes higher levels in lungs and kidneys compared to the non-targeted ones. The subsequent substitution of Tf by lactoferrin (Lf) yielded dendriplexes with higher accumulation and transfection efficiency in brain, undoubtedly optimizing the formulation, although the precise mechanism remains unclear [219,220].

Other promising targeting moieties for enhance BBB delivery include Angiopep (LRP1 receptor), RVG29 (acetylcholine receptors of neuronal and brain capillary endothelial cells), and leptin-derived peptide [221–223]. Taking this idea a little bit further, Kuang and co-workers recently published an elegant nanostructure for the double targeting of glioma tumor gene therapy [224]. The authors relied on the over-expression of Tf-receptors in brain capillary endothelial cells and glioma cells [214–217], but used a T7 peptide for targeting for which binding and transport is not hindered but boosted in the presence of competing endogenous Tf [225]. Using the T7 peptide grafted to poly-l-lysine (DGL) dendrimers to deliver psiLuc (50 µg DNA per animal) efficiently silenced luciferase expression in mice bearing orthotopic human gliomas (U87-luciferase cells) more efficiently than for the non-targeted system potentially due to the three-fold increased uptake [224].

6.10 Conclusions

Since their first use as transfection agents around 20 years ago, and due to their favorable balance of transfection efficiency and toxicity, dendrimers have emerged as one of the most promising class of compounds in the area of synthetic vectors for gene therapy.

Their unique architecture facilitates chemical modification to modulate their physical chemistry and thus various aspects of the formulation, such as cytotoxicity, gene condensation capacity, “stealth” properties, or targeting specifics organs. This versatility has been extensively demonstrated in a broad range of *in vitro* studies. However, under the more complex conditions of *in vivo* application specific challenges, e.g., related to colloidal stability, biodistribution and tissue specificity remain to be resolved in order for such systems to move from the laboratory to clinical use.

One of the key challenges in the field remains the difficult correlation of dendrimer chemical structure to the actual performance of the synthetic vector *in vivo*; chemical modifications of dendrimer molecules intended to modulate a specific property of the system often induce a range of further effects that can be difficult to predict. The challenge invariably increases as the complexity of the test system increases, i.e., from *in vitro* to *in vivo* conditions. For example, the control of the dendriplexes biodistribution stills is difficult to control, probably due to the multiple interactions of the nanostructure with the components of the crowded serum soup after their administration. In this sense, specific composition of the protein corona may affect the final fate of the dendriplexes.

To promote tissue specificity some target motifs have been covalently linked to the dendrimer. While these modifications have been shown to successfully modify the biodistribution pattern of the nanostructures, this has not always also increased the efficacy of the targeted dendriplex formulations.

These results highlight the importance of the understanding of the colloidal behavior of these promising nanostructures under *in vivo* conditions by monitoring the effect of their physicochemical properties on their interaction with the physiological milieu after administration.

Systematic rational development of such synthetic vector systems can only progress satisfactorily if it is matched by the development of tools and methods allowing us to understand the individual steps and mechanistic basis of the transfection process so that eventually a QSAR-based understanding of these systems can emerge and guide further development.

References

[1] Kukowska-Latallo JF, et al. Efficient transfer of genetic material into mammalian cells using Starburst polyamidoamine dendrimers. Proc Natl Acad Sci U S A 1996;93(10):4897–902.

[2] Haensler J, Szoka FC. Polyamidoamine cascade polymers mediate efficient transfection of cells in culture. Bioconjug Chem 1993;4(5):372–9.

[3] Yamagata M, et al. Structural advantage of dendritic poly(l-lysine) for gene delivery into cells. Bioorg Med Chem 2007;15(1):526–32.

[4] Fant K, et al. DNA condensation by PAMAM dendrimers: self-assembly characteristics and effect on transcription. Biochemistry 2008;47(6):1732–40.

[5] Bielinska AU, Kukowska-Latallo JF, Baker Jr. JR. The interaction of plasmid DNA with polyamidoamine dendrimers: mechanism of complex formation and analysis of alterations induced in nuclease sensitivity and transcriptional activity of the complexed DNA. Biochim Biophys Acta 1997;1353(2):180–90.

[6] Tang MX, Szoka FC. The influence of polymer structure on the interactions of cationic polymers with DNA and morphology of the resulting complexes. Gene Ther 1997;4(8):823–32.

[7] De la Fuente M, et al. A nano-enabled ITCH RNAi-driven cancer-specific booster strategy for the combination therapy of pancreatic cancer. Nanomedicine NBM 2014, submitted.

[8] Wu J, Huang W, He Z. Dendrimers as carriers for siRNA delivery and gene silencing: a review. ScientificWorldJournal 2013;2013:630654.

[9] Raviña M, et al. Knocking down gene expression with dendritic vectors. Mini Rev Med Chem 2010;10(1):73–86.

[10] Hollins AJ, et al. Evaluation of generation 2 and 3 poly(propylenimine) dendrimers for the potential cellular delivery of antisense oligonucleotides targeting the epidermal growth factor receptor. Pharm Res 2004;21(3):458–66.

[11] Wang P, et al. Generation 4 polyamidoamine dendrimers is a novel candidate of nano-carrier for gene delivery agents in breast cancer treatment. Cancer Lett 2010;298(1):34–49.

[12] Marano RJ, et al. Dendrimer delivery of an anti-VEGF oligonucleotide into the eye: a long-term study into inhibition of laser-induced CNV, distribution, uptake and toxicity. Gene Ther 2005;12(21):1544–50.

[13] Tomalia DA. Birth of a new macromolecular architecture: dendrimers as quantized building blocks for nanoscale synthetic organic chemistry. Aldrichimica Acta 2004;37(2):39–57.

[14] Dufes C, Uchegbu IF, Schatzlein AG. Dendrimers in gene delivery. Adv Drug Deliv Rev 2005;57(15):2177–202.

[15] Svenson S, Tomalia DA. Dendrimers in biomedical applications - Reflections on the field. Adv Drug Deliv Rev 2005;57(15):2106–29.

[16] Sakthivel T, Florence AT. Dendrimers and Dendrons: facets of Pharmaceutical Nanotechnology. Drug Delivery Technology 2003;3:73–8.

[17] Dutta T, et al. Dendrimer nanocarriers as versatile vectors in gene delivery. Nanomedicine-Nanotechnology Biology and Medicine 2010;6(1):25–34.

[18] Parekh HS. The advance of dendrimers - A versatile targeting-platform for gene/drug delivery. Curr Pharm Des 2007;13(27):2837–50.

[19] Singha K, Namgung R, Kim WJ. Polymers in small-interfering RNA delivery. Nucleic Acid Ther 2011;21(3):133–47.

[20] Yoo H, Juliano RL. Enhanced delivery of antisense oligonucleotides with fluorophore-conjugated PAMAM dendrimers. Nucleic Acids Res 2000;28(21):4225–31.

[21] Kihara F, et al. Effects of structure of polyamidoamine dendrimer on gene transfer efficiency of the dendrimer conjugate with α-cyclodextrin. Bioconjug Chem 2002;13(6):1211–9.

[22] Kihara F, et al. In vitro and in vivo gene transfer by an optimized α-cyclodextrin conjugate with polyamidoamine dendrimer. Bioconjug Chem 2003;14(2):342–50.

[23] Godbey WT, Wu KK, Mikos AG. Tracking the intracellular path of poly(ethylenimine)/DNA complexes for gene delivery. Proc Natl Acad Sci U S A 1999;96(9):5177–81.

[24] Posadas I, et al. Highly efficient transfection of rat cortical neurons using carbosilane dendrimers unveils a neuroprotective Role for HIF-1α in early chemical hypoxia-mediated neurotoxicity. Pharm Res 2009;26(5):1181–91.

[25] Lee H, Larson RG. Membrane Pore Formation Induced by Acetylated and Polyethylene Glycol-Conjugated Polyamidoamine Dendrimers. J Phys Chem C 2011;115(13):5316–22.

[26] Fant K, et al. Effects of PEGylation and acetylation of PAMAM dendrimers on DNA binding, cytotoxicity and in vitro transfection efficiency. Mol Pharmaceutics 2010;7(5):1734–46.

[27] Choi JS, et al. Enhanced transfection efficiency of PAMAM dendrimer by surface modification with L-arginine. J Control Release 2004;99(3):445–56.

[28] Männistö M, et al. Structure-activity relationships of poly(L-lysines): effects of pegylation and molecular shape on physicochemical and biological properties in gene delivery. J Control Release 2002;83(1):169–82.

[29] Patil ML, et al. Internally cationic polyamidoamine PAMAM-OH dendrimers for siRNA delivery: effect of the degree of Quaternization and cancer targeting. Biomacromolecules 2009;10(2):258–66.

[30] Han L, et al. Tat-BMPs-PAMAM conjugates enhance therapeutic effect of small interference RNA on U251 glioma cells in vitro and in vivo. Hum Gene Ther 2010;21(4):417–26.

[31] Nam HY, et al. Biodegradable PAMAM ester for enhanced transfection efficiency with low cytotoxicity. Biomaterials 2009;30(4):665–73.

[32] Schatzlein AG, et al. Preferential liver gene expression with polypropylenimine dendrimers. J Control Release 2005;101(1-3 SPEC. ISS):247–58.

[33] Tomalia DA, Reyna LA, Svenson S. Dendrimers as multi-purpose nanodevices for oncology drug delivery and diagnostic imaging. Biochem Soc Trans 2007;35:61–7.

[34] Chisholm EJ, et al. Cancer-specific transgene expression mediated by systemic injection of nanoparticles. Cancer Res 2009;69(6):2655–62.

[35] Shah V, et al. Targeted nanomedicine for suppression of CD44 and simultaneous cell death induction in ovarian cancer: an optimal delivery of siRNA and anticancer drug. Clin Cancer Res 2013;19(22):6193–204.

[36] Aggarwal P, et al. Nanoparticle interaction with plasma proteins as it relates to particle biodistribution, biocompatibility and therapeutic efficacy. Adv Drug Deliv Rev 2009;61(6):428–37.

[37] Åkesson A, et al. The protein corona of dendrimers: PAMAM binds and activates complement proteins in human plasma in a generation dependent manner. RSC Advances 2012;2(30):11245–8.

[38] Monopoli MP, et al. Biomolecular coronas provide the biological identity of nanosized materials. Nat Nanotechnol 2012;7(12):779–86.

[39] Monopoli MP, et al. Physical-Chemical aspects of protein corona: relevance to in vitro and in vivo biological impacts of nanoparticles. J Am Chem Soc 2011;133(8):2525–34.

[40] Fischer D, et al. In vitro cytotoxicity testing of polycations: influence of polymer structure on cell viability and hemolysis. Biomaterials 2003;24(7):1121–31.

[41] Bielinska AU, et al. DNA complexing with polyamidoamine dendrimers: implications for transfection. Bioconjug Chem 1999;10(5):843–50.

[42] Poxon SW, et al. Dendrimer delivery of oligonucleotides. Drug Deliv 1996;3(4):255–61.

[43] Wu J, et al. Polycationic dendrimers interact with RNA molecules: polyamine dendrimers inhibit the catalytic activity of Candida ribozymes. Chem Commun (Camb) 2005;3:313–5.

[44] Malik N, Evagorou eg, Duncan R. Dendrimer-platinate: a novel approach to cancer chemotherapy. Anticancer Drugs 1999;10(8):767–76.

[45] Gebhart CL, Kabanov AV. Evaluation of polyplexes as gene transfer agents. J Control Release 2001;73(2-3):401–16.

[46] Taratula O, et al. Poly(propyleneimine) dendrimers as potential siRNA delivery nanocarrier: from structure to function. International Journal of Nanotechnology 2011;8(1-2):36–52.

[47] Zinselmeyer BH, et al. The lower-generation polypropylenimine dendrimers are effective gene-transfer agents. Pharm Res 2002;19(7):960–7.

[48] Koppu S, et al. Tumor regression after systemic administration of a novel tumor-targeted gene delivery system carrying a therapeutic plasmid DNA. J Control Release 2010;143(2):215–21.

[49] Dufès C, et al. Synthetic anticancer gene medicine exploits intrinsic antitumor activity of cationic vector to cure established tumors. Cancer Res 2005;65(18):8079–84.

[50] Denkewalter RG, Kolc J, Lukasavage WJ. Macromolecular highly branched homogeneous compound based on lysine units; 1981, Google Patents.

[51] Bikram M, et al. Long-circulating DNA-complexed biodegradable multiblock copolymers for gene delivery: degradation profiles and evidence of dysopsonization. J Control Release 2005;103(1):221–33.

[52] Yang S, et al. Cellular uptake of self-assembled cationic peptide-DNA complexes: multifunctional role of the enhancer chloroquine. J Control Release 2009;135(2):159–65.

[53] Marano RJ, et al. Inhibition of in vitro VEGF expression and choroidal neovascularization by synthetic dendrimer peptide mediated delivery of a sense oligonucleotide. Exp Eye Res 2004;79(4):525–35.

[54] Weber N, et al. Characterization of carbosilane dendrimers as effective carriers of siRNA to HIV-infected lymphocytes. J Control Release: 2008;132(1):55–64.

[55] Bermejo JF, et al. Water-soluble carbosilane dendrimers: synthesis biocompatibility and complexation with oligonucleotides; evaluation for medical applications. Chemistry 2007;13(2):483–95.
[56] Gras R, et al. Changes in gene expression pattern of human primary macrophages induced by carbosilane dendrimer 2G-NN16. Pharm Res 2009;26(3):577–86.
[57] Cloninger MJ. Biological applications of dendrimers. Curr Opin Chem Biol 2002;6(6):742–8.
[58] Wada K, et al. Improvement of gene delivery mediated by mannosylated dendrimer/alpha-cyclodextrin conjugates. J Control Release 2005;104(2):397–413.
[59] Nishikawa M, et al. Pharmacokinetics and in vivo gene transfer of plasmid DNA complexed with mannosylated poly(L-lysine) in mice. J Drug Target 2000;8(1):29–38.
[60] Arima H, Motoyama K, Higashi T. Sugar-appended polyamidoamine dendrimer conjugates with cyclodextrins as cell-specific non-viral vectors. Adv Drug Deliv Rev 2013;65(9):1204–14.
[61] Nanjwade BK, et al. Dendrimers: emerging polymers for drug-delivery systems. Eur J Pharm Sci 2009;38(3):185–96.
[62] Bai S, Thomas C, Ahsan F. Dendrimers as a carrier for pulmonary delivery of enoxaparin, a low-molecular weight heparin. J Pharm Sci 2007;96(8):2090–106.
[63] Esfand R, Tomalia DA. Poly(amidoamine) (PAMAM) dendrimers: from biomimicry to drug delivery and biomedical applications. Drug Discov Today 2001;6(8):427–36.
[64] Tang MX, Redemann CT, Szoka FC. In vitro gene delivery by degraded polyamidoamine dendrimers. Bioconjug Chem 1996;7(6):703–14.
[65] Navarro G, de Ilarduya CT. Activated and non-activated PAMAM dendrimers for gene delivery in vitro and in vivo. Nanomedicine-Nanotechnology Biology and Medicine 2009;5(3):287–97.
[66] Dennig J, Duncan E. Gene transfer into eukaryotic cells using activated polyamidoamine dendrimers. Reviews in Molecular Biotechnology 2002;90:339–47.
[67] Navarro G, et al. Low generation PAMAM dendrimer and CpG free plasmids allow targeted and extended transgene expression in tumors after systemic delivery. J Control Release 2010;.
[68] Zintchenko A, et al. Drug nanocarriers labeled with near-infrared-emitting quantum dots (quantoplexes): imaging fast dynamics of distribution in living animals. Mol Ther 2009;17(11):1849–56.
[69] de Bruin KG, et al. Dynamics of photoinduced endosomal release of polyplexes. J Control Release 2008;130(2):175–82.
[70] Glover DJ, et al. The efficiency of nuclear plasmid DNA delivery is a critical determinant of transgene expression at the single cell level. J Gene Med 2010;12(1):77–85.
[71] Pollard H, et al. Polyethylenimine but not cationic lipids promotes transgene delivery to the nucleus in mammalian cells. J Biol Chem 1998;273(13):7507–11.
[72] de Jong G, et al. Efficient in-vitro transfer of a 60-Mb mammalian artificial chromosome into murine and hamster cells using cationic lipids and dendrimers. Chromosome Res 2001;9(6):475–85.
[73] Tomalia DA, Majoros I. Dendrimeric supramolecular and supramacromolecular assemblies (Reprinted from Supramolecular Polymers, Marcel Dekker, Inc.: New York, pg 359-434, 2002). Journal of Macromolecular Science-Polymer Reviews 2003;C43(3):411–77.
[74] Chen W, Turro NJ, Tomalia DA. Using ethidium bromide to probe the interactions between DNA and dendrimers. Langmuir 2000;16(1):15–9.
[75] Santander-Ortega MJ, Uchegbu IF, Schätzlein AG. Dendrimer-based gene delivery systems: administration routes and in vivo evaluation; 2012, p. 329-354.

[76] Santander-Ortega MJ, et al. Dendrimers for drug delivery. In: Gordon R, editor. Encyclopedia of Biophysics. Berlin Heidelberg: Springer; 2013. p. 441–51.

[77] Tomalia DA, et al. A new class of polymers - starburst-dendritic macromolecules. Polymer Journal 1985;17(1):117–32.

[78] Rietveld IB, et al. Location of the outer shell and influence of pH on carboxylic acid-functionalized poly(propylenei mine) dendrimers. Macromolecules 2001;34(23):8380–3.

[79] Jones JW, et al. Crowned dendrimers: pH-responsive pseudorotaxane formation. J Org Chem 2003;68(6):2385–9.

[80] Hecht S, Frechet JM. Dendritic encapsulation of function: applying nature's site isolation principle from biomimetics to materials science. Angew Chem Int Ed Engl 2001;40(1):74–91.

[81] Chai M, et al. Structure and conformation of DAB dendrimers in solution via multidimensional NMR techniques. J Am Chem Soc 2001;123(20):4670–8.

[82] Carnerup AM, et al. Condensation of DNA using poly(amido amine) dendrimers: effect of salt concentration on aggregate morphology. Soft Matter 2011;7(2):760–8.

[83] Carnerup AM, et al. Watching DNA condensation induced by poly(amido amine) dendrimers with time-resolved Cryo-TEM. Langmuir 2009;25(21):12466–70.

[84] Ottaviani MF, et al. Formation of supramolecular structures between DNA and starburst dendrimers studied by EPR, CD, UV, and melting profiles. Macromolecules 2000;33(21):7842–51.

[85] Santander-Ortega MJ, et al. Optimisation of synthetic vector systems for cancer gene therapy – the role of the excess of cationic dendrimer under physiological conditions. Current topics in medical chemistry; 2014, in press.

[86] Leroueil PR, et al. Nanoparticle interaction with biological membranes: does nanotechnology present a janus face? Acc Chem Res 2007;40(5):335–42.

[87] Santander-Ortega MJ, et al. Hydration forces as a tool for the optimization of core-shell nanoparticle vectors for cancer gene therapy. Soft Matter 2012;8(48):12080–92.

[88] Majoros IJ, et al. Acetylation of poly(amidoamine) dendrimers. Macromolecules 2003;36(15):5526–9.

[89] Lee JH, et al. Polyplexes assembled with internally quaternized PAMAM-OH dendrimer and plasmid DNA have a neutral surface and gene delivery potency. Bioconjug Chem 2003;14(6):1214–21.

[90] Kolhatkar RB, et al. Surface acetylation of polyamidoamine (PAMAM) dendrimers decreases cytotoxicity while maintaining membrane permeability. Bioconjug Chem 2007;18(6):2054–60.

[91] Stasko NA, et al. Cytotoxicity of polypropylenimine dendrimer conjugates on cultured endothelial cells. Biomacromolecules 2007;8(12):3853–9.

[92] Woodle MC, Newman MS, Martin FJ. Liposome leakage and blood circulation: comparison of adsorbed block copolymers with covalent attachment of PEG. Int J Pharm 1992;88(1-3):327–34.

[93] Allen TM, Cullis PR. Drug delivery systems: entering the mainstream. Science 2004;303(5665):1818–22.

[94] Moghimi SM, et al. Effect of splenic congestion associated with haemolytic anaemia on filtration of 'spleen-homing' microspheres. Clin Sci (Lond) 1993;84(6):605–9.

[95] Moghimi SM, et al. An investigation of the filtration capacity and the fate of large filtered sterically-stabilized microspheres in rat spleen. Biochim Biophys Acta 1993;1157(3):233–40.

[96] Kaminskas LM, et al. The impact of molecular weight and PEG chain length on the systemic pharmacokinetics of PEGylated poly L-lysine dendrimers. Mol Pharm 2008;5(3):449–63.

[97] Kim Y, Klutz AM, Jacobson KA. Systematic investigation of polyamidoamine dendrimers surface-modified with poly(ethylene glycol) for drug delivery applications: synthesis, characterization, and evaluation of cytotoxicity. Bioconjug Chem 2008;19(8):1660–72.

[98] Tack F, et al. Modified poly(propylene imine) dendrimers as effective transfection agents for catalytic DNA enzymes (DNAzymes). J Drug Target 2006;14(2):69–86.

[99] Braun CS, et al. Structure/function relationships of polyamidoamine/DNA dendrimers as gene delivery vehicles. J Pharm Sci 2005;94(2):423–36.

[100] Chaltin P, et al. Delivery of antisense oligonucleotides using cholesterol-modified sense dendrimers and cationic lipids. Bioconjug Chem 2005;16(4):827–36.

[101] Bijsterbosch MK, et al. bis-cholesteryl-conjugated phosphorothioate oligodeoxynucleotides are highly selectively taken up by the liver. Journal of Pharmacology and Experimental Therapeutics 2002;302(2):619–26.

[102] Dung TH, et al. Preparation and evaluation of cholesteryl PAMAM dendrimers as nano delivery agents for antisense oligonucleotides. Colloids and Surfaces a-Physicochemical and Engineering Aspects 2008;313:273–7.

[103] Kono K, et al. Transfection activity of polyamidoamine dendrimers having hydrophobic amino acid residues in the periphery. Bioconjug Chem 2005;16(1):208–14.

[104] Biswas S, et al. Lipid modified triblock PAMAM-based nanocarriers for siRNA drug co-delivery. Biomaterials 2013;34(4):1289–301.

[105] Arima H, et al. Enhancement of gene expression by polyamidoamine dendrimer conjugates with α-, β-, and γ-cyclodextrins. Bioconjug Chem 2001;12(4):476–84.

[106] Tsutsumi T, et al. Evaluation of polyamidoamine dendrimer/α-cyclodextrin conjugate (generation 3, G3) as a novel carrier for small interfering RNA (siRNA). J Control Release 2007;119(3):349–59.

[107] Boeneman K, et al. Selecting improved peptidyl motifs for cytosolic delivery of disparate protein and nanoparticle materials. ACS Nano 2013;7(5):3778–96.

[108] Liu C, et al. Arginine-terminated generation 4 PAMAM dendrimer as an effective nanovector for functional siRNA delivery in vitro and in vivo. Bioconjug Chem 2014;25(3):521–32.

[109] Kim TI, et al. Arginine-conjugated polypropylenimine dendrimer as a non-toxic and efficient gene delivery carrier. Biomaterials 2007;28(11):2061–7.

[110] Kim TI, et al. Synthesis and characterization of a novel arginine-grafted dendritic block copolymer for gene delivery and study of its cellular uptake pathway leading to transfection. Bioconjug Chem 2007;18(2):309–17.

[111] Liu XX, et al. PAMAM dendrimers mediate siRNA delivery to target Hsp27 and produce potent antiproliferative effects on prostate cancer cells. ChemMedChem 2009;4(8):1302–10.

[112] Shen XC, et al. Importance of size-to-charge ratio in construction of stable and uniform nanoscale RNA/dendrimer complexes. Org Biomol Chem 2007;5(22):3674–81.

[113] Zhou J, et al. PAMAM dendrimers for efficient siRNA delivery and potent gene silencing. Chem Commun (Camb) 2006;22:2362–4.

[114] Rodrigo AC, et al. Efficient, non-toxic hybrid PPV-PAMAM dendrimer as a gene carrier for neuronal cells. Biomacromolecules 2011;12(4):1205–13.

[115] Kim SH, et al. VEGF siRNA delivery system using arginine-grafted bioreducible poly(disulfide amine). Mol Pharm 2009;6(3):718–26.

[116] Kim SH, et al. Reductive degradation behavior of bioreducible poly(disulfide amine) for enhancing SiRNA efficiency. Macromol Biosci 2010;10(8):898–905.

[117] Dunn SS, et al. Reductively responsive siRNA-conjugated hydrogel nanoparticles for gene silencing. J Am Chem Soc 2012;134(17):7423–30.

[118] Baigude H, et al. In vivo delivery of RNAi by reducible interfering nanoparticles (iNOPs). ACS Medicinal Chemistry Letters 2013;4(8):720–3.

[119] Baigude H, et al. Design and creation of new nanomaterials for therapeutic RNAi. ACS Chem Biol 2007;2(4):237–41.

[120] Su J, et al. Silencing microRNA by interfering nanoparticles in mice. Nucleic Acids Res 2011;39(6).

[121] An S, et al. Brain-targeting delivery for RNAi neuroprotection against cerebral ischemia reperfusion injury. Biomaterials 2013;34(35):8949–59.

[122] Sumer B, Gao J. Theranostic nanomedicine for cancer. Nanomedicine (Lond) 2008;3(2):137–40.

[123] Taratula O, et al. Multifunctional nanomedicine platform for cancer specific delivery of siRNA by superparamagnetic iron oxide nanoparticles-dendrimer complexes. Curr Drug Deliv 2011;8(1):59–69.

[124] Novoa-Carballal R, Pergushov DV, Muller AHE. Interpolyelectrolyte complexes based on hyaluronic acid-block-poly(ethylene glycol) and poly-L-lysine. Soft Matter 2013;9(16):4297–303.

[125] van Duijvenbode RC, Borkovec M, Koper GJM. Acid-base properties of poly(propylene imine) dendrimers. Polymer 1998;39(12):2657–64.

[126] Peng SF, et al. Effects of the nanostructure of dendrimer/DNA complexes on their endocytosis and gene expression. Biomaterials 2010;31(21):5660–70.

[127] Kabanov VA, et al. Interaction of Astramol poly(propyleneimine) dendrimers with linear polyanions. Macromolecules 1999;32(6):1904–9.

[128] Kabanov AV, et al. Soluble stoichiometric complexes from poly(N-ethyl-4-vinylpyridinium) cations and poly(ethylene oxide) block-polymethacrylate anions. Macromolecules 1996;29(21):6797–802.

[129] Roessler BJ, et al. Substituted beta-cyclodextrins interact with PAMAM dendrimer-DNA complexes and modify transfection efficiency. Biochem Biophys Res Commun 2001;283(1):124–9.

[130] Su CJ, et al. Columnar mesophases of the complexes of in DNA with low-generation poly(amido amine) dendrimers. Biomacromolecules 2009;10(4):773–83.

[131] Evans HM, et al. Structural polymorphism of DNA-dendrimer complexes. Phys Rev Lett 2003;91(7).

[132] Boeckle S, et al. Purification of polyethylenimine polyplexes highlights the role of free polycations in gene transfer. J Gene Med 2004;6(10):1102–11.

[133] Finsinger D, et al. Protective copolymers for nonviral gene vectors: synthesis, vector characterization and application in gene delivery. Gene Ther 2000;7(14):1183–92.

[134] Clamme JP, Azoulay J, Mely Y. Monitoring of the formation and dissociation of polyethylenimine/DNA complexes by two photon fluorescence correlation spectroscopy. Biophys J 2003;84(3):1960–8.

[135] Yue Y, et al. Revisit complexation between DNA and polyethylenimine — Effect of uncomplexed chains free in the solution mixture on gene transfection. J Control Release 2011;155(1):67–76.

[136] Bolton K. Gene delivery with amphiphilic lower generation polypropilenimine dendrimer in pharmaceutical sciences. Glasgow: University of Strathclide; 2008.

[137] Eichman JD, et al. The use of PAMAM dendrimers in the efficient transfer of genetic material into cells. Pharmaceutical Science & Technology Today 2000;3(7):232–45.

[138] Kaneshiro TL, Lu ZR. Targeted intracellular codelivery of chemotherapeutics and nucleic acid with a well-defined dendrimer-based nanoglobular carrier. Biomaterials 2009;30(29):5660–6.

[139] Jain K, et al. Dendrimer toxicity: let's meet the challenge. Int J Pharm 2010;394(1-2):122–42.

[140] Liu CX, Zhang N. Pharmaceutical strategies enhancing cell penetration efficiencies of non-viral gene delivery systems. Curr Gene Ther 2009;9(4):267–90.
[141] Conner SD, Schmid SL. Regulated portals of entry into the cell. Nature 2003;422(6927):37–44.
[142] Perumal OP, et al. The effect of surface functionality on cellular trafficking of dendrimers. Biomaterials 2008;29(24-25):3469–76.
[143] Parton RG, Simons K. The multiple faces of caveolae. Nat Rev Mol Cell Biol 2007;8(3):185–94.
[144] Simons K, Ikonen E. Functional rafts in cell membranes. Nature 1997;387(6633):569–72.
[145] Norkin LC, Wolfrom SA, Stuart ES. Association of caveolin with Chlamydia trachomatis inclusions at early and late stages of infection. Exp Cell Res 2001;266(2):229–38.
[146] Pelkmans L, Kartenbeck J, Helenius A. Caveolar endocytosis of simian virus 40 reveals a new two-step vesicular-transport pathway to the ER. Nat Cell Biol 2001;3(5):473–83.
[147] Peyron P, et al. Nonopsonic phagocytosis of Mycobacterium kansasii by human neutrophils depends on cholesterol and is mediated by CR3 associated with glycosylphosphatidylinositol-anchored proteins. J Immunol 2000;165(9):5186–91.
[148] Varma R, Mayor S. GPI-anchored proteins are organized in submicron domains at the cell surface. Nature 1998;394(6695):798–801.
[149] Manunta M, et al. Gene delivery by dendrimers operates via a cholesterol dependent pathway. Nucleic Acids Res 2004;32(9):2730–9.
[150] Manunta M, et al. Gene delivery by dendrimers operates via different pathways in different cells, but is enhanced by the presence of caveolin. J Immunol Methods 2006;314(1-2):134–46.
[151] Schitzer JE, et al. Filipin-sensitive caveolae-mediated transport in endothelium: reduced transcytosis, scavenger endocytosis and capillary permeability of select macromolecules. J Cell Biol 1994;127:1217–32.
[152] Damke H, et al. Clathrin-independent pinocytosis is induced in cells overexpressing a temperature-sensitive mutant of dynamin. J Cell Biol 1995;131(1):69–80.
[153] Nishikawa M, Huang L. Nonviral vectors in the new millennium: delivery barriers in gene transfer. Hum Gene Ther 2001;12(8):861–70.
[154] Juliano R, et al. Mechanisms and strategies for effective delivery of antisense and siRNA oligonucleotides. Nucleic Acids Res 2008;36(12):4158–71.
[155] Sonawane ND, Szoka FC, Verkman AS. Chloride accumulation and swelling in endosomes enhances DNA transfer by polyamine-DNA polyplexes. J Biol Chem 2003;278(45):44826–31.
[156] Nel AE, et al. Understanding biophysicochemical interactions at the nano-bio interface. Nat Mater 2009;8(7):543–57.
[157] Behr JP. Gene-transfer with synthetic cationic amphiphiles - prospects for gene-therapy. Bioconjug Chem 1994;5(5):382–9.
[158] Cho YW, Kim JD, Park K. Polycation gene delivery systems: escape from endosomes to cytosol. J Pharm Pharmacol 2003;55(6):721–34.
[159] Mack GS. MicroRNA gets down to business. Nat Biotechnol 2007;25(6):631–8.
[160] Leong KW, Grigsby CL. Balancing protection and release of DNA: tools to address a bottleneck of non-viral gene delivery. J R Soc Interface 2010;7:S67–82.
[161] Schaffer DV, et al. Vector unpacking as a potential barrier for receptor-mediated polyplex gene delivery. Biotechnol Bioeng 2000;67(5):598–606.
[162] Thomas M, Klibanov AM. Non-viral gene therapy: polycation-mediated DNA delivery. Appl Microbiol Biotechnol 2003;62(1):27–34.

[163] Huth S, et al. Interaction of polyamine gene vectors with RNA leads to the dissociation of plasmid DNA-carrier complexes. J Gene Med 2006;8(12):1416–24.

[164] Zhang Y, Wang Z, Gemeinhart RA. Progress in microRNA delivery. J Control Release 2013;172(3):962–74.

[165] Lechardeur D, et al. Metabolic instability of plasmid DNA in the cytosol: a potential barrier to gene transfer. Gene Ther 1999;6(4):482–97.

[166] Kreiss P, et al. Plasmid DNA size does not affect the physicochemical properties of lipoplexes but modulates gene transfer efficiency. Nucleic Acids Res 1999;27(19):3792–8.

[167] Lam AP, Dean DA. Progress and prospects: nuclear import of nonviral vectors. Gene Ther 2010;17(4):439–47.

[168] Pichon C, Billiet L, Midoux P. Chemical vectors for gene delivery: uptake and intracellular trafficking. Curr Opin Biotechnol 2010;21(5):640–55.

[169] Featherstone C, Darby MK, Gerace L. A monoclonal antibody against the nuclear pore complex inhibits nucleocytoplasmic transport of protein and RNA in vivo. J Cell Biol 1988;107(4):1289–97.

[170] Dowty ME, et al. Plasmid DNA entry into postmitotic nuclei of primary rat myotubes. Proc Natl Acad Sci U S A 1995;92(10):4572–6.

[171] Kosugi S, et al. Six classes of nuclear localization signals specific to different binding grooves of importin alpha. J Biol Chem 2009;284(1):478–85.

[172] Wagstaff KM, Jans DA. Nucleocytoplasmic transport of DNA: enhancing non-viral gene transfer. Biochem J 2007;406(2):185–202.

[173] Pouton CW, Seymour LW. Key issues in non-viral gene delivery. Adv Drug Deliv Rev 2001;46(1-3):187–203.

[174] Brunner S, et al. Cell cycle dependence of gene transfer by lipoplex polyplex and recombinant adenovirus. Gene Ther 2000;7(5):401–7.

[175] Wilke M, et al. Efficacy of a peptide-based gene delivery system depends on mitotic activity. Gene Ther 1996;3(12):1133–42.

[176] Monopoli MP, et al. Formation and characterization of the nanoparticle-protein corona. Methods Mol Biol 2013;1025:137–55.

[177] Walczyk D, et al. What the cell "sees" in bionanoscience. J Am Chem Soc 2010;132(16):5761–8.

[178] Dell'Orco D, et al. Modeling the time evolution of the nanoparticle-protein corona in a body fluid. PLoS One 2010;5(6):e10949.

[179] Santander-Ortega MJ, Bastos-Gonzalez D, Ortega-Vinuesa JL. Electrophoretic mobility and colloidal stability of PLGA particles coated with IgG. Colloids Surf B Biointerfaces 2007;60(1):80–8.

[180] Santander-Ortega MJ, et al. Insulin-loaded PLGA nanoparticles for oral administration: an in vitro physico-chemical characterization. J Biomed Nanotechnol 2009;5(1):45–53.

[181] Tirado-Miranda M, et al. Infuence of surface characteristics on fast-aggregating protein-coated polymer colloids. Trends in Colloid and Interface Science Xv 2001;118:285–9.

[182] Kukowska-Latallo JF, et al. Intravascular and endobronchial DNA delivery to murine lung tissue using a novel, nonviral vector. Hum Gene Ther 2000;11(10):1385–95.

[183] Ortega-Vinuesa JL, Hidalgo-Alvarez R. Sequential adsorption of F(ab')(2) and BSA on negatively and positively charged polystyrene latexes. Biotechnol Bioeng 1995;47(6):633–9.

[184] Hellstrand E, et al. Complete high-density lipoproteins in nanoparticle corona. FEBS J 2009;276(12):3372–81.

[185] Gulyaev AE, et al. Significant transport of doxorubicin into the brain with polysorbate 80-coated nanoparticles. Pharm Res 1999;16(10):1564–9.

[186] Dagher G, et al. HDL-mediated cholesterol uptake and targeting to lipid droplets in adipocytes. J Lipid Res 2003;44(10):1811–20.
[187] Zhang Y, et al. The inhibition of endocytosis affects HDL-lipid uptake mediated by the human scavenger receptor class B type I. Mol Membr Biol 2007;24(5-6):442–54.
[188] Kim HR, et al. Low-density lipoprotein receptor-mediated endocytosis of PEGylated nanoparticles in rat brain endothelial cells. Cell Mol Life Sci 2007;64(3):356–64.
[189] Santander-Ortega MJ, et al. Colloidal stability of pluronic F68-coated PLGA nanoparticles: a variety of stabilisation mechanisms. J Colloid Interface Sci 2006;302(2):522–9.
[190] Dobrovolskaia MA, McNeil SE. Immunological properties of engineered nanomaterials. Nat Nanotechnol 2007;2(8):469–78.
[191] Dash BC, et al. The influence of size and charge of chitosan/polyglutamic acid hollow spheres on cellular internalization, viability and blood compatibility. Biomaterials 2010;31(32):8188–97.
[192] Dobrovolskaia MA, et al. Interaction of colloidal gold nanoparticles with human blood: effects on particle size and analysis of plasma protein binding profiles. Nanomedicine 2009;5(2):106–17.
[193] Fortina P, et al. Nanobiotechnology: the promise and reality of new approaches to molecular recognition. Trends Biotechnol 2005;23(4):168–73.
[194] Roberts JC, Bhalgat MK, Zera RT. Preliminary biological evaluation of polyamidoamine (PAMAM) Starburst dendrimers. J Biomed Mater Res 1996;30(1):53–65.
[195] Malik N, et al. Dendrimers: relationship between structure and biocompatibility in vitro, and preliminary studies on the biodistribution of ¹²⁵I-labelled polyamidoamine dendrimers in vivo. J Control Release 2000;65(1-2):133–48.
[196] Wilbur DS, et al. Biotin reagents for antibody pretargeting. 3. Synthesis, radioiodination, and evaluation of biotinylated starburst dendrimers. Bioconjug Chem 1998;9(6):813–25.
[197] Matsumura Y, Maeda H. A new concept for macromolecular therapeutics in cancer chemotherapy: mechanism of tumoritropic accumulation of proteins and the antitumor agent smancs. Cancer Res 1986;46(12 Pt 1):6387–92.
[198] Yew NS, et al. High and sustained transgene expression in vivo from plasmid vectors containing a hybrid ubiquitin promoter. Mol Ther 2001;4(1):75–82.
[199] de Wolf HK, et al. Plasmid CpG depletion improves degree and duration of tumor gene expression after intravenous administration of polyplexes. Pharm Res 2008;25(7):1654–62.
[200] Herweijer H, et al. Time course of gene expression after plasmid DNA gene transfer to the liver. J Gene Med 2001;3(3):280–91.
[201] Qi R, et al. PEG-conjugated PAMAM dendrimers mediate efficient intramuscular gene expression. AAPS J 2009;11(3):395–405.
[202] Mignani S, et al. Expand classical drug administration ways by emerging routes using dendrimer drug delivery systems: a concise overview. Adv Drug Deliv Rev 2013;65(10):1316–30.
[203] Hudde T, et al. Activated polyamidoamine dendrimers, a non-viral vector for gene transfer to the corneal endothelium. Gene Ther 1999;6(5):939–43.
[204] Nakanishi H, et al. Nonviral genetic transfer of Fas ligand induced significant growth suppression and apoptotic tumor cell death in prostate cancer in vivo. Gene Ther 2003;10(5):434–42.
[205] Kang C, et al. Evaluation of folate-PAMAM for the delivery of antisense oligonucleotides to rat C6 glioma cells in vitro and in vivo. J Biomed Mater Res 2010;93(2):585–94.
[206] Maruyama-Tabata H, et al. Effective suicide gene therapy in vivo by EBV-based plasmid vector coupled with polyamidoamine dendrimer. Gene Ther 2000;7(1):53–60.

[207] Vincent L, et al. Efficacy of dendrimer-mediated angiostatin and TIMP-2 gene delivery on inhibition of tumor growth and angiogenesis: in vitro and in vivo studies. Int J Cancer 2003;105(3):419–29.
[208] Bell HS, et al. A p53-derived apoptotic peptide derepresses p73 to cause tumor regression in vivo. J Clin Invest 2007;117(4):1008–18.
[209] Taratula O, et al. Surface-engineered targeted PPI dendrimer for efficient intracellular and intratumoral siRNA delivery. J Control Release 2009;140(3):284–93.
[210] Calzolari A, et al. Transferrin receptor 2 is frequently expressed in human cancer cell lines. Blood Cells Mol Dis 2007;39(1):82–91.
[211] Pardridge WM, Eisenberg J, Yang J. Human blood-brain barrier transferrin receptor. Metabolism 1987;36(9):892–5.
[212] Aktas Y, et al. Development and brain delivery of chitosan-PEG nanoparticles functionalized with the monoclonal antibody OX26. Bioconjug Chem 2005;16(6):1503–11.
[213] De La Fuente M, et al. Chapter 7.3 Drug delivery strategies: nanostructures for improved brain delivery, in nanostructured biomaterials for overcoming biological barriers. The Royal Society of Chemistry 2012;392–432.
[214] Ji B, et al. Pharmacokinetics and brain uptake of lactoferrin in rats. Life Sci 2006;78(8):851–5.
[215] Pang ZQ, et al. Brain delivery and cellular internalization mechanisms for transferrin conjugated biodegradable polymersomes. Int J Pharm 2011;415(1-2):284–92.
[216] Zhang PC, et al. Transferrin-conjugated polyphosphoester hybrid micelle loading paclitaxel for brain-targeting delivery: synthesis, preparation and in vivo evaluation. J Control Release 2012;159(3):429–34.
[217] Calzolari A, et al. Transferrin receptor 2 is frequently and highly expressed in glioblastomas. Transl Oncol 2010;3(2):123–34.
[218] Huang RQ, et al. Efficient gene delivery targeted to the brain using a transferrin- conjugated polyethyleneglycol-modified polyamidoamine dendrimer. FASEB J 2007;21(4):1117–25.
[219] Huang RQ, et al. The use of lactoferrin as a ligand for targeting the polyamidoamine-based gene delivery system to the brain. Biomaterials 2008;29(2):238–46.
[220] Huang R, et al. Brain-targeting mechanisms of lactoferrin-modified DNA-loaded nanoparticles. J Cereb Blood Flow Metab 2009;29(12):1914–23.
[221] Ke W, et al. Gene delivery targeted to the brain using an Angiopep-conjugated polyethyleneglycol-modified polyamidoamine dendrimer. Biomaterials 2009;30(36): 6976–85.
[222] Liu Y, et al. Brain-targeting gene delivery and cellular internalization mechanisms for modified rabies virus glycoprotein RVG29 nanoparticles. Biomaterials 2009;30(25):4195–202.
[223] Liu Y, et al. A leptin derived 30-amino-acid peptide modified pegylated poly-l-lysine dendrigraft for brain targeted gene delivery. Biomaterials 2010;31(19):5246–57.
[224] Kuang Y, et al. T7 peptide-functionalized nanoparticles utilizing RNA interference for glioma dual targeting. Int J Pharm 2013;454(1):11–20.
[225] Oh S, et al. Synthesis and anti-cancer activity of covalent conjugates of artemisinin and a transferrin-receptor targeting peptide. Cancer Lett 2009;274(1):33–9.
[226] Ohkuma S, Poole B. Fluorescence probe measurement of the intralysosomal pH in living cells and the perturbation of pH by various agents. Proc Natl Acad Sci U S A 1978;75(7):3327–31.

Polymeric nanoparticles for gene delivery

7

T. Werfel, C. Duvall
Vanderbilt University, Nashville, TN, USA

7.1 Introduction

Gene therapies have emerged as a part of the revolution in cell and molecular biology over the second half of the twentieth century. While viral gene therapies represent the gold standard in terms of efficiency, therapeutic use of viruses remains shadowed by concerns of safety. Due to the tremendous opportunity for gene therapies to impact medicine, there is great motivation to work toward safer and targeted viral vectors and to also carry out parallel efforts toward engineering more efficient non-viral systems that can achieve safe, efficacious gene therapy in humans. This chapter will focus on the primary barriers that have limited clinical translation of non-viral gene therapies and outline recent advances in polymeric nanoparticles that have the potential to overcome these challenges. Many of the recent breakthroughs in non-viral gene therapy are a result of advances in materials science, chemistry, and engineering that have enabled the synthesis/fabrication of more homogeneous materials with an array of well-defined and tunable functionalities, in addition to the evolution of technologies for more rigorous characterization of chemical composition and biological interactions of nanomaterials. Taken together, rapid advances in materials synthesis and characterization, combined with tremendous progress in our ability to manipulate the genome and molecular-level and nanoscale phenomena, are predicted to yield clinically-relevant breakthroughs in non-viral gene therapies that provide new and improved treatments for some of our most challenging diseases. This chapter overviews advances in the design and testing of polymeric nanoparticles that continue to motivate our quest for clinical translation of non-viral gene therapies for both gene expression and silencing.

Nucleic acid-based drugs are sought for the replacement of missing, mutated, or deficient genes and for the suppression of gene expression by RNA interference (RNAi). Common classes of nucleic acid-based drugs include plasmid DNA (pDNA), messenger RNA (mRNA), short interfering RNA (siRNA), short hairpin RNA (shRNA), microRNA (miRNA), miRNA inhibitors, peptide nucleic acids (PNA), RNA-based adjuvants, and clustered regularly interspaced short palindromic repeats/Cas (CRISPR/Cas) gene editing systems. This chapter will focus on the more heavily investigated delivery strategies for siRNA and pDNAs, but note that strategies are similar for other nucleic acid-based drug classes not herein discussed. miRNA, miRNA inhibitor, RNA-adjuvant and PNA delivery strategies typically coincide with siRNA

Polymers and Nanomaterials for Gene Therapy. http://dx.doi.org/10.1016/B978-0-08-100520-0.00007-2
© 2016 Elsevier Ltd. All rights reserved.

delivery, while mRNA (although trafficking to the nucleus is not needed), shRNA and CRISPR/Cas systems typically coincide with pDNA. Gene replacement/addition is the more mature field but comes with the extra challenges associated with delivering pDNA into the nucleus and the limitation that plasmids often do not stably integrate and thus do not have high potential for curing chronic genetic diseases. By contrast, RNAi by siRNA only requires delivery into the cytosol where the RNA-induced silencing complex (RISC) machinery is located. The activated RISC complex contains the antisense siRNA strand and mediates the recognition and subsequent enzymatic degradation of complementary mRNA. Both approaches require some common and disparate delivery considerations based on the desired intracellular pharmacokinetics. For instance, both require accumulation at the tissue compartment of interest and must cross the outer cell membrane. However, RNAi by siRNA will only require a carrier with endosomal escape mechanisms while gene augmentation/replacement achieved through delivery of pDNA requires intracellular trafficking into nucleus.

In general, three major delivery barriers are considered when developing polymeric nanoparticles for gene therapy: 1) preferential biodistribution to the site of action, 2) cellular internalization by the targeted cell type, and 3) trafficking to and unpackaging within the intracellular compartment where the nucleic acid cargo will be active (Figure 7.1). Viruses have evolved mechanisms to efficiently overcome these barriers and are the inspiration for the design of biomimetic polymer nanoparticles. Continued breakthroughs in the elucidation of the mechanisms of action of viruses, combined with rapid advances in synthetic fabrication and characterization of virus-mimetic polymeric nanoparticles, suggests that more safe and effective non-viral gene therapies will continue to emerge.

Five classes of polymer nanoparticles are herein reviewed with the goal of highlighting techniques for synthesis and characterization and the gene therapy applications for each. The classes of materials to be discussed are micelles/polyplexes, cross-linked micelles, polymersomes, and microgels/nanogels.

7.2 Micelles

"Block copolymers with amphiphilic character, having a large solubility difference between hydrophilic and hydrophobic segments, are known to assemble in an aqueous milieu into polymeric micelles with a mesoscopic size range" [1].

Micelles are self-assembling amphiphilic block copolymers with a core-shell (hydrophilic-hydrophobic) architecture. They are typically assembled in aqueous solutions by methods such as dialysis, pH adjustment, and dropwise addition of concentrated organic solutions. The formation of core-shell architecture is driven by entropic processes, resulting in segregation of a hydrophobic core from solution by a hydrophilic outer corona. This is an attractive architecture for drug delivery applications because it is possible to load hydrophobic small molecule therapeutics into the hydrophobic micelle core. The hydrophilic outer shell is often composed of PEG and serves as an aide for increased blood circulation and reduced cytotoxicity. In many gene therapy applications, the outer corona or the core are endowed with cationic

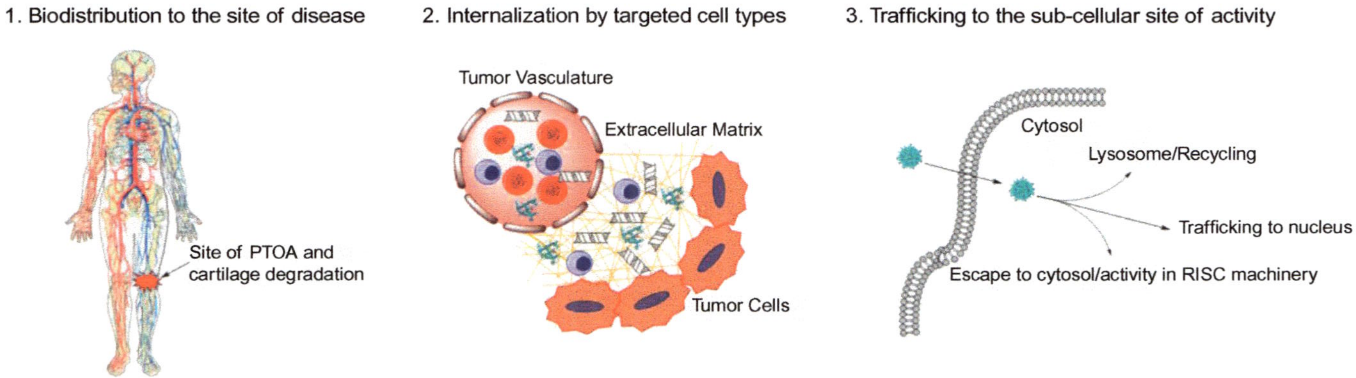

Figure 7.1 Schematic of the three major systemic and intracellular barriers to delivery of gene therapies. 1) The drug should be able to accumulate at the site of pathology, whether passively or by active targeting. 2) The drug should escape the circulation, navigate the ECM, and be internalized by the targeted cell types. 3) The drug should be trafficked to and released within the sub-cellular compartment where it is active.

functionalities that enable the electrostatic packaging of nucleic acids. Pre-assembled micelles with a cationic corona can be used to load nucleic acids, forming "micelleplexes." In other approaches, a diblock polymer with one core-forming, cationic block and one neutral hydrophilic block (typically PEG) is assembled into PEGylated polyplexes, also known as polyion complex (PIC) micelles, upon mixing with anionic nucleic acid cargo. In this chapter, we will utilize the "micelle" as an umbrella term to describe both of these sub-classes of delivery systems. Physicochemical attributes such as size, shape, chemical composition, and charge are all important factors when considering the design of micelles as therapeutic carriers because they have profound effect on their effectiveness in overcoming *in vivo* delivery barriers. This section reviews researched approaches for modulating the physicochemical properties of micellar gene therapies to approach the delivery efficiency of viral gene vectors.

The earliest approaches to using micelles as non-viral gene vectors employed cationic, amine-rich polymers such as poly(ethyleneimine) (PEI), poly(2-(dimethylamino)ethyl methacrylate) (PDMAEMA), and poly(L-lysine) (PLys). The amines of this class of cationic polymers are used for three purposes: 1) condensation/nuclease protection of negatively charged nucleic acids, 2) enhanced cellular uptake, and 3) endosomolytic escape. The positively charged amines form electrostatic interactions with and very efficiently condense nucleic acids. They also drive cellular uptake through mechanisms that are not yet fully elucidated, but which are likely mediated through anionic, heparin sulfate proteoglycans present on cell surfaces [2]. After internalization, polymers with secondary and tertiary amines with pKa values at or just below pH 7 drive endosomolytic escape through the proton-sponge effect [3,4].

Polyplexes formed of solely of cationic homopolymers have shown promise as *in vitro* tranfection agents but have many translational issues [5]. Their cationic character causes cytotoxicity and limits their utility for systemic delivery. The cationic surface charge causes aggregation with serum proteins and red blood cells, generally thought to be the reason for their eventual disproportionate accumulation in the capillary beds of the lungs, liver, and spleen and rapid clearance by the reticuloendothelial system (RES). In extreme cases, this can lead to rapid blockage of pulmonary capillaries and acute mortality in animal models. Therefore, cationic polymers are typically synthesized as diblock copolymers with hydrophilic polymers such as PEG and used to form PIC micelles [1]. Kataoka's group pioneered the development of these micelles and their application as gene vectors. They used PEG-PLys as a diblock copolymer to condense DNAs and form PIC micelles that utilize PEGylation to shield their cationic character. This approach has served as the basis for numerous subsequent designs that utilize polyion complex formation to produce stable, nucleic acid-loaded micelles.

Other "micelleplex" approaches have leveraged block copolymers that pre-assemble into stable micelles in the absence of polyanionic nucleic acids. These stable micelles incorporate a cationic outer shell that can be utilized to condense nucleic acids into their corona. Stayton et al. developed a diblock polymer capable of forming stable micelles in the absence of any nucleic acids and condensing siRNA onto their surfaces, while the micelle core provided active endosome disruption behavior [6]. These micelles consist of a 100% PDMAEMA block and a random terpolymer block of 50% butyl methacrylate (BMA), 25% DMAEMA, and 25% propylacrylic acid (PAA) (Figure 7.2a). When a

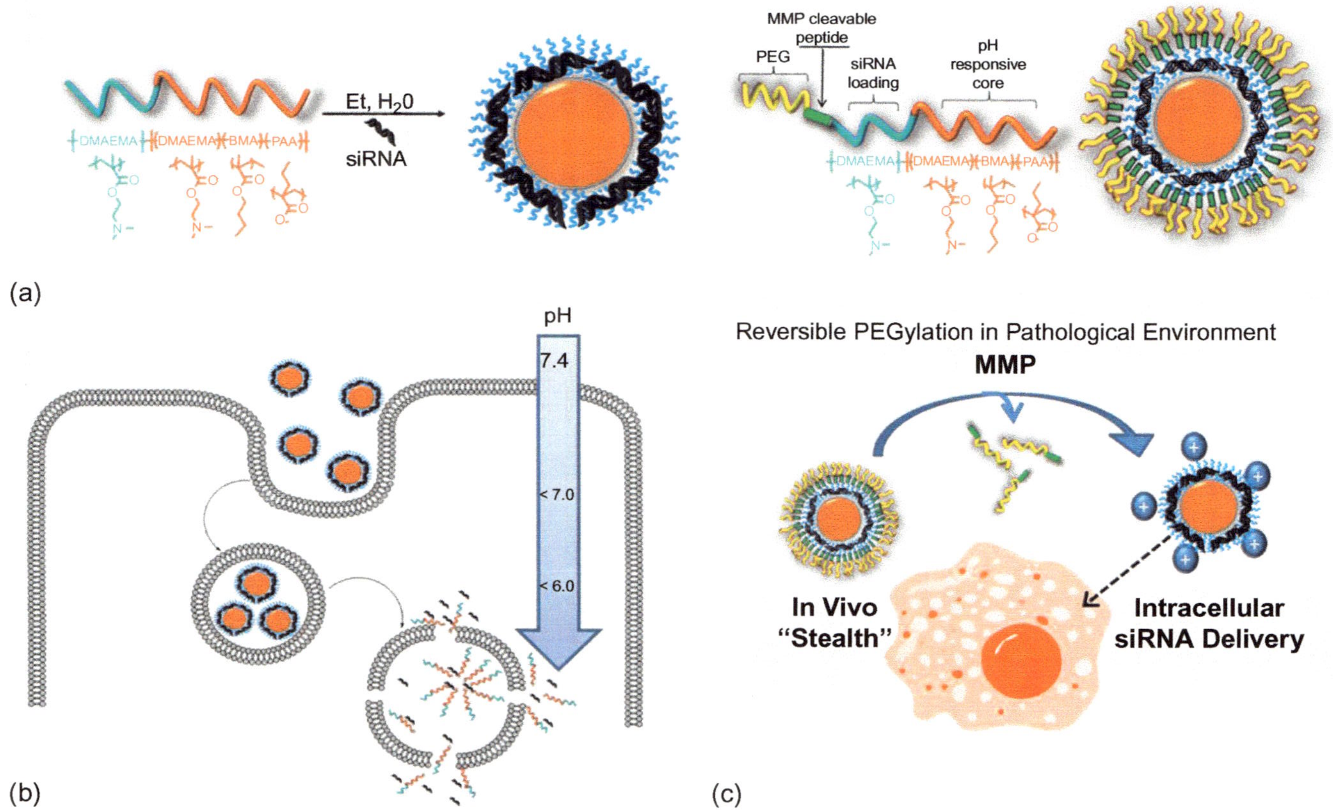

Figure 7.2 Endosomolytic Micelles for siRNA packaging and delivery [17,18]**.** a) siRNA loading and micellization in aqueous solution. B) Cellular internalization of the cationic nanocarriers and pH-responsive endolysosomal escape of siRNA. c) PEGylation of the cationic nanocarriers for *in vivo* stealth and MMP-triggered, reversible PEGylation for preferential siRNA delivery into pathological sites.

concentrated stock of the polymer in ethanol is added dropwise into excess PBS, micellar nanoparticles (~50 nm diameter) form with a PDMAEMA shell and p(DMAEMA-co-BMA-co-PAA) core. The hydrophobic BMA and electrostatic stabilization between DMAEMA and PAA provide core stability at physiological pH. The highly cationic PDMAEMA block provides stable siRNA loading into the outer micelle shell. The core stabilization is disrupted at pH values representative of the endolysosomal pathway (~5-6) due to concurrent protonation of the secondary amines of DMAEMA and carboxylic acid of PAA, leading to net cationic charge and electrostatic repulsion/destabilization of the core. The exposed p(DMAEMA-co-BMA-co-PAA) core-forming block exhibits pH-dependent lipid bilayer membrane disruptive activity, driving endosomolysis and escape of siRNA into the cytosol (Figure 7.2b).

While they are very effective for *in vitro* delivery and potentially for topical/local delivery *in vivo*, these micelleplexes possess highly cationic surface charge, resulting in similar limitations as the cationic homopolymers described above with regard to systemic delivery applications *in vivo*. To overcome this shortcoming, Stayton et al. have extended the use of the endosomolytic p(DMAEMA-co-BMA-co-PAA) micelle core-forming composition by combining it with a charge neutral corona containing sites for siRNA conjugation (rather than electrostatic loading) [7]. The corona of these charge neutral polymeric micelles consists of neutral, hydrophilic poly[N-(2-hydroxypropyl)methacrylamide-co-N-(2-(pyridine-2-yldisulfanyl)ethyl)methacrylamide)] (p[HPMA-co-PDSMA]). This block promotes aqueous stability and provides pyridyl disulfide functionalities amenable to conjugation of thiol-functionalized siRNA. This approach shows mRNA knockdown and excellent biocompatibility and represents a potential alternative approach to PEGylation of cationic micelles, although the efficiency of siRNA conjugation is still much lower than electrostatic condensation.

Approaches that promote micelle surface charge neutrality, such as PEGylation, covalent siRNA loading as described above, and others also have limitations. By shielding of the cationic surface charge, the mechanism for cellular uptake of the micelles is also significantly neutralized. Attempts at modulating the effects of PEGylation are currently under investigation. Groups are elucidating what effect different lengths and densities of PEG on surfaces of micelles have on uptake and *in vivo* pharmacokinetics [8]. Two promising approaches for overcoming reductions in bioactivity due to PEGylation are: 1) the addition of targeting moieties on micelle surfaces that engage actively-internalized cell surface receptors (rather than depending on non-specific electrostatic interactions with the cell surface for uptake) and 2) reversible PEGylation.

Micelle surface functionalization with targeting moieties can be beneficial for both PEGylated and non-PEGylated micelles. Stayton et al. employed this approach to their cationic micelles by conjugating folate for cancer cell targeting [9]. They were able to exhibit folate receptor specific interaction of the folate conjugated micelles as well as *in vitro* gene knockdown, though this system is also not optimized for intravenous delivery due to high residual surface charge. More commonly, targeting moieties are added to the surface of PEGylated micelles. Kataoka et al. attached targeting molecules to their PEGylated PIC micelles for various applications [10]. They attached lactose for targeting the ASGP receptor of HepG2 cells, showing a significant increase in cellular transfection efficiency compared with micelles lacking lactose [11]. They also employed the cyclic RGD peptide as a means to target $\alpha_v\beta_3$ integrin receptors. This integrin is expressed on many cell types that play an important role in angiogenesis, and is a potentially useful means to target anti-angiogenic tumor therapies. In these studies, this targeting approach improved PIC micelle tumor delivery of pDNA encoding suicide genes [12]. Hydrophobic stabilization of the core in PIC micelles has also been employed in order to achieve better pharmacokinetics *in vivo* [13]. By balancing cationic DMAEMA with hydrophobic BMA within the core of PEG-P(DMAEMA-co-BMA) PIC micelles, significantly higher endosomal escape and luciferase protein level knockdown was achieved *in vitro*. Further, knockdown of the model gene PPIB was achieved *in vivo* without any active targeting moiety.

Similar targeting approaches have been taken for more traditional (non-PIC) micelles as well. For example, Xiong et al. recently coined a "virus-mimetic" targeted micelle for siRNA delivery [14]. They targeted PEG-poly(ε-capralactone) (PEG-PCL) micelles by decorating the PEG shell with a combination of RGD targeting peptide and TAT cell-penetrating peptide. They observed an increase in cellular uptake for micelles with the targeting moieties, seeing the largest increase for the combined targeting with RGD and TAT peptides. Increased cellular uptake also led to increased levels of P-glycoprotein (P-gp) knockdown at the mRNA and protein levels, re-sensitizing the cancer cells to doxorubicin treatment. A follow-up study was conducted where doxorubicin (Dox) was co-delivered with siRNA within the RGD- and TAT-targeted PEG-PCL micelles [15]. The co-delivery technique improved the efficacy of Dox in multi-drug resistant (MDR) MDA-MB-435 tumor cells over-expressing P-gp. Cell viability *in vitro* was reduced two-three fold by co-delivery of MDR-1 siRNA and Dox in RGD/TAT targeted micelles verses untargeted micelles or targeted micelles with scrambled siRNA. As for most studies in this area to date, these are preliminary results and require more investigation in preclinical animal trials. These results do, however, provide proof of concept and promise of future success of this targeting approach for increased therapeutic efficacy *in vivo*.

The second approach that has recently evolved as a means for overcoming reduced activity due to micelle PEGylation is to use reversible/sheddable conjugation chemistry for micelle PEG attachment. By this approach, micelles will contain a cellular transfection moiety, most commonly cationic charge, which is shielded by PEG. PEG shields the micelles from uptake until it is removed within the pathological environment of interest. Giorgio et al. pioneered a form of enzyme activation of quantum dot nanoparticles (QD-NPs) coined as "proximity-activated" targeting [16]. With this approach, QD-NPs are only activated in proximity of the pathological environment where matrix-metalloproteinase (MMP) enzymes are expressed in higher concentrations than in circulation and normal tissue environments. A modification of this approach has been applied for reversible PEGylation of micelles for siRNA delivery [17]. In this approach, PDMAEMA-P(DMAEMA-co-BMA-co-PAA) cationic micelles were PEGylated through PEG attachment to the micelle surface using an MMP-7 cleavable peptide (Figure 7.2c). PEGylation shields the cationic pDMAEMA shell utilized for siRNA loading, providing stealth in circulation. This PEG shielding of the cationic layer is reversed in pathological environments of high MMP-7 expression (e.g., aggressive metastatic tumors), exposing cationic character of the micelles and driving cellular uptake. The surface charge of responsive micelles was increased 2.5-3 fold after exposure to MMP-7, indicating the potential of this design to shed PEGylation in pathological microenvironments. In the preliminary studies, MMP-7 responsive micelles showed significant increases in knockdown of the model enzyme luciferase in R221A-Luc mammary tumor cells following MMP-7 activation.

The use of biodegradable micelles is a separate approach that has been explored, predominantly as a means to lower toxicity profiles of micelles. Poly(beta-amino esters) (PBAEs) are a particularly promising class of cationic, biodegradable micelles showing better transfection efficiency *in vitro* than LF2K [19,20]. Langer and Green et al. developed a high-throughput system for synthesizing and screening combinatorial libraries of PBAEs to discover polymers that provide high cell transfection efficiency with low toxicity profiles [21]. Initially, they identified an ideal PBAE polymer candidate through high-throughput screening. A library of end-group functionalities was then screened to identify which would give the greatest increase in transfection efficiency while maintaining cytocompatibility. It was found that PBAEs with primary alkyl di-amines were more effective than PBAEs with di-amine PEG spacers. Most importantly, they reported transfection efficiencies comparable to adenoviruses and orders of magnitude above PBAEs without end-group functionalities. Further studies of similar systems (i.e., poly(amine-co-esters)) have shown a continuum of success for cationic, biodegradable systems in pre-clinical studies and suggest that these classes of non-viral gene delivery systems are moving closer to replicating virus-like efficiency and becoming a clinical reality [22].

To this point, most of the micellar systems discussed were of fully synthetic origin. Many micellar delivery systems employ amino acid-based monomers as well. The PEG-P(Lys) derived PIC micelles discussed earlier were one of the first delivery systems to employ "natural" monomers. Other micelles have been developed since using natural source monomers such as histidine, chitosan, β-cyclodextrins, and glycoamidoamines that show promise as non-viral gene vectors.

Poly(histidine) (P(His)) has a secondary amine containing imidazole group with its pKa around 6.0, appropriate for the proton sponge effect in late endosomes where the pH is shifted to below 6.0. P(His) is typically combined with P(Lys) in either a grafted or highly branched copolymer. The primary amines of lysine are fully protonated at physiological pH and serve to efficiently condense nucleic acids. The lower pKa secondary amines of P(His) provide an endosomal escape mechanism by the proton sponge effect [23]. However, it was shown that the addition of endosomal disruption agent, chloroquine, significantly improved the transfection efficiency of these micelles, indicating that they are still prone to entrapment in the endolysosomal pathway [24]. Other approaches to hybrid micelles using P(His) have also been investigated such as a promising chitosan-P(His) hybrid, where the histidine residues are incorporated into a chitosan backbone in order to enhance endosomal escape and transfection efficiency [25].

Chitosan, a polysaccharide composed of N-acetyl glucosamine units bonded via $\beta(1 \rightarrow 4)$ glycosidic bonds, has been extensively investigated as a saccharide polymer-based delivery vector. Chitosan is a "green" approach because it is a biodegradable, renewable resource derived from chitin. Howard et al. conducted a series of successful experiments in which they delivered siRNA to knockdown enhanced green fluorescence protein (eGFP). They achieved high levels of eGFP knockdown in H1299 human lung carcinoma cells and murine peritoneal macrophages (77.9% and 89.3%, respectively) [26]. In their *in vivo* studies, the siRNA loaded chitosan micelles achieved 37% and 43% reduction in eGFP in bronchiolar epithelial cells in transgenic eGFP mice after nasal administration as compared to mismatched and untreated controls, respectively. Although promising results have been reported, chitosan micelles are limited by poor solubility in physiological buffers and poor escape from the endolysosomal pathway. Therefore, several variants such as chitosan-PEI hybrids and the chitosan-P(His) hybrid mentioned above have been developed in an attempt to increase solubility and endosomal escape [27].

Reineke and Davis have been prolific publishers on natural source-based micelles, reporting some of the most convincing results that non-viral gene vectors could have profound clinical impacts. The Reineke lab developed a class of glycopolycations, the poly(glicoamidoamines) (PGAAs) [28]. Initially, a library of PGAAs was made through condensation reactions of carbohydrates with oligoamine co-monomers. PGAAs were varied based on parameters such as carbohydrate size, hydroxyl number and stereochemistry, amine number, and presence of heterocyclic groups. The library was screened for gene delivery and optimized formulations were identified that facilitate efficient DNA packaging and intracellular delivery. The Reineke group also investigated the effectiveness of trehalose-based "click" polymers and are optimizing aqueous RAFT synthesized formulations for *in vivo* performance [29,30]. Through these studies, they have found formulations of diblock glycopolymers that effectively condense pDNA and siRNA while promoting stability under physiological salt and serum conditions. Moreover, the diblock glycopolymer formulations were made even more stable in physiological conditions when surface functionalized with poly(trehalose) as an alternative to PEGylation [31]. The poly(trehalose) functionalized diblock glycopolymer micelles showed rapid internalization into glioblastoma cells and tunable rates of siRNA delivery.

The Davis lab has pioneered the use of β-cyclodextrin (β-CD) based polymers for non-viral gene delivery. Cationic β-cyclodextrin-based polymers (βCDPs) were synthesized by the condensation of diamino-cyclodextrin monomers with diimidate co-monomers and were capable of forming polyplexes with nucleic acids [32]. βCDPs are unique in that their large, interior cavities are capable of forming inclusion complexes with hydrophobic moieties that can then be functionalized with PEG or targeting ligands [33,34]. The Davis lab was the first to translate a targeted βCDP-based micelle into the clinic for RNA interference (RNAi) human trials. They reported on the first targeted delivery of siRNA in humans using a βCDP, PEGylated for increased circulation and targeted by way of a human transferrin (Tf) targeting ligand [35]. The success of this approach was realized just a year later when the Davis lab reported the first evidence of RNAi in humans using targeted polymeric NPs [36]. In the study, a reduction of the targeted mRNA and the protein level expression was observed as compared to "pre-dosing tissue." The presence of predicted mRNA cleavage fragments further demonstrated the siRNA-mediated degradation of mRNA leading to RNAi. This initial report is a profound accomplishment, verifying the potential of siRNA and other forms of RNAi (discovered in 1998, won Nobel Prize in Physiology and Medicine in 2006[37]) as potent and specific therapeutics that could transform the treatment of a variety of hard to drug pathologies in the future.

7.3 Cross-linked micelles

"Supramolecular self-assembly techniques have provided a versatile means by which to selectively assemble polymer molecules into well-defined three dimensional core–shell nanostructures. The covalent stabilization and tailoring of these dynamic nanostructures can be achieved using a range of chemistries within the assembly to afford robust functional nanoparticles" [38].

Micelles, as described above, are supramolecular self-assemblies of amphiphilic block copolymers. Their self-assembly is driven by the hydrophobic effect – an entropic phenomenon where water molecules move from their ordered architecture along the polymer (water "cage") and into solution. Self-assembled micelles can suffer from lack of stability in biologic environments. Often, micellar systems are susceptible to dynamic changes such as temperature, salts, and concentration. Cross-linking has been employed as a strategy to overcome micelle instability by introducing covalent linkages between polymer chains that make up the micelles. Cross-linking can be used to offer greater stability to micelles by reinforcing the weaker intermolecular forces that drive micelle self-assembly [38].

Two major categories of cross-linking are employed for the stabilization of micelles. Core cross-linking is the process by which cross-linkable groups are introduced onto the hydrophobic segment of the block copolymer of the micelle. The cross-linking reaction is performed after micellization, resulting in a core cross-linked micelle with increased stability over its micellar precursor. Alternatively, cross-linkable groups can be incorporated onto the hydrophilic portion of the polymer that forms the micelle shell. In this case, the cross-linking reactions can be similarly performed post-assembly, but the result is a shell cross-linked micelle. Both strategies have been investigated for increasing micellar stability and used for a variety of applications (therapeutics, molecular imaging agents, nano/micro-electronics, etc.) [38]. Moreover, each strategy comes with its advantages and limitations. The choice of core versus shell cross-linking strategies is often based on the eventual application. However, shell cross-linked micelles are considered by many to have greater potential as delivery agents than their core cross-linked counterparts due to their higher core mobility, increased diversity of core composition, and improved encapsulation efficiency by the membrane-like, cross-linked shell layer.

Several chemistries have been employed for micelle core and shell cross-linking. Core cross-linking is commonly achieved by the copolymerization of monomers containing cross-linkable groups during the polymerization of the hydrophobic polymer block. The introduction of cross-linking reagents or external stimuli (such as photo/UV-irradiation) can be used to drive the cross-linking reactions within the core following polymer synthesis and micelle self-assembly. Recently, induction of core cross-linking by external cues such as pH and temperature have been investigated as alternative strategies [39,40]. Radical polymerization techniques were first reported for the cross-linking of the shell of micelles and condensation reactions were reported soon after [41,42]. However, the most common shell cross-linking technique is achieved by introducing cross-linking reagents. A diamine reagent can cross-link carboxylic acid groups of the polymer backbone in the presence of a carbodiimide activator in aqueous solution. Alkylation is an alternative cross-linking strategy for cross-linking shells comprised of hydrophilic monomers, such as *N,N*-dimethylaminoethyl methacrylate, that do not contain carboxylic acid functionalities (Figure 7.3) [43,44].

Selectively achieving intramicellar and not intermicellar coupling is a critical criteria for producing consistent batches of cross-linked micelles. Intramicellar couplings lead to the desired shell or core cross-linked micelles while intermicellar couplings lead to the formation of large, covalently bound aggregates. Two major approaches are taken to overcome this potential limitation. By one approach, cross-linking is simply done at extremely dilute concentrations. The extreme dilution greatly reduces the probability of intermicellar cross-linking in solution and typically leads to consistent intramicellar cross-linking. Translation to an industrial scale is a limitation of this approach, and therefore, cross-linking of ABC copolymers has been introduced as an alternative. AB copolymers consist of a hydrophobic-hydrophilic diblock architecture. ABC copolymers consist of an additional intermediate "B" segment of polymer that contains the cross-linking moieties. For example, Armes et al. developed an ABC copolymer consisting of a poly(propylene oxide) (PPO) core, poly(2-(dimethylamino) ethyl methacrylate) (DMAEMA) intermediate layer, and methoxy capped poly(oligo(ethylene glycol) methacrylate) (OEGMA) shell (Figure 7.4) [45]. The DMAEMA intermediate layer residues were cross-linked using the bifunctional quaternizing agent, 1,2-bis(2-iodoethoxy)ethane (BIEE). The introduction of this intermediate layer for coupling allows for cross-linking at higher concentrations, and offers a more desirable approach for promoting intramicellar cross-linking over intermicellar.

Although the two approaches described above have been used for developing cross-linked micelles with good functionality, more effective cross-linking strategies are still sought after. The traditional cross-linking reactions are inherently inefficient and hard to quantify. Orthogonal cross-linking chemistries have been more recently investigated as a more efficient alternative. Namely, "click" chemistry has been an increasingly popular approach. One of the most used "click" chemistries employs the copper(I)-catalyzed Huisgen 1,3-dipolar cycloaddition reaction between alkynes and azides. The reaction is highly efficient, selective, and results in good compatibility and orthogonality with a variety of biological molecules. Hawker et al. demonstrated the potential for this approach by incorporating click-reactive functional groups in either the core or shell of micelles in aqueous solution [46]. Using azide/alkyne click-functionalized fluorescent dyes, they showed the selective reactivity of these groups to complementary azide/alkyne groups within their micelle architectures. Further work has followed up on this initial report, verifying "click" chemistry as a powerful cross-linking approach and investigating its utility for the surface-functionalization of micelles [47,48]. Reversible or environmentally-responsive cross-linking, such as Katoaka's disulfide cross-linked PIC micelles [49], is also a more recent evolution and has been developed to ensure that micelles can unpackage their cargo (i.e., so that micelle overstabilization doesn't hinder drug release/bioactivity).

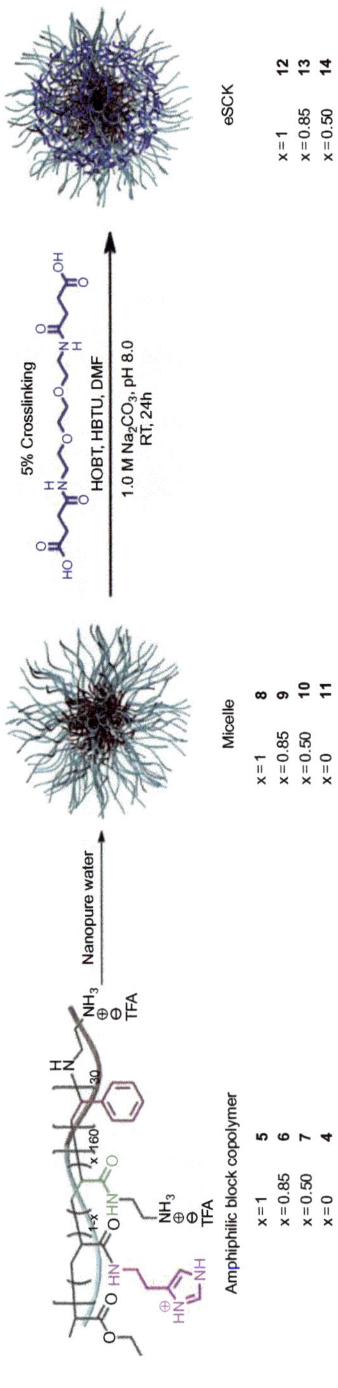

Figure 7.3 Schematic of polymerization and cross-linking of shell cross-linked micelles. Adapted with permission from [44]. Copyright 2012, Elsevier.

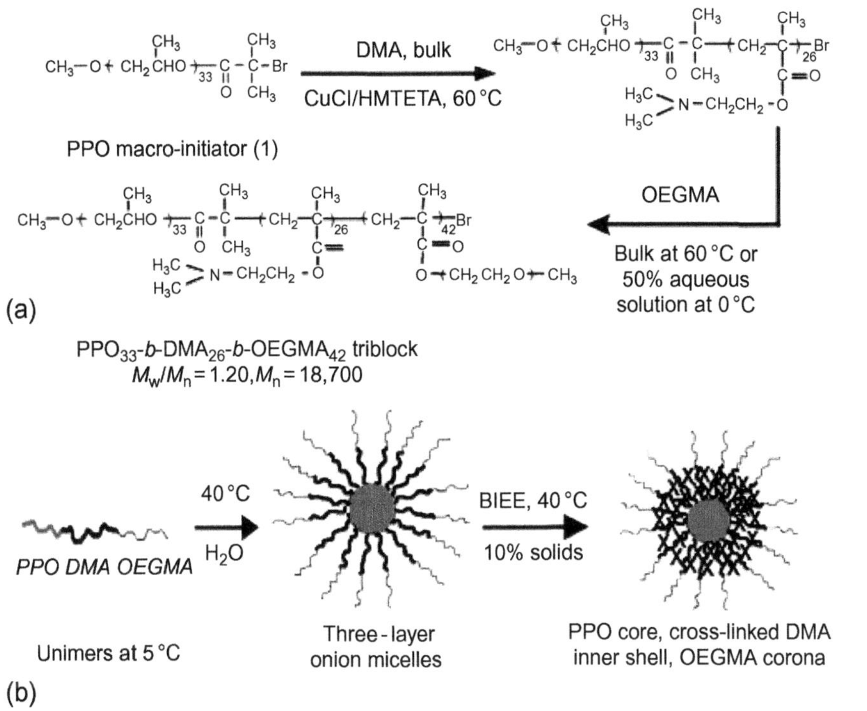

Figure 7.4 Schematic of Intramicellar cross-linking by ABC triblock copolymers. Adapted with permission from [45]. Copyright 2001, American Chemical Society.

Wooley et al. pioneered synthesis of shell cross-linked micelles for packaging nucleic acids [50]. McCormick and Armes et al. investigated another shell cross-linked system, which they described as for the release of "bioactive agents" [39]. Their design incorporated a reversible cross-linking moiety into the B-block of their ABC triblock copolymer to achieve controlled release. The thermoresponsive and reducible nature of these SCL micelles made the drug-release profile responsive to changes in environmental conditions. Other groups focused on further optimizing shell cross-linked micelle chemistry for more efficient encapsulation and improved biocompatibility for gene delivery. Stenzel et al. thoroughly investigated cationic polymers made by RAFT polymerization for their application as shell cross-linked micelles for gene delivery [51]. They showed that shell cross-linked micelles were more biocompatible than their cationic non-cross-linked micelle counterparts and that cross-linked micelles retained desirable levels of transfection. Wooley's group published a set of articles also demonstrating the effective transfection of mammalian cells by shell cross-linked micelles [52,53]. Their initial study highlighted a shell cross-linked micelle composed of poly(acrylamidoethyl amine)-*b*-polystyrene (PAEA-*b*-PS). They demonstrated the ability to condense oligonucleotides and pDNA and to effectively transfect mammalian HeLa cells, indicating that a 6:1 N/P ratio provided the highest transfection efficiency. The follow-up study optimized transfection efficiency and cytotoxicity by investigating various relative quantities of primary versus tertiary amines and exploring a range of N/P ratios used for nucleic acid condensation. They showed that cationic shell cross-linked nanoparticles (cSCKs) with 25% primary amines and 75% tertiary amines (cSCK-pa_{25}-ta_{75}) transfected pDNA into HeLa cells best at a 20:1 N/P ratio. Alternatively, cSCK-pa_{50}-ta_{50} micelles at a 10:1 N/P ratio were best for transfecting oligodeoxynucleotides (ODNs). Importantly, they also demonstrated that the incorporation of tertiary amines into the shell of micelles was an effective strategy for maintaining transfection efficiency while reducing cytotoxicity.

Building upon these initial proof of concept and optimization studies, other groups have recently further investigated the effectiveness of both shell and core cross-linked micelles as gene delivery vectors. Kataoka's group employed their environmentally responsive, disulfide core cross-linked micelles to this end [54]. The PIC micelles were formed from iminothiolane-modified PEG-*b*-PLL (PEG-*b*-(PLL-IM)). This polymer efficiently condensed siRNA and formed a cross-linked core through linkages between iminothiolane groups. They showed that reductive conditions (such as those within the cytosol) destabilize the core of these micelles and lead to increased gene silencing compared to a non-reducible analog. Finally, they also demonstrated that tuning of the ratio of PEG-b-(PLL-IM) to siRNA was imperative for achieving gene silencing activity.

Wooley et al. recently investigated additional shell-cross linked micellar designs for siRNA and peptide nucleic acid (PNA) delivery. Interestingly, their first report in this vein was on the packaging and efficient delivery of PNAs [55]. PNAs are synthetic hybrid molecules comprising a polypeptide backbone but with nitrogenous base R-groups that enables Watson-Crick base pairing with nucleic acids. PNAs are a promising class of therapeutics because they are not susceptible to degradation by nucleases that are classically linked to the *in vivo* degradation of gene therapies. Yet, PNAs still suffer from short circulation half-lives, low transfection efficiency, and a lack of endolysosomal escape mechanisms. Therefore, the Wooley group investigated the packaging of PNAs into their previously developed cSCKs. They took two approaches to packaging PNAs: 1) electrostatic complexation of PNA-ODN hybrid, and 2) bioreducible, covalent disulfide linkage. They showed that both versions of the cSCKs were as effective at transfection as LF2K and had higher cell viabilities. They also showed that cSCKs were more effective than previously developed Arg-mediated methods of PNA delivery. This was attributed to cSCK facilitation of PNA endosomal escape, while PNAs alone suffered from entrapment within endosomes. Next, the Wooley lab investigated the cSCKs for their ability to enhance the delivery of siRNA [44]. In particular, they studied the effect of altering the cSCKs buffering capacity on siRNA binding, transfection, endosomal escape, and cytotoxicity. The buffering capacity of cSCKs was tuned by altering the amount of histamine verse primary amines in the shell cross-linking region. The addition of histamine was shown to reduce the siRNA encapsulation efficiency as well as cellular binding. However, cytotoxicity was significantly reduced, and the increased buffering capacity led to an increase in transfection efficiency.

Theranostic nanomaterials represent another recent development for shell cross-linked micelles capable of both delivery and diagnostic sensing. The Wooley lab recently developed a hierarchically assembled theranostic material from the cSCK platform described above [56]. The material is capable of delivering siRNA as a therapeutic and can be radiolabeled for potential diagnostic applications. The hierarchical assembly is initiated by preparing anionic, shell cross-linked nanorods of the amphiphilic block copolymer poly(acrylic acid)-*b*-poly(*p*-hydroxystyrene) (PAA-*b*-P*p*HS) in nanopure water. Cationic shell cross-linked nanoparticles are then added to coat the nanorod surfaces. These theranostic, hierarchically assembled nanostructures were demonstrated to condense siRNA, transfect several cell lines *in vitro*, and have sufficient radiolabeling yield. The nanostructures condensed siRNA at lower N/P ratios and achieved higher intracellular delivery of Cy3-labeled siRNA compared to LF2K or cSCKs alone. The hierarchically assembled theranostic nanostructures also showed higher radiolabeling efficiency compared to cSCKs, suggesting their potential for use with PET imaging. A follow-up study by the Wooley group incorporated paclitaxel, a hydrophobic anticancer drug into the nanostructures and also introduced a targeting ligand onto the surface of cSCKs [57]. Multifunctional approaches such as the hierarchically assembled theranostic nanostructures may expect to be met with significant regulatory barriers for clinical use yet represent key advances in basic science that are adding to our understanding of how nanomaterials interact with biological systems.

7.4 Polymersomes

"Polymersomes are polymer-based vesicular shells that form upon hydration of amphiphilic block copolymers" [58].

Discher et al. developed a novel class of polymeric nanoparticles at the turn of the twenty-first century termed "polymersomes." This term represented a polymer-based "liposome" because they adopt a vesicular structure similar to previously-described liposomes but were composed of amphiphilic diblock copolymers rather than lipids (Figure 7.5) [59]. Polymersomes create unique capabilities relative to micellar structures, which are typically utilized to load hydrophobic drugs into their core, because they can encapsulate a wide variety of therapeutics, including entrapment of larger, water-soluble biologics such as nucleic acids and proteins, within the aqueous interior. Like other polymer-based nanoparticles, one can also incorporate additional polymer functionality (i.e., responsiveness to environmental stimuli) to improve delivery. For example, recent developments on controlled disassembly, degradation, targeting, and endolysosomal escape have further increased the potential of polymersomes as nanocarriers for therapeutics. These developments are discussed herein with a focus on applications in gene therapy.

Polymersomes have a large range of tunable properties based on the polymer molecular weight (MW), composition, and chemistry, a key advantage relative to classical liposomes. Using a relatively inert, synthetic diblock copolymer, poly(ethelyne glycol)-polybutadiene (PEG-PBD), the effect of the molecular packing factor (a measure of the ratio of hydrophobic to hydrophilic sections of an amphiphilic copolymer) on morphology of copolymer aggregates has been elucidated. By changing the length of the hydrophilic block, the resulting aggregates could be tuned to form either spherical micelles, worm-like micelles, or polymersomes [60]. For hydrophobic:hydrophilic

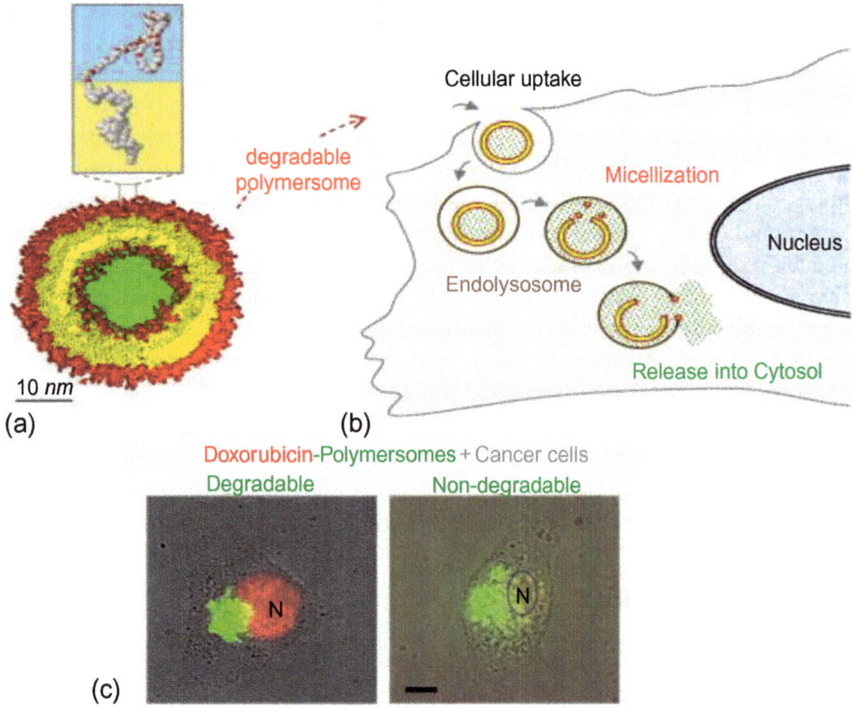

Figure 7.5 Polymersomes. a) Polymersome assembly and structure in aqueous milieu. b) Cellular uptake and endolysosomal escape. c) Fluorescent microscopy imaging of doxorubicin delivery from polymersomes.
Adapted with permission from [58]. Copyright 2009, Elsevier.

ratios that form polymersomes, the vesicle interior diameter correlates with the total MW of the block copolymer [61]. Further, it is only for lower MW copolymers that the polymer bilayer exhibits mid-planes of low density ("methyl troughs") reminiscent of lipid bilayers. Higher MW copolymers form a single thick shell of homogeneous density [62,63]. Membrane fluidity generally decreases with increasing MW, most drastically once an entanglement threshold is crossed. Polymersomes have generally been described as having mechanical properties somewhere between lipid vesicles (optimized for fluidity) and viral capsids (evolved to possess robustness and stability) [58].

Polymersome activity *in vivo* is largely dependent upon the ability to tune the material chemistry to create ideal pharmacokinetic properties. The earliest polymersomes were formed from inert polymers, such as PEG-PBD and poly(ethylene gycol)-polyethyleneethylene (PEG-PEE). These polymersomes exhibit long circulation half-lives, being 100% PEGylated to provide serum stability. However, the need for controlled release to enable effective *in vivo* bioactivity has motivated the recent development of biodegradable and environmentally- and stimuli-responsive polymersome chemistries. The bioactivity of therapeutics delivered through nanocarriers are dependent upon the ability of the carrier to overcome physiological barriers such as rapid clearance through the RES, transfection across the plasma membrane, escape from endolysosomes, and appropriate intracellular trafficking. Groups have, accordingly, focused on mechanisms to overcome these barriers and increase the efficiency of polymersomes. External cues such as temperature and UV light, as well as internal cues such as pH and redox changes have been investigated [64].

Temperature and UV-light exposure are popular examples of using externally applied cues to stimulate endolysosomal escape and release of therapeutics from polymersomes. Kros et al. and Jiang et al. explored two different UV-sensitive polymersomes. Kros et al. investigated a system based on the polymer PEG-poly(methylphenylsilane) [65]. This polymersome is capable of encapsulating hydrophilic drugs, and the Si-Si bond is cleaved upon exposure to UV-light, shifting polarity of the hydrophobic core and triggering disassembly of and cargo release from the polymersomes. Jiang et al. investigated a system based on the polymer PEG-(Malachite Green-CN) (PEG-MG-CN) [66]. A similar mechanism of core disruption is employed by PEG-MG-CN where UV-light induces the loss of cyanide anions from MG and the resultant PEG-MG disassembles and releases its hydrophilic cargo. UV exposure presents many challenges *in vivo* and is therefore more likely to be used for *in vitro* or *ex vivo* applications. As an alternative, temperature-responsive motifs have been investigated that could be of more utility *in vivo*. The polymer, poly(N-isopropylacrylamide) (PNIPAm), is a prototypical lower critical solution temperature (LCST) polymer that has been pursued for thermally responsive polymersomes [67,68]. PNIPAm solubility is temperature-dependent ($T > 32\,°C$ – hydrophobic; $T < 32\,°C$ – hydrophilic). As an example, the diblock polymer PEG-PNIPAm is amphiphilic at temperatures above 32 °C and can self-assemble into polymersomes when the appropriate relative block lengths of PEG and PNIPAm are used. As the temperature drops below 32 °C, amphiphilicity is lost and the polymersomes disassemble, providing a controlled release mechanism for the encapsulated drugs (such as doxorubicin). As is to be expected, it is necessary to use these carriers in combination with some form of hypothermic therapy to stimulate the temperature-dependent response, which is below normal physiologic temperature (37 °C) [68].

Leveraging endogenous, physiologic cues for polymersome disassembly are desirable, relative to externally actuated systems, because they are not as limited by accessibility for applications deeper within the body. For example, changes in pH and redox conditions have been investigated as ways to internally trigger the release of therapeutics within the desired cellular microenvironment. Polymersomes experience a change in pH during endocytosis, from extracellular pH of 7.4 to ~5.0 in late endolysosomes. Triggered release can be achieved with polymer chemistries, which are sensitive to this change in pH. Two primary approaches have been taken to develop polymersomes that respond to pH change: 1) leveraging changes in rate of polyester hydrolysis with decreasing pH [69,70] and 2) incorporation of acidic/basic monomers whose protonation state effects polymer water solubility [71]. The first approach is less desirable as degrading polymersomes already non-specifically release therapeutics prior to the decrease in pH (lower pH only accelerates the process). Alternatively, designs based on a shift in polymer hydrophobicity in response to pH more stably encapsulate therapeutics, and have more switchlike release behavior. Poly(2-vinylpyridine)-PEG (P2VP-PEG) [71] and poly(2-(methacryloyloxy)ethyl phosphorylcholine)-*b*-poly(2-(diisopropylamino(ethyl methacrylate) (PMPC-*b*-PDPA) [72] are two examples of diblock polymers that have been utilized to fabricate pH responsive polymersomes that release their cargo within the pH of the endolysosomal pathway. Redox conditions within the endolysosomal pathway can also be leveraged for controlled release. Hubbell et al. developed reduction [73] and oxidation [74,75] responsive polymersomes. They developed a triblock amphiphile composed of PEG-poly(propylene-sulfide)-PEG (PEG-PPS-PEG) that self assembles into polymersomes. Upon exposure to reactive-oxygen species (ROS) such as hydrogen peroxide (H_2O_2), PPS (initially hydrophobic) is oxidized to more polar poly(sulfoxides) and poly(sulfones), triggering polymersome disruption. To employ a reduction-responsive mechanism, they incorporated a disulfide linkage into the diblock version of the previous polymer (PEG-SS-PPS). The disulfide bond is broken in reducing environments such as by high concentrations of intracellular glutathione [76], driving polymersome disassembly and release of therapeutics.

Over the last 10 years, polymersome design principles have begun to be applied for development of carriers that overcome non-viral gene therapy barriers such as low transfection efficiency, rapid degradation, transient expression, and off-target toxicity. Korobko et al. were one of the first to apply polymersomes to effectively deliver genes *in vitro* [77,78]. They used a cationic and amphiphilic diblock copolymer p(butadiene)-*b*-p(*N*-methyl-4-vinyl pyridinium) (PBD-P4VPQI) to encapsulate DNA in the vesicle lumen as well as being electrostatically condensed with the cationic P4VPQI. The DNA was efficiently encapsulated, and controlled release was achieved by osmotic pressure (proton sponge-like effect driven by cationic P4VPQI monomer) triggered disruption of the polymersomes. They further showed that the polymersomes could encapsulate pDNA, thus overcoming a shortfall of many previous polymersome designs that could only encapsulate very small genetic materials. Though promising, the downfall of this design was that the overly cationic character of the polymersomes drives non-specific uptake and cytotoxicity, making this design less applicable for *in vivo* use. Brown et al. developed seminal poly(amino acid) (poly(AA)) based polymersomes. Their design utilizes amphiphilic triblock copolymers of methoxy-(PEG), palmitic acid, and either poly-L-lysine (PLL) or poly-L-orthinine (PO) [79]. These triblocks self-assemble into polymersomes. Importantly, the cationic charge of this system can be tuned by the ratio of polymer to DNA (polymer:DNA). Effective *in vivo* delivery to the lungs and liver were shown when the charge was tuned to be overall slightly anionic, but significant cytotoxicity was seen *in vitro* at very low concentrations [80]. Thus, more biocompatible designs have been a focus for continued polymersome development.

PEG-polylactic acid (PEG-PLA or "OLA") polymersomes were a promising initial design for balancing cellular transfection efficiency and cytotoxicity, which are often competing considerations [81]. The PEGylated surface increases blood circulation half-life, decreases cytotoxicity, and doesn't affect nucleic acid encapsulation. The OLA polymersomes (OLA-Psomes) were effectively applied to deliver siRNA *in vitro* at comparable levels to the commercial transfection reagent Lipofectamine 2000 (LF2K). The OLA-Psomes + siRNA and LF2K + siRNA showed about a 33% knockdown effect while free siRNA showed no knockdown. The OLA-Psomes are promising for further development because their PEGylated surfaces make them more translatable *in vivo* relative to highly cation lipoplexes formed with LF2K and similar reagents.

More recently, several designs have been investigated with the goal of increasing circulation half-lives and balancing the efficiency of therapeutic delivery and off-target cytotoxicity. Like all classes of nanoparticles, PEGylation is heavily investigated for increasing the circulation half-lives of polymersomes. It is generally understood that PEGylation of nanoparticles decreases their interaction with plasma proteins in circulation and decreases the immune response to the carrier. In order to address the problem of inefficient and non-specific delivery, groups have also begun employing active targeting mechanisms. Development of multifunctional polymersomes with long circulation time and efficient uptake through active targeting may enable discover of gene therapies that approach the efficiency of viral vectors.

Several approaches to targeting have been investigated and are still in development. Initially, conventional receptor ligands and antibodies have been sought for active targeting [82]. Non-specific cell penetrating peptides (CPPs) such as the HIV-derived Tat peptide (TATp) have also been utilized for increasing cell uptake. As an example, Christian et al. conjugated TATp to PEG-PBD polymersomes to target dendritic cells [83]. Antibodies have the potential to be utilized as a more specific targeting mechanism. Lecommandoux et al. recently functionalized magnetic polymersomes with an antibody (trastuzumab) in order to target and image overexpressing HER-2 cells in a metastatic breast cancer model [84]. Biotin and polyguanacyclic acid have also been used to achieve active targeting with polymersomes [85,86]. All of these mechanisms have been shown to be effective *in vitro*, but none have yet been well characterized *in vivo*. It will be crucial to see the translation from *in vitro* studies to preclinical *in vivo* studies and to observe how well these modified polymersomes perform *in vivo* relative to other classes of carriers.

Other targeting peptides have also been explored by Kokkoli et al. for polymersome-based gene therapies. Initially, they pursued a PEG-poly(γ-methyl-ε-caprolactone) (PEG-PMCL) polymersome containing a vinyl sulfone electrophile in the corona [87]. The electrophile reacted preferentially with thiol-containing peptides for active targeting. They also utilized a bio-orthogonal "click" chemistry approach to attach targeting peptides [88]. In this case, polymersomes were modified with a fibronectin mimetic peptide (PR_b) to target to areas of higher $\alpha_5\beta_1$ integrin (various cancer cells). The polymersomes were found to significantly outperform non-targeted analogs in terms of cytotoxicity and delivery efficiency. Recently, this group showed enhanced delivery of siRNA to highly expressing $\alpha_5\beta_1$ integrin breast cancer cells using PR_b targeting [89]. In this study, PEG-PBD polymersomes surface-functionalized with PR_b by click chemistry were used to encapsulate and deliver siRNA. Down-regulation of Orai3 (calcium channel protein) was shown to reduce breast cancer cell viability in previous studies and was therefore chosen as a target for siRNA knockdown in this study. The PR_b functionalized polymersomes inhibited Orai3 significantly better than unfunctionalized polymersomes in T47D breast cancer cells in vitro. Although commercial transfection agents had higher levels of knockdown, the polymersomes represent a more translatable system with regards to potential *in vivo* biocompatibility. This, and other, studies serve as a promising proof of concept for application of polymersomes for targeted gene delivery, motivating continued preclinical development of this approach.

7.5 Microgels/nanogels

"Hydrophilic polymers are the center of research emphasis in nanotechnology because of their perceived "intelligence." They can be used as thin films, scaffolds, or nanoparticles in a wide range of biomedical and biological applications" [90].

Over the last few decades, advances in materials engineering have enabled synthesis of defined networks of hydrophilic polymers with finely controlled physicochemical and mechanical properties. These networks of hydrophilic polymers can be physically or chemically cross-linked to yield "hydrogels." Hydrogels have a high affinity for water because of their hydrophilic nature, and their cross-linked architecture results in swelling. Hydrogel chemistry can be utilized for loading and controlled release of biologics such as proteins, nucleic acids, and cells. Defined functionalities can also be engineered into hydrogels to control host cell infiltration into and response to hydrogels following implantation *in vivo*. Their high water content and similarity in structure to native extracellular matrices makes hydrogels especially useful for tissue engineering, but they are currently under investigation for a broad variety of biomedical applications including drug delivery, biosensing, microarrays, and diagnostic imaging [90].

Hydrogels are conventionally synthesized at the macro/tissue scale, but new methods are amenable to creating hydrogels of varied size and shape for specialized applications such as gene therapy (Figure 7.6). Three of the most important parameters for characterizing hydrogel bulk structure are the swollen state polymer volume fraction, molecular weight of polymers between cross-links, and the mesh size (consult this recent review for mathematical models of these parameters [90]). While bulk hydrogels

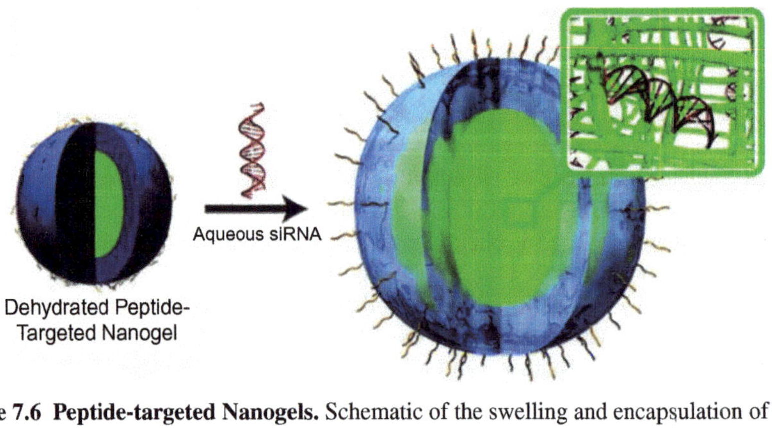

Figure 7.6 Peptide-targeted Nanogels. Schematic of the swelling and encapsulation of siRNA in aqueous milieu.
Adapted with permission from [91]. Copyright 2011, American Chemical Society.

are often utilized as tissue engineering implants, microgels and nanogels offer flexibility that is often harnessed for drug delivery applications. Microgels are typically in the hundreds of nanometer to micrometer scale. Preparation into thin films is the most common approach for their application in drug delivery [92]. Recent advances in fabrication techniques have also allowed the production of nanogels, nano-sized networks of cross-linked hydrophilic polymers. Nanogels are created using specialized fabrication approaches and are typically administered similarly to the other categories of nanoparticles discussed in this chapter.

Like other classes of gene therapy nanoparticles, hydrogels can be derived from natural and synthetic sources. Neutral synthetic polymers commonly utilized to generate hydrogels are PEG, poly(hydroxyethyl methacrylate) (PHEMA), and poly(vinyl alcohol) (PVA), with PEG being the most heavily investigated. Multi-arm PEG precursors with various reactive end chemistries are commercially available that can be used to form cross-linked hydrogel networks that are resistant to protein adsorption and do not generate significant cytotoxicity or excessive immune response [93]. A strength of basic PEG hydrogels is that they block protein or cell adhesion and are largely non-degradable. Thus, PEG provides a "blank slate," enabling engineering of the hydrogels to possess carefully tuned properties such as degradation and cell adhesion. PEG hydrogel cross-linking is achieved using end functional groups such as silane, acrylate, thiols, and groups that participate in ionic and hydrogen bonding. Often, peptides are engineered into hydrogels, both to provide degradation mechanisms (e.g., MMP dependent) that mimic native ECM and to provide sites for cell adhesion (e.g., RGD peptide) [94–96]. Like PEG, PHEMA is extremely stable in water and has robust mechanical properties. PHEMA is especially well-known for its application in contact lenses, but it is also an attractive material for drug delivery [97]. A variety of modifications have been employed to enhance the properties of PHEMA hydrogels (e.g., copolymerization with other monomers has been utilized to tune degradability, swelling, and mechanical properties). PVA hydrogels are generated by repeated freeze-thaw cycles or through covalent cross-linking. Physical cross-linking is also possible with PVA and is of particular utility in drug delivery as it is more amenable to achieving biodegradability [98,99].

Biological hydrogels formed from naturally derived polymers or biohybrids of both natural and synthetic materials are also a topic of heavy investigation. Commonly used natural polymers include collagen, hyaluronic acid (HA), fibrin, alginate, agarose, dextran, and chitosan [100]. These natural polymers are derived from mammalian extracellular matrix (collagen), tissues (HA), or from marine algae sources (alginate and agarose), and they are advantageous for their low levels of cytotoxicity. Collagen provides an especially conducive cell-material interface because it is a natural matrix for cellular growth and inherently possesses a multitude of cell-signaling domains that are difficult to replicate using synthetic methods. However, natural polymers like collagen also suffer from poor mechanical properties, and cross-linking strategies are commonly employed to improve its mechanical strength [101]. Alginate and chitosan are linear polysaccharides that can be cross-linked to form hydrogels. Alginate generates gels upon the formation of ionic bridges in the presence of divalent cations. It degrades slowly over time, and the mechanical integrity of the alginate hydrogels is concurrently lost. Natural polymers typically possess inherent degradation mechanisms, which is advantageous for most *in vivo* applications. For example, collagen is naturally degraded by collagenases and HA is degraded by cells through the release of enzymes like hyaluronidase [90].

Biohybrid hydrogel systems attempt to synergize the advantages of synthetic and biologic materials by combining them into a single platform that enable one to retain synthetic control while also integrating biologic signals that enhance biocompatibility and overall performance. A common approach is the addition of degradable or biologically active peptides into the matrix of synthetic hydrogels [102,103]. Another approach is the addition of enzymes or genetically-engineered proteins into the mesh space of hydrogels [104,105]. Both approaches are used to enhance cell adhesion and invasion into the interior of the gels as well as to provide the possibility for cell-mediated degradation. Growth factors can also be included in hydrogels by covalent attachments and regulate cellular function and migration within the gels [106,107]. Biologics such as peptides, chitosan, alginate, agarose, and dextran can also be copolymerized with synthetic monomers [108–110]. Polypeptide approaches such as the elastin-like class of polypeptides are particularly popular and can be designed and synthesized using genetic engineering protocols [111]. Expense and synthetic complexity are potential limitations for polypeptide-based technologies, but recent advances are making their production more scalable and cost efficient [112].

7.6 Microgels

Microscopic scale hydrogels, or microgels, are useful for applications in drug delivery and the controlled release of bioactive therapeutics. Mucoadhesive microgels for oral and non-oral drug delivery to areas such as the nasal and pulmonary cavities are under particularly heavy investigation [113,114]. Hydrophilic microgels composed of mucoadhesive polymers (e.g., hydroxypropylcellulose, chitosan, carbopol, carboxymethylcellulose, hyaluronic acid, polyacrylic acid) naturally adhere to mucus layers, a property which makes them a natural fit for these applications [113]. Recent research has also explored functionalizing the surface of microgels with anionic or thiolated moieties, or both. These targeting moieties could allow for better interpenetration into and adhesion within mucus layers [115,116]. These mucoadhesive microgels can be used to encapsulate a variety of therapeutics and serve as a drug depot for the sustained release of therapeutics. A few particularly exciting formulations have been developed for oral delivery of therapeutics [117]. For example, an oral polymer microgel drug delivery system was optimized with varying amounts of poly(acrylic acid) (PAA) and Pluronic® [118]. Nine Pluronic® copolymers were grafted to PAA with a range of molecular weights and varying poly(propylene oxide) (PPO) and PEG content, with or without cross-linking. The longest Pluronic® polymers with grafted PAA attachments resulted in the microgels with the strongest mucoadhesive properties. Another mucoadhesive microgel system incorporated cysteine-modified PAA polymers [119]. The thiol moieties of cysteine were shown to form disulfide bonds and intermolecular cross-links with mucus material and increase the mucoadhesive properties 14-fold over unmodified microgels. Microgels show promise for mucoadhesive applications, and these recent studies suggest that microgels will make an impact on oral and pulmonary drug delivery.

The Lyon group has innovated functional microgel-based drug delivery systems. They initially published on the synthesis and characterization of environmentally responsive core-shell morphology microgels [120]. They tested two compositions: 1) PNIPAm-co-acrylic acid (AAc) core/PNIPAm shell and 2) PNIPAm core/PNIPAm-co-AAc shell. It was found that the location (core verses shell) of PNIPAm and AAc have an effect on the temperature (PNIPAm) and pH (AAc) dependent swelling. Importantly, it was found that there were only small differences in the effect on pH between the two core-shell particles, indicating that the spatial organization of temperature responsive PNIPAm dominates the behavior in this system. They also reported an application of PNIPAm-co-AAc microgels in which they loaded microgel thin films with Dox and thermally regulated its loading and release [121]. In this study, they used a spin coating, layer-by-layer (scLBL) approach to prepare thin films of anionic PNIPAm-co-AAc alternating with layers of cationic poly(allylamine hydrochloride) (PAH). They showed that doxorubicin could be loaded into 10, 20, and 30 layer thick microgel films and that release kinetics of doxorubicin from the films could be controlled using temperature. They were further able to tune the swelling kinetics of core-shell microgels by the incorporation of the hydrophobic monomer butyl methacrylate (BMA) into the shell [122]. This showed the ability to control the kinetics of microgel de-swelling simply by modifying the periphery of particles. Alternatively, the incorporation of hydrophilic PEG moieties into PNIPAm core-shell microgels was studied with the goal of reducing protein adsorption into the gels [123]. Their results confirmed that protein adsorption was reduced by the incorporation of PEG chains, especially when PEG was located within the microgel shell. They further showed that once PNIPAm undergoes its phase transition from hydrophobic to hydrophilic, PEG grafts can interpenetrate or flip with the PNIPAm layer. They also showed early evidence for the ability to target microgel therapies using folate-conjugated PNIPAm microgels [124]. In sum, their studies on an array of PNIPAm-based, thermo-responsive microgels have provided the field with significant mechanistic insights into the synthesis, characterization, and functionality of microgel drug delivery platforms.

Peppas and Elbert have also deeply influenced the development of microgels as drug delivery systems. Peppas laid much of the groundwork for how we understand swelling and drug release from hydrogel networks [125–127]. More recently, his group has investigated the incorporation of biological molecules such as insulin into microgels/microparticles [128]. They have showed highly efficient encapsulation of insulin within p(methacrylic acid-*g*-ethylene glycol) (P(MAA-*g*-EG)) microgels with preferential release in neutral/basic media over acidic media, with the goal of protecting insulin during passage through acidic environments within the body (i.e. stomach). They tested the encapsulation of other hydrophilic model drugs, namely theophylline, vancomycin, and fluorescently labeled dextrans. Insulin showed the highest loading efficiency into their P(MAA-g-EG) formulation. One could envision this technology as an oral carrier for insulin and also imagine the possibility of tailoring this formulation to increase encapsulation of other therapeutics for oral delivery. Elbert's group has developed many PEG-based materials for the coating of biomedical materials. In work with Hubbell, he developed approaches for coating biomaterials for increasing biocompatibility by reducing the adsorption of serum proteins and cell adhesion [129–133]. Elbert's group more recently pioneered a microgel-based system aimed at reducing late-term thrombosis on drug-eluting stents [134]. In this approach, glass or poly(ethylene terephthalate) substrates were dip-coated with PEG-octavinylsulfone(OVA)-BSA microgel conjugates. The difference in nucleophilicity of thiols verse primary amines was used to avoid macrogelation and then induce rapid covalent attachment of microgels to the glass surface under diluted conditions. The PEG-OVA-BSA coated surfaces reduced protein adsorption. Cellular adhesion and migration were inhibited by the microgel formulation and promoted by a control, fibrinogen-capped coating. The authors proposed that this microgel-based coating could serve as a useful surface modification for blood-contacting devices through resistance of protein and cell adhesion.

Only very recently have microgels been developed with the capability of encapsulating gene therapies and nucleic acid based therapeutics. Moreover, the delivery of nucleic acids from hydrogels is strongly biased to nanogel formulations over microgels. To our knowledge, two approaches have been taken in delivering nucleic acids from microgels: direct incorporation and incorporation of nucleic acid containing nanoparticles into microgels.

De Smedt et al. introduced a microgel system by which they could induce the time-controlled release of siRNA [135]. They started by testing two microgel preparation techniques. They found that it was ideal to load siRNA into neutral dextran microgels prior to cross-linking and gel formation, while the copolymerization of cationic DMAEMA monomers allowed for post-gelation siRNA loading. Time-controlled siRNA release was achieved from the microgels by varying the density of cross-linking incorporated during gelation. Microgels with higher orders of cross-linking led to much slower release profiles. The authors argue that this is particularly desirable for siRNA delivery because of the transient nature of its activity. Therefore, the functional impact of gene knockdown should be increased if siRNA delivery can be sustained over time periods that span its transient activity profile. The authors also allude to the major setback of attempting siRNA delivery from a microgel platform; many microgels are too large for cellular internalization. To address this shortcoming, they filtered out larger microgels to obtain a mean population size between 2 and 3 um before applying them as cell treatments. Queiroz et al. co-loaded a similar system with pDNA against a gene for p53 and the model anti-cancer drug Dox [136]. They showed effective encapsulation of both and observed modest sustained release for up to 5 days. They propose application of this delivery system for local chemotherapy/gene delivery approaches where the therapeutics would diffuse from the microgels and into the local tumor microenvironment.

The second microgel-based gene therapy approach has involved loading of nucleic acid loaded-nanoparticles into the microgels. This stategy has been attempted recently by both Peppas and Roy. The Peppas group loaded cationic nanoparticles into anionic p(methacrylic acid-*co*-*N*-vinylpyrrolidone) (P(MAA-co-NVP)) microgels and performed chemical characterization of the system [137]. They demonstrated that swelling of the system was dependent upon both pH and the cross-linking density of the microgels. Although they used a protein and small molecule as models for loading and release studies, one could easily envision using the cationic nanoparticles to condense siRNA and incorporating siRNA loaded cationic nanoparticles into anionic microgels. Roy et al. developed a similar but more functional system for pulmonary delivery of biologic drugs [138]. Their system incorporates an enzymatically degraded peptide sequence into the backbone of the microgel as a cross-linker. After cross-linking, the microgel is also loaded with either nanoparticles, biologic drugs, or biologic drugs within nanoparticles. The "nano-inside-micro" formulation exhibited desirable properties for pulmonary delivery: it was mucoadhesive, enzymatically degraded, protected from clearance by alveolar macrophages, and induced better penetration into/through pulmonary mucus membrane versus a microgel alone. The authors argue that the formulation can be versatile and tailored to the delivery of DNA, siRNA, and proteins. Moreover, they argue that the system is able to encapsulate biologic-loaded nanoparticles in the range of 20–200 nm. The "nano-inside-micro" approach is certainly a step toward overcoming the physical size limitation of delivering nucleic acids from a microgel platform. It should be noted that similar systems have also been developed as scaffolds and also porous silicon microparticles, where drug-loaded nanoparticles are incorporated into polymer or other "bulk" scaffolds for applications in controlled/sustained release and tissue engineering [139–145].

7.7 Nanogels

Nanogels have been significantly more developed for gene therapies relative to their microscale counterparts. Size is the major distinction between the two classes of hydrogels and is the motivating factor for choosing nanogels as gene delivery vectors over microgels. Nanogels, or hydrogel nanoparticles, are typically in a similar size range to the other classes of nanoparticles discussed up to this point (20–200 nm). Systemic delivery by intravenous injection of therapeutics into the blood circulation is the preferable gene delivery strategy for many potential applications (e.g., dispersed cancers). Therefore, the size range of nanogels is preferable over the size of microgels for most gene delivery applications as microgels risk getting lodged within and blocking the capillaries of the systemic circulation. Moreover, the polydispersity of microgel samples is typically larger than that of nanogels. The ability to better control the polydispersity in nanogels is highly desirable, ensuring more consistency and reduced toxicity of treatments *in vivo*.

While microgels are typically polymerized by emulsion or precipitation polymerizations (although controlled polymerizations are becoming more popular), nanogels are typically prepared by controlled polymerizations and cross-linking reactions [146]. The controlled polymerization techniques such as RAFT, ATRP, and ROMP are needed to achieve better control over size distribution in nanogels. This, in turn, leads to the production of consistent gene vectors that can encapsulate nucleic acids efficiently and provide some tuning of their release kinetics based on swelling, biodegradability, diffusion, or environmental responsiveness.

Vinogradov et al. initially reported nanogel formulations for delivering oligodeoxynucleotides (ODNs) [146–149]. Their formulation consisted of a cross-linkable copolymer of PEG-PEI prepared via an emulsification/solvent evaporation technique. Gel permeation chromatography (GPC) was used to fractionate the resulting nanogels and separate the smallest size range for use in the studies (~120 nm). The ODN-loaded nanogels were condensed to a smaller size (~80 nm) and had a decreased zeta potential. These nanogels facilitated better uptake of ODNs in multidrug resistant (MDR) carcinoma cells compared to ODNs free in solution. Moreover, ODNs encapsulated in the nanogels demonstrated increased inhibition of P-gp efflux pumps in MDR cell lines when compared with free ODNs. This approach was adapted for a follow-up study aimed at delivering ODNs to the brain. The previously reported nanogel delivery system was tested for its ability to transport across the blood-brain barrier (BBB) using polarized monolayers of bovine brain microvessel endothelial cells as an *in vitro* model. They demonstrated that nanogel-ODN formulations facilitated increased transport across the BBB and that surface modification of the nanogels with transferrin and insulin further increased the transport efficiency. The accumulation of ODN in the brain was increased 15-fold one hour after intravenous injections and decreased two-fold in the liver and spleen. Delivery of nucleoside analog 5'triphosphates was also attempted in the same nanogel system. 5-fluoroadenosine arabinoside (FATP) was electrostatically complexed into the nanogels and delivered *in vitro*. The nanogels protected 90% of encapsulated FATP from nuclease degradation and folate-targeted nanogels resulted in a 10-fold increase in transfection of human breast carcinoma MCF-7 cells.

Several other research groups have followed up on Vinogradov's initial findings and have effectively developed nanogel formulations for siRNA and pDNA delivery. One of these applications utilized co-delivery of pH-responsive nanogels with PIC micelles loaded with pDNA [150]. Co-delivery of the nanogels and PIC micelles led to increased transfection efficiency compared to delivery by PIC micelles alone, suggesting the importance of the pH-responsive nanogels in this strategy. With this approach, it was assumed that the pH-responsive nanogels are co-localized with PIC micelles within endolysosomal vesicles and that the nanogels introduce the proton sponge buffering capacity needed to promote PIC micelle endolysosomal escape and trafficking to the nucleus.

The most common nanogel gene therapy strategy has been synthesis of surface PEGylated nanogel carriers with cross-linked, cationic cores amenable to nucleic acid loading. For example, nanogels consisting of a cross-linked PDEAMA core and PEGylated surface were investigated for their ability to deliver siRNA and compared with delivery from a PEG-PDEAMA PIC micelle analog [151]. The nanogel/siRNA complexes showed enhanced salt and serum stability compared with their PIC micelle analogs. Importantly, the nanogel/siRNA complexes greatly enhanced gene silencing over the PIC micelles and demonstrated a more pronounced tendency for endosomal escape as observed by confocal microscopy. This system was further studied with the addition of quaternary ammonium groups for enhanced siRNA binding and nanogel stability, and the effect of tertiary verse quaternary amine groups in the core was analyzed [152]. The introduction of quaternary ammonium groups was shown to improve siRNA binding at lower N/P ratios as well as increase nanogel stability against polyanion displacement reagents. However, the zeta-potential of these nanogels was over +20 mV, potentially limiting the use of this strategy *in vivo*, especially for intravenous delivery applications. Interestingly, it was found that quaternization increased uptake of nanogel/siRNA complexes but showed no effect on the uptake of the unloaded nanogels. The buffering capacity of the nanogels was reduced with increasing quaternization, and it was concluded that the ratio of quaternary ammonium groups to tertiary amine groups should be optimized in this kind of system since they have competing effects on cellular uptake and proton sponge-based endosomal escape. Another effective nanogel gene delivery system was made by conjugating branched PEI onto preformed gelatin nanoparticles [153]. The nanogels showed less toxicity than PEI alone and significantly higher gene silencing than LF2K *in vitro*. The nanogels also effectively escaped the endolysosomal pathway, and increasing the N/P ratio from 10 to 30 in the formulation doubled the intracellular uptake. Again, the zeta-potential of these nanogels, up to +40 mV, causes concerns for *in vivo* delivery applications. In sum, these initial preclinical studies on nanogel/nucleic acid formulations show great promise but need further development to become translatable into *in vivo* models and the clinic.

A series of nanogels has also been synthesized from natural polymeric sources. The gelatin-PEI hybrid system described above represents one example. Hyaluronic acid (HA) nanogels were also synthesized in the size range of 200 to 500 nm and used to encapsulate siRNA by an inverse water-in-oil emulsion [154]. The siRNA was physically entrapped within the nanogels after cross-linking by this method. The HA/siRNA complexes effectively targeted HA receptor positive cells (HCT-116 cells). The nanogels showed selective uptake by HCT-116 cells via receptor mediated endocytosis, but addition of glutathione was necessary to reduce the nanogel cross-links and induce siRNA release. De Smedt et al. also investigated a nanogel analog to their previously described dextran-based microgels [155]. Dextran-based nanogels with [2-(methacryloyloxy)-ethyl]trimethylammonium chloride (TMAEMA) and 2-aminoethyl methacrylate (AEMA) cationic cross-linkers were used to electrostatically condense siRNA. The nanogels were then co-delivered alongside amphiphilic photosensitizers (photochemical internalization (PCI)) in order to destabilize endosomal vesicles and prolong the effects of gene silencing. The De Smedt group followed up this study by investigating a PEGylated model of the dextran-nanogel, concluding that the non-PEGylated system was effective *in vitro* but would be too cationic for effective *in vivo* delivery [156]. The PEGylation technique was used, as has been discussed with many carriers, to improve circulation and minimize aggregation with serum proteins after intravenous injections. Multiple methods of PEGylation were tested, with the authors concluding that covalent attachment of PEG to the nanogels was the most effective fabrication technique. Importantly, PEGylated nanogels were also internalized by the cell types used in this study, and gene silencing was achieved. Therefore, the modification of cationic dextran nanogels with PEG was an effective method for increasing the translatability of the system without removing its functionality. Biodegradable, carbohydrate-based nanogels are another promising platform from natural sources. The chemistry of a carbohydrate-based nanogel system was recently optimized and tested for its efficacy for pDNA delivery to hepatocytes *in vitro* [157,158]. An optimized design was ultimately identified with transfection efficiency comparable to branched PEI model systems, and the addition of small amounts PEI into the carbohydrate-based nanogels further increased delivery of pDNA to hepatocytes. The biodegradability of this system was an advantage to many other nanogels and was shown to promote controlled release of pDNA.

Multi-functional nanogels designed to overcome multiple *in vivo* delivery barriers have shown particular promise during preclinical investigations. Lyon's peptide-functionalized nanogels serve as a good example [159]. They developed PNIPAm-based core/shell nanogels made through seeded precipitation polymerization. The nanogels were then surface functionalized with a peptide for targeting the EphA2 receptor of ovarian carcinoma cell types. The nanogels protected siRNA and achieved targeted gene knockdown of the EGF receptor. A chemosensitization analysis was conducted in a follow up study by co-delivering docetaxel with the siRNA-loaded, targeted nanogels [160]. EphA2 positive Hey cells showed decreased EGFR levels and cell viability after co-delivery of the nanogels with docetaxel while EphA2 negative SK-OV-3 cells were not chemosensitized by the nanogels. This study suggests that targeted co-delivery of siRNA is a promising strategy for increasing the efficacy of established chemotherapies. DeSimone et al. developed a multifunctional, reductively-responsive nanogel for siRNA delivery (Figure 7.7) [161]. Initially, siRNA was incorporated into the nanogels by electrostatic condensation, but siRNA loading and function was increased by polymerizing a siRNA prodrug with reducible, disulfide linkages directly into the PEGylated nanogels. The nanogels released the siRNA cargo under reducing conditions (indicative of the cytosol) while remaining stable in other physiological environments. Moreover, the amine content of the nanogels was optimized in order to achieve gene silencing while maintaining cytocompatibility.

Roy et al. directly investigated the function of nanogel shape on cellular internalization [162]. Based on the knowledge that nanorods are internalized at a higher rate than spherical nanoparticles, Roy's group extended this field by then comparing the internalization of nanorods versus nanodiscs. They found that mammalian epithelial and immune cells preferentially internalize nanodiscs over nanorods. They also showed that larger-sized nanorods and nanodiscs were internalized more than smaller sized rods and discs, although the opposite has been seen with nanospheres. They concluded that nanoparticle internalization should be thought of as a manifestation of the following shape- and size-dependent parameters: particle surface-to-cell membrane contact area, strain energy for membrane deformation, and local particle concentration at the cell membrane. These results suggest the need for more optimal carriers that leverage all, or as many as possible, of the above functionalities in concert.

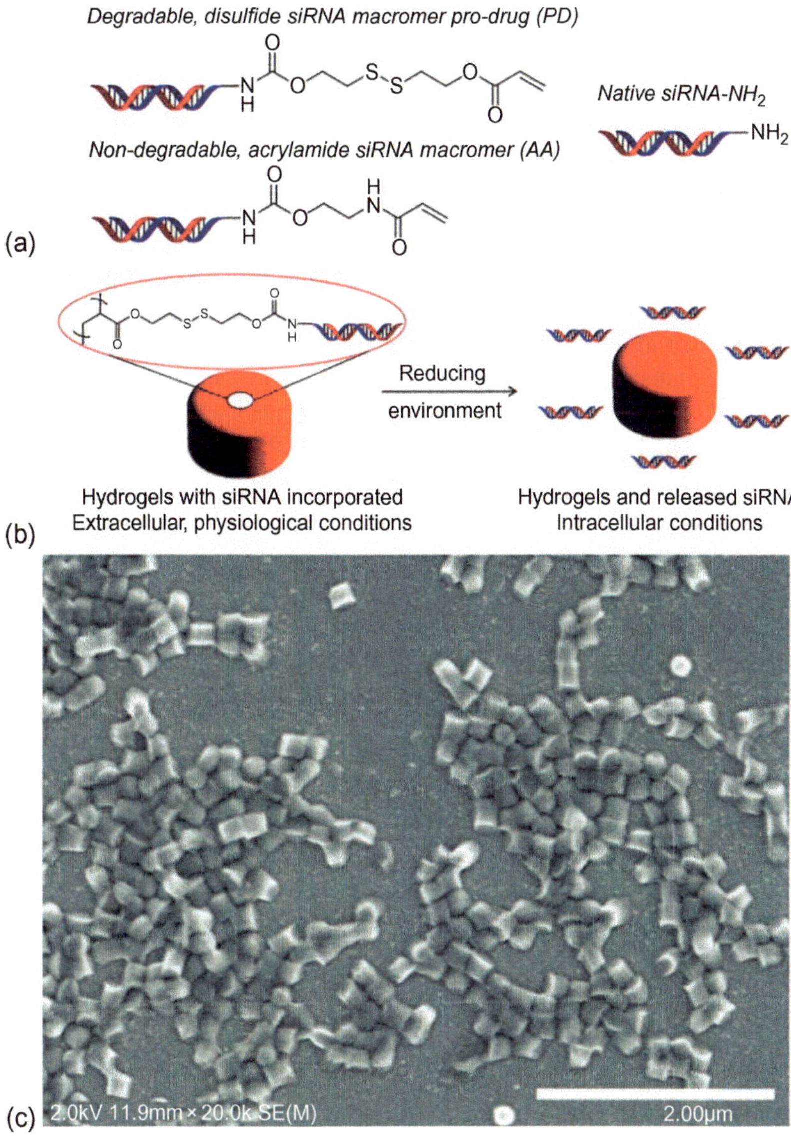

Figure 7.7 Reductively responsive nanogels for siRNA delivery. a) siRNA macromer pro-drugs used for nanogel synthesis. b) Schematic of the hydrogels and their response to reducing environments. c) SEM of the reductively-responsive nanogel architecture.
Adapted with permission from [161]. Copyright 2012, American Chemical Society.

7.8 Outlook

Two to three decades of research into gene therapy has revealed both the great potential and technical challenges of implementing successful therapies. Initial successes such as the first clinical trial for viral-based gene therapy against adenosine deaminase severe combined immunodeficiency (ADA-SCID) affirm the great clinical potential of gene therapies and motivate further investigation [163–165]. However, several studies in gene therapy have been halted, and the clinical trials have been canceled due to safety concerns and lack of efficacy [166–168]. These setbacks provide motivation for continued development of new, multifunctional nanocarriers that meet both safety and efficacy thresholds. Advances in our ability to synthesize nanocarriers using twenty-first century rational design principles are affording rapid advances towards this goal. For example, liposomal formulations of DNA vaccines have shown increased uptake of pDNA by cells and increased immunogenicity of the vaccines in clinical trials [169,170]. Over half of the clinical trials for RNA interference by siRNA currently employ formulations of liposomes or polymeric conjugates (Table 7.1) [171]. We refer readers to the recent review on worldwide clinical trials up to year 2012 for more detailed information on over 1800 gene therapy clinical trials that have been attempted to date [172].

The recent advances in polymeric nanoparticles have begun yielding multi-functional, virus-mimetic vectors that offer increasingly effective spatiotemporal control over biodistribution, controlled release, and intracellular delivery *in vivo*. Current formulations require tuning to achieve application-dependent optimization. For example, formulations for systemic delivery (i.e., intravenous injection) should maximize circulation time and accumulation at the site of interest while minimizing opsonization by the RES and off-target or non-specific cellular uptake. By contrast, local delivery formulations should shoot for rapid cell recognition and uptake while minimizing diffusion from the site of interest. In all cases, strategies for increasing the therapeutic index of the genetic agents are crucial for developing more efficacious systems with fewer side effects.

It is important to note that rapid progress in the understanding of both cellular and molecular biology has led to the ability to create rationally designed delivery vehicles based on the new knowledge of the cellular and molecular environments. It is expected that this progress will only grow in scope and continue to accelerate discovery by enabling more effective design of gene vectors. For example, nanoparticle diffusion through tissue and cellular internalization is still poorly understood. Elucidation of the mechanisms involved in cellular internalization and endosomal escape of nanoparticles is also critical. It is not fully understood how particle size, sterics, targeting, cargo chemistry, and shape influence internalization and subsequent intracellular trafficking, nor which properties are most important. Moreover, it is not yet understood which properties are most important for different cell and tissue types. Better elucidation of these structure-function properties will provide engineers with the ability to design more rational systems with universally better characteristics as well as application-specific advantages. The recent advances in top-down synthesis approaches suggest we will ultimately be able to manipulate understood pathways in order to influence uptake in only desired cell types through specified internalization routes.

Table 7.1 **RNAi-based drugs in clinical trials**

Drug	Target	Delivery system	Disease	Phase	Status	Company	ClinicalTrials.gov identifier
DPC, dynamic polyconjugate; LNP, lipid nanoparticle; NP, nanoparticle; shRNA, short hairpin RNA							
ALN–VSP02	KSP and VEGF	LNP	Solid tumours	I	Completed	Alnylam Pharmaceuticals	NCT01158079
siRNA–EphA2–DOPC	EphA2	LNP	Advanced cancers	I	Recruiting	MD Anderson Cancer Center	NCT01591356
Atu027	PKN3	LNP	Solid tumours	I	Completed	Silence Therapeutics	NCT00938574
TKM–080301	PLK1	LNP	Cancer	I	Recruiting	Tekmira Pharmaceutical	NCT01262235
TKM–100201	VP24, VP35, Zaire Ebola L-polymerase	LNP	Ebola-virus infection	I	Recruiting	Tekmira Pharmaceutical	NCT01518881
ALN–RSV01	RSV nucleocapsid	Naked siRNA	Respiratory syncytial virus infections	II	Completed	Alnylam Pharmaceuticals	NCT00658086
PRO-040201	ApoB	LNP	Hypercholesterolaemia	I	Terminated	Tekmira Pharmaceutical	NCT00927459
ALN–PCS02	PCSK9	LNP	Hypercholesterolaemia	I	Completed	Alnylam Pharmaceuticals	NCT01437059
ALN–TTR02	TTR	LNP	Transthyretin-mediated amyloidosis	II	Recruiting	Alnylam Pharmaceuticals	NCT01617967

Continued

Table 7.1 Continued

Drug	Target	Delivery system	Disease	Phase	Status	Company	ClinicalTrials.gov identifier
CALAA-01	RRM2	Cyclodextrin NP	Solid tumours	I	Active	Calando Pharmaceuticals	NCT00689065
TD101	K6a (N171K mutation)	Naked siRNA	Pachyonychia congenita	I	Completed	Pachyonychia Congenita Project	NCT00716014
AGN211745	VEGFR1	Naked siRNA	Age-related macular degeneration, choroidal neovascularization	II	Terminated	Allergan	NCT00395057
QPI-1007	CASP2	Naked siRNA	Optic atrophy, non-arteritic anterior ischaemic optic neuropathy	I	Completed	Quark Pharmaceuticals	NCT01064505
I5NP	p53	Naked siRNA	Kidney injury, acute renal failure	I	Completed	Quark Pharmaceuticals	NCT00554359
			Delayed graft function, complications of kidney transplant	I, II	Recruiting	Quark Pharmaceuticals	NCT00802347
PF-655 (PF-04523655)	RTP801 (Proprietary target)	Naked siRNA	Choroidal neovascularization, diabetic retinopathy, diabetic macular oedema	II	Active	Quark Pharmaceuticals	NCT01445899
siG12D LODER	KRAS	LODER polymer	Pancreatic cancer	II	Recruiting	Silenseed	NCT01676259

Bevasiranib	VEGF	Naked siRNA	Diabetic macular oedema, macular degeneration	II	Completed	Opko Health	NCT00306904
SYL1001	TRPV1	Naked siRNA	Ocular pain, dry-eye syndrome	I, II	Recruiting	Sylentis	NCT01776658
SYL040012	ADRB2	Naked siRNA	Ocular hypertension, open-angle glaucoma	II	Recruiting	Sylentis	NCT01739244
CEQ508	CTNNB1	*Escherichia coli*-carrying shRNA	Familial adenomatous polyposis	I, II	Recruiting	Marina Biotech	Unknown
RXi-109	CTGF	Self-delivering RNAi compound	Cicatrix scar prevention	I	Recruiting	RXi Pharmaceuticals	NCT01780077
ALN–TTRsc	TTR	siRNA–GalNAc conjugate	Transthyretin-mediated amyloidosis	I	Recruiting	Alnylam Pharmaceuticals	NCT01814839
ARC-520	Conserved regions of HBV	DPC	HBV	I	Recruiting	Arrowhead Research	NCT01872065

Adapted with permission from [171]. *Copyright 2013, Nature.*

This chapter covered a multitude of nanocarriers with significant promise for overcoming the delivery barriers that have crippled the advancement of gene therapies into the clinic. The benefits of nucleic acid protection, increased circulation, active and passive targeting, evasion of immune system, cellular internalization, and endolysosomal escape, all hallmarks of virus functionality, have been demonstrated in preclinical models and are being realized in ongoing clinical trials. Further investigation is needed in order to fully understand the relative importance of each of these design considerations and to develop multi-functional nanocarriers that address these delivery concerns as comprehensively as possible. Through the advancement of nanoparticle fabrication techniques, further understanding of physiological mechanisms, and investigation of biomaterial-biological system interactions, methods of gene delivery will continue to advance toward safe and efficacious clinical gene therapies.

References

[1] Kataoka K, Harada A, Nagasaki Y. Block copolymer micelles for drug delivery: design, characterization and biological significance. Adv Drug Deliv Rev 2001;47(1): 113–31.

[2] Poon GM, Gariepy J. Cell-surface proteoglycans as molecular portals for cationic peptide and polymer entry into cells. Biochem Soc Trans 2007;35(Pt 4):788–93.

[3] Behr J. The proton sponge, a means to enter cells viruses never thought of. M S-Med Sci 1996;12(1):56–8.

[4] Erbacher P, Remy JS, Behr JP. Gene transfer with synthetic virus-like particles via the integrin-mediated endocytosis pathway. Gene Ther 1999;6(1):138–45.

[5] Verbaan F, et al. Intravenous fate of poly(2-(dimethylamino)ethyl methacrylate)-based polyplexes. Eur J Pharm Sci 2003;20(4-5):419–27.

[6] Convertine AJ, et al. Development of a novel endosomolytic diblock copolymer for siRNA delivery. J Control Release 2009;133(3):221–9.

[7] Lundy BB, et al. Neutral polymeric micelles for RNA delivery. Bioconjug Chem 2013;24(3):398–407.

[8] Omedes Pujol M, et al. Determination of key structure-activity relationships in siRNA delivery with a mixed micelle system. J Control Release 2013;172(3):939–45.

[9] Benoit DS, et al. Synthesis of folate-functionalized RAFT polymers for targeted siRNA delivery. Biomacromolecules 2011;12(7):2708–14.

[10] Osada K, Christie RJ, Kataoka K. Polymeric micelles from poly(ethylene glycol)-poly(amino acid) block copolymer for drug and gene delivery. J R Soc Interface 2009;6(Suppl 3):S325–39.

[11] Wakebayashi D, et al. Lactose-conjugated polyion complex micelles incorporating plasmid DNA as a targetable gene vector system: their preparation and gene transfecting efficiency against cultured HepG2 cells. J Control Release 2004;95(3):653–64.

[12] Oba M, et al. Cyclic RGD Peptide-Conjugated Polyplex Micelles as a Targetable Gene Delivery System Directed to Cells Possessing αvβ3 and αvβ5 Integrins. Bioconjug Chem 2007;18(5):1415–23.

[13] Nelson CE, et al. Balancing Cationic and Hydrophobic Content of PEGylated siRNA Polyplexes Enhances Endosome Escape, Stability, Blood Circulation Time, and Bioactivity in Vivo. ACS Nano 2013;7(10):8870–80.

[14] Xiong XB, Uludag H, Lavasanifar A. Virus-mimetic polymeric micelles for targeted siRNA delivery. Biomaterials 2010;31(22):5886–93.
[15] Xiong XB, Lavasanifar A. Traceable multifunctional micellar nanocarriers for cancer-targeted co-delivery of MDR-1 siRNA and doxorubicin. ACS Nano 2011;5(6):5202–13.
[16] Smith R, Sewell SL, Giorgio TD. Proximity-activated nanoparticles: in vitro performance of specific structural modification by enzymatic cleavage. Int J Nanomedicine 2008;3(1):95–103.
[17] Li H, et al. Matrix Metalloproteinase Responsive, Proximity-Activated Polymeric Nanoparticles for siRNA Delivery. Adv Funct Mater 2013;23(24):3040–52.
[18] Convertine AJ, et al. pH-Responsive Polymeric Micelle Carriers for siRNA Drugs. Biomacromolecules 2010;11(11):2904–11.
[19] Lynn DM, Langer R. Degradable Poly(β-amino esters): Synthesis, Characterization, and Self-Assembly with Plasmid DNA. J Am Chem Soc 2000;122(44):10761–8.
[20] Akinc A, et al. Synthesis of poly(beta-amino ester)s optimized for highly effective gene delivery. Bioconjug Chem 2003;14(5):979–88.
[21] Green JJ, et al. Combinatorial Modification of Degradable Polymers Enables Transfection of Human Cells Comparable to Adenovirus. Adv Mater 2007;19(19):2836–42.
[22] Zhou J, et al. Biodegradable poly(amine-co-ester) terpolymers for targeted gene delivery. Nat Mater 2012;11(1):82–90.
[23] Midoux P, Monsigny M. Efficient gene transfer by histidylated polylysine/pDNA complexes. Bioconjug Chem 1999;10(3):406–11.
[24] Benns JM, et al. pH-sensitive cationic polymer gene delivery vehicle: N-Ac-poly(L-histidine)-graft-poly(L-lysine) comb shaped polymer. Bioconjug Chem 2000;11(5): 637–45.
[25] Moreira C, et al. Improving chitosan-mediated gene transfer by the introduction of intracellular buffering moieties into the chitosan backbone. Acta Biomater 2009;5(8):2995–3006.
[26] Howard KA, et al. RNA interference in vitro and in vivo using a novel chitosan/siRNA nanoparticle system. Mol Ther 2006;14(4):476–84.
[27] Jiang HL, et al. Efficient gene delivery using chitosan-polyethylenimine hybrid systems. Biomed Mater 2008;3(2):025013.
[28] Reineke TM. Poly(glycoamidoamine)s: Cationic glycopolymers for DNA delivery. Journal of Polymer Science Part A: Polymer Chemistry 2006;44(24):6895–908.
[29] Srinivasachari S, et al. Trehalose Click Polymers Inhibit Nanoparticle Aggregation and Promote pDNA Delivery in Serum. J Am Chem Soc 2006;128(25):8176–84.
[30] Smith AE, et al. Diblock Glycopolymers Promote Colloidal Stability of Polyplexes and Effective pDNA and siRNA Delivery under Physiological Salt and Serum Conditions. Biomacromolecules 2011;12(8):3015–22.
[31] Sizovs A, et al. Poly(trehalose): Sugar-Coated Nanocomplexes Promote Stabilization and Effective Polyplex-Mediated siRNA Delivery. J Am Chem Soc 2013;135(41):15417–24.
[32] Hwang SJ, Bellocq NC, Davis ME. Effects of Structure of β-Cyclodextrin-Containing Polymers on Gene Delivery. Bioconjug Chem 2001;12(2):280–90.
[33] Pun SH, Davis ME. Development of a Nonviral Gene Delivery Vehicle for Systemic Application. Bioconjug Chem 2002;13(3):630–9.
[34] Bellocq NC, et al. Transferrin-Containing, Cyclodextrin Polymer-Based Particles for Tumor-Targeted Gene Delivery. Bioconjug Chem 2003;14(6):1122–32.
[35] Davis ME. The first targeted delivery of siRNA in humans via a self-assembling, cyclodextrin polymer-based nanoparticle: from concept to clinic. Mol Pharm 2009;6(3):659–68.

[36] Davis ME, et al. Evidence of RNAi in humans from systemically administered siRNA via targeted nanoparticles. Nature 2010;464(7291):1067–70.

[37] Fire A, et al. Potent and specific genetic interference by double-stranded RNA in Caenorhabditis elegans. Nature 1998;391(6669):806–11.

[38] O'Reilly RK, Hawker CJ, Wooley KL. Cross-linked block copolymer micelles: functional nanostructures of great potential and versatility. Chem Soc Rev 2006;35(11):1068–83.

[39] Li Y, et al. Synthesis of Reversible Shell Cross-Linked Micelles for Controlled Release of Bioactive Agents†. Macromolecules 2006;39(8):2726–8.

[40] Sumerlin BS, et al. Aqueous solution properties of pH-responsive AB diblock acrylamido–styrenic copolymers synthesized via aqueous reversible addition–fragmentation chain transfer. Journal of Polymer Science Part A: Polymer Chemistry 2004;42(7):1724–34.

[41] Thurmond KB, Kowalewski T, Wooley KL. Water-Soluble Knedel-like Structures: The Preparation of Shell-Cross-Linked Small Particles. J Am Chem Soc 1996;118(30):7239–40.

[42] Thurmond KB, Kowalewski T, Wooley KL. Shell Cross-Linked Knedels: A Synthetic Study of the Factors Affecting the Dimensions and Properties of Amphiphilic Core-Shell Nanospheres. J Am Chem Soc 1997;119(28):6656–65.

[43] Pilon LN, et al. Synthesis and Characterization of Shell Cross-Linked Micelles with Hydroxy-Functional Coronas: A Pragmatic Alternative to Dendrimers? Langmuir 2005;21(9):3808–13.

[44] Shrestha R, et al. Endosomal escape and siRNA delivery with cationic shell cross-linked knedel-like nanoparticles with tunable buffering capacities. Biomaterials 2012;33(33):8557–68.

[45] Liu S, Armes SP. The facile one-pot synthesis of shell cross-linked micelles in aqueous solution at high solids. J Am Chem Soc 2001;123(40):9910–1.

[46] O'Reilly RK, et al. Functionalization of Micelles and Shell Cross-linked Nanoparticles Using Click Chemistry. Chem Mater 2005;17(24):5976–88.

[47] Jiang X, et al. Fabrication of Two Types of Shell-Cross-Linked Micelles with "Inverted" Structures in Aqueous Solution from Schizophrenic Water-Soluble ABC Triblock Copolymer via Click Chemistry. Langmuir 2009;25(4):2046–54.

[48] Yu SS, et al. Macrophage-Specific RNA Interference Targeting via "Click". Mannosylated Polymeric Micelles Molecular Pharmaceutics 2013;10(3):975–87.

[49] Kakizawa Y, Harada A, Kataoka K. Environment-Sensitive Stabilization of Core – Shell Structured Polyion Complex Micelle by Reversible Cross-Linking of the Core through Disulfide Bond. J Am Chem Soc 1999;121(48):11247–8.

[50] Thurmond 2nd KB, et al. Packaging of DNA by shell crosslinked nanoparticles. Nucleic Acids Res 1999;27(14):2966–71.

[51] Zhang L, et al. Shell-Cross-Linked Micelles Containing Cationic Polymers Synthesized via the RAFT Process: Toward a More Biocompatible Gene Delivery System. Biomacromolecules 2007;8(9):2890–901.

[52] Zhang K, et al. Cationic shell-crosslinked knedel-like nanoparticles for highly efficient gene and oligonucleotide transfection of mammalian cells. Biomaterials 2009;30(5):968–77.

[53] Zhang K, et al. Structure-activity relationships of cationic shell-crosslinked knedel-like nanoparticles: shell composition and transfection efficiency/cytotoxicity. Biomaterials 2010;31(7):1805–13.

[54] Matsumoto S, et al. Environment-Responsive Block Copolymer Micelles with a Disulfide Cross-Linked Core for Enhanced siRNA Delivery. Biomacromolecules 2008;10(1):119–27.

[55] Fang H, et al. Cationic shell-cross-linked knedel-like (cSCK) nanoparticles for highly efficient PNA delivery. Mol Pharm 2009;6(2):615–26.
[56] Shrestha R, et al. Hierarchically assembled theranostic nanostructures for siRNA delivery and imaging applications. J Am Chem Soc 2012;134(42):17362–5.
[57] Elsabahy M, et al. Multifunctional hierarchically assembled nanostructures as complex stage-wise dual-delivery systems for coincidental yet differential trafficking of siRNA and paclitaxel. Nano Lett 2013;13(5):2172–81.
[58] Christian DA, et al. Polymersome carriers: from self-assembly to siRNA and protein therapeutics. Eur J Pharm Biopharm 2009;71(3):463–74.
[59] Discher BM, et al. Polymersomes: Tough Vesicles Made from Diblock Copolymers. Science 1999;284(5417):1143–6.
[60] Won Y-Y, Davis HT, Bates FS. Giant Wormlike Rubber Micelles. Science 1999;283(5404):960–3.
[61] Aranda-Espinoza H, et al. Electromechanical limits of polymersomes. Phys Rev Lett 2001;87(20):208301.
[62] Srinivas G, Discher DE, Klein ML. Self-assembly and properties of diblock copolymers by coarse-grain molecular dynamics. Nat Mater 2004;3(9):638–44.
[63] Srinivas G, Discher DE, Klein ML. Key roles for chain flexibility in block copolymer membranes that contain pores or make tubes. Nano Lett 2005;5(12):2343–9.
[64] Rijcken CJ, et al. Triggered destabilisation of polymeric micelles and vesicles by changing polymers polarity: an attractive tool for drug delivery. J Control Release 2007;120(3):131–48.
[65] Kros A, et al. Silane-based hybrids for biomedical applications. J Adhes Sci Tech 2002;16(2):143–55.
[66] Jiang Y, et al. Reversible self-organization of a UV-responsive PEG-terminated malachite green derivative: vesicle formation and photoinduced disassembly. Langmuir 2007;23(7):4029–34.
[67] Li Y, Lokitz BS, McCormick CL. Thermally Responsive Vesicles and Their Structural "Locking" through Polyelectrolyte Complex Formation. Angew Chem 2006;118(35):5924–7.
[68] Qin S, et al. Temperature-Controlled Assembly and Release from Polymer Vesicles of Poly(ethylene oxide)-block-poly(N-isopropylacrylamide). Adv Mater 2006;18(21):2905–9.
[69] Ahmed F, Discher DE. Self-porating polymersomes of PEG-PLA and PEG-PCL: hydrolysis-triggered controlled release vesicles. J Control Release 2004;96(1):37–53.
[70] Ahmed F, et al. Shrinkage of a rapidly growing tumor by drug-loaded polymersomes: pH-triggered release through copolymer degradation. Mol Pharm 2006;3(3):340–50.
[71] Borchert U, et al. pH-induced release from P2VP-PEO block copolymer vesicles. Langmuir 2006;22(13):5843–7.
[72] Du J, et al. pH-Sensitive Vesicles Based on a Biocompatible Zwitterionic Diblock Copolymer. J Am Chem Soc 2005;127(51):17982–3.
[73] Cerritelli S, Velluto D, Hubbell JA. PEG-SS-PPS: reduction-sensitive disulfide block copolymer vesicles for intracellular drug delivery. Biomacromolecules 2007;8(6):1966–72.
[74] Napoli A, et al. Glucose-oxidase based self-destructing polymeric vesicles. Langmuir 2004;20(9):3487–91.
[75] Napoli A, et al. Oxidation-responsive polymeric vesicles. Nat Mater 2004;3(3):183–9.
[76] Cheng R, et al. Glutathione-responsive nano-vehicles as a promising platform for targeted intracellular drug and gene delivery. J Control Release 2011;152(1):2–12.
[77] Korobko AV, Backendorf C, van der Maarel JR. Plasmid DNA encapsulation within cationic diblock copolymer vesicles for gene delivery. J Phys Chem B 2006;110(30):14550–6.

[78] Korobko AV, Jesse W, van der Maarel JRC. Encapsulation of DNA by Cationic Diblock Copolymer Vesicles. Langmuir 2004;21(1):34–42.

[79] Brown MD, et al. Preliminary Characterization of Novel Amino Acid Based Polymeric Vesicles as Gene and Drug Delivery Agents. Bioconjug Chem 2000;11(6):880–91.

[80] Brown MD, et al. In vitro and in vivo gene transfer with poly(amino acid) vesicles. J Control Release 2003;93(2):193–211.

[81] Kim Y, et al. Polymersome delivery of siRNA and antisense oligonucleotides. J Control Release 2009;134(2):132–40.

[82] Torchilin VP, et al. p-Nitrophenylcarbonyl-PEG-PE-liposomes: fast and simple attachment of specific ligands, including monoclonal antibodies, to distal ends of PEG chains via p-nitrophenylcarbonyl groups. Biochim Biophys Acta 2001;1511(2):397–411.

[83] Christian NA, et al. Tat-functionalized near-infrared emissive polymersomes for dendritic cell labeling. Bioconjug Chem 2007;18(1):31–40.

[84] Pourtau L, et al. Antibody-Functionalized Magnetic Polymersomes: In vivo Targeting and Imaging of Bone Metastases using High Resolution MRI. Adv Healthc Mater 2013;2(11):1420–4.

[85] Dalhaimer P, Bermudez H, Discher DE. Biopolymer mimicry with polymeric wormlike micelles: Molecular weight scaled flexibility, locked-in curvature, and coexisting microphases. J Polym Sci B 2004;42(1):168–76.

[86] Broz P, et al. Cell targeting by a generic receptor-targeted polymer nanocontainer platform. J Control Release 2005;102(2):475–88.

[87] Petersen MA, et al. Synthesis and characterization of reactive PEO-PMCL polymersomes. Polymer Chem 2010;1(8):1281–90.

[88] Pangburn TO, Bates FS, Kokkoli E. Polymersomes functionalized via "click" chemistry with the fibronectin mimetic peptides PR_b and GRGDSP for targeted delivery to cells with different levels of [small alpha]5[small beta]1 expression. Soft Matter 2012;8(16):4449–61.

[89] Pangburn TO, et al. Targeted Polymersome Delivery of siRNA Induces Cell Death of Breast Cancer Cells Dependent upon Orai3 Protein Expression. Langmuir 2012;28(35):12816–30.

[90] Peppas NA, et al. Hydrogels in Biology and Medicine: From Molecular Principles to Bionanotechnology. Adv Mater 2006;18(11):1345–60.

[91] Smith MH, Lyon LA. Multifunctional Nanogels for siRNA Delivery. Acc Chem Res 2011;45(7):985–93.

[92] Lyon LA, et al. Thermoresponsive microgel-based materials. Chem Soc Rev 2009;38(4):865–74.

[93] Buddy Ratner AH, Schoen Frederick, Lemons Jack. Bomaterials Science: An Introduction to Materials in Medicine. Amsterdam, Netherlands: Elsevier; 2012.

[94] Whitesides GM, et al. Soft lithography in biology and biochemistry. Annu Rev Biomed Eng 2001;3:335–73.

[95] West JL, Hubbell JA. Polymeric biomaterials with degradation sites for proteases involved in cell migration. Macromolecules 1999;32(1):241–4.

[96] Hern DL, Hubbell JA. Incorporation of adhesion peptides into nonadhesive hydrogels useful for tissue resurfacing. J Biomed Mater Res 1998;39(2):266–76.

[97] Kidane A, Szabocsik JM, Park K. Accelerated study on lysozyme deposition on poly(HEMA) contact lenses. Biomaterials 1998;19(22):2051–5.

[98] Peppas NA, Merrill EW. Development of semicrystalline poly(vinyl alcohol) hydrogels for biomedical applications. J Biomed Mater Res 1977;11(3):423–34.

[99] Martens PJ, Bryant SJ, Anseth KS. Tailoring the degradation of hydrogels formed from multivinyl poly(ethylene glycol) and poly(vinyl alcohol) macromers for cartilage tissue engineering. Biomacromolecules 2003;4(2):283–92.

[100] Drury JL, Mooney DJ. Hydrogels for tissue engineering: scaffold design variables and applications. Biomaterials 2003;24(24):4337–51.

[101] Lee CR, Grodzinsky AJ, Spector M. The effects of cross-linking of collagen-glycosaminoglycan scaffolds on compressive stiffness, chondrocyte-mediated contraction, proliferation and biosynthesis. Biomaterials 2001;22(23):3145–54.

[102] Lutolf MP, et al. Synthetic matrix metalloproteinase-sensitive hydrogels for the conduction of tissue regeneration: engineering cell-invasion characteristics. Proc Natl Acad Sci U S A 2003;100(9):5413–8.

[103] Burdick JA, Khademhosseini A, Langer R. Fabrication of Gradient Hydrogels Using a Microfluidics/Photopolymerization Process. Langmuir 2004;20(13):5153–6.

[104] Podual K, Doyle 3rd FJ, Peppas NA. Glucose-sensitivity of glucose oxidase-containing cationic copolymer hydrogels having poly(ethylene glycol) grafts. J Control Release 2000;67(1):9–17.

[105] Ehrick JD, et al. Genetically engineered protein in hydrogels tailors stimuli-responsive characteristics. Nat Mater 2005;4(4):298–302.

[106] Mann BK, et al. Smooth muscle cell growth in photopolymerized hydrogels with cell adhesive and proteolytically degradable domains: synthetic ECM analogs for tissue engineering. Biomaterials 2001;22(22):3045–51.

[107] Suzuki Y, et al. Alginate hydrogel linked with synthetic oligopeptide derived from BMP-2 allows ectopic osteoinduction in vivo. J Biomed Mater Res 2000;50(3):405–9.

[108] Zhang S. Fabrication of novel biomaterials through molecular self-assembly. Nat Biotechnol 2003;21(10):1171–8.

[109] Petka WA, et al. Reversible hydrogels from self-assembling artificial proteins. Science 1998;281(5375):389–92.

[110] Wang C, Stewart RJ, Kopecek J. Hybrid hydrogels assembled from synthetic polymers and coiled-coil protein domains. Nature 1999;397(6718):417–20.

[111] MacEwan, S.R. and A. Chilkoti, Applications of elastin-like polypeptides in drug delivery. Journal of Controlled Release, (0).

[112] Amiram M, et al. A highly parallel method for synthesizing DNA repeats enables the discovery of 'smart' protein polymers. Nat Mater 2011;10(2):141–8.

[113] Alpar HO, et al. Biodegradable mucoadhesive particulates for nasal and pulmonary antigen and DNA delivery. Adv Drug Deliv Rev 2005;57(3):411–30.

[114] Vinogradov SV. Colloidal microgels in drug delivery applications. Curr Pharm Des 2006;12(36):4703–12.

[115] Peppas NA, Huang Y. Nanoscale technology of mucoadhesive interactions. Adv Drug Deliv Rev 2004;56(11):1675–87.

[116] Chowdary KP, Rao YS. Mucoadhesive microspheres for controlled drug delivery. Biol Pharm Bull 2004;27(11):1717–24.

[117] Peppas NA, Wood KM, Blanchette JO. Hydrogels for oral delivery of therapeutic proteins. Expert Opin Biol Ther 2004;4(6):881–7.

[118] Bromberg L, et al. Bioadhesive properties and rheology of polyether-modified poly(acrylic acid) hydrogels. Int J Pharm 2004;282(1-2):45–60.

[119] Krauland AH, Bernkop-Schnurch A. Thiomers: development and in vitro evaluation of a peroral microparticulate peptide delivery system. Eur J Pharm Biopharm 2004;57(2):181–7.

[120] Jones CD, Lyon LA. Synthesis and Characterization of Multiresponsive Core – Shell Microgels. Macromolecules 2000;33(22):8301–6.

[121] Serpe MJ, et al. Doxorubicin Uptake and Release from Microgel Thin Films. Biomacromolecules 2004;6(1):408–13.
[122] Gan D, Lyon LA. Tunable swelling kinetics in core–shell hydrogel nanoparticles. J Am Chem Soc 2001;123(31):7511–7.
[123] Gan D, Lyon LA. Synthesis and Protein Adsorption Resistance of PEG-Modified Poly(N-isopropylacrylamide) Core/Shell Microgels. Macromolecules 2002;35(26):9634–9.
[124] Nayak S, et al. Folate-Mediated Cell Targeting and Cytotoxicity Using Thermoresponsive Microgels. J Am Chem Soc 2004;126(33):10258–9.
[125] Peppas NA, Franson NM. The swelling interface number as a criterion for prediction of diffusional solute release mechanisms in swellable polymers. Journal of Polymer Science: Polymer Physics Edition 1983;21(6):983–97.
[126] Ritger PL, Peppas NA. A simple equation for description of solute release II. Fickian and anomalous release from swellable devices. Journal of Controlled Release 1987;5(1):37–42.
[127] Ritger PL, Peppas NA. A simple equation for description of solute release I. Fickian and non-fickian release from non-swellable devices in the form of slabs, spheres, cylinders or discs. J Control Release 1987;5(1):23–36.
[128] Morishita M, et al. Elucidation of the mechanism of incorporation of insulin in controlled release systems based on complexation polymers. J Control Release 2002;81(1–2):25–32.
[129] Elbert DL, Hubbell JA. Surface Treatments of Polymers for Biocompatibility. Annual Review of Materials Science 1996;26(1):365–94.
[130] Elbert DL, Hubbell JA. Reduction of fibrous adhesion formation by a copolymer possessing an affinity for anionic surfaces. J Biomed Mater Res 1998;42(1):55–65.
[131] Elbert DL, Herbert CB, Hubbell JA. Thin Polymer Layers Formed by Polyelectrolyte Multilayer Techniques on Biological Surfaces. Langmuir 1999;15(16):5355–62.
[132] Kenausis GL, et al. Poly(l-lysine)-g-Poly(ethylene glycol) Layers on Metal Oxide Surfaces: Attachment Mechanism and Effects of Polymer Architecture on Resistance to Protein Adsorption†. J Phys Chem B 2000;104(14):3298–309.
[133] Huang N-P, et al. Poly(l-lysine)-g-poly(ethylene glycol) Layers on Metal Oxide Surfaces: Surface-Analytical Characterization and Resistance to Serum and Fibrinogen Adsorption. Langmuir 2000;17(2):489–98.
[134] Scott EA, et al. Protein adsorption and cell adhesion on nanoscale bioactive coatings formed from poly(ethylene glycol) and albumin microgels. Biomaterials 2008;29(34):4481–93.
[135] Raemdonck K, et al. Dextran Microgels for Time-Controlled Delivery of siRNA. Adv Funct Mater 2008;18(7):993–1001.
[136] Costa D, et al. Plasmid DNA microgels for a therapeutical strategy combining the delivery of genes and anticancer drugs. Macromol Biosci 2012;12(9):1243–52.
[137] Knipe JM, Chen F, Peppas NA. Multiresponsive polyanionic microgels with inverse pH responsive behavior by encapsulation of polycationic nanogels. J Appl Polym Sci 2014;131(7) p. n/a-n/a.
[138] Wanakule P, et al. Nano-inside-micro: Disease-responsive microgels with encapsulated nanoparticles for intracellular drug delivery to the deep lung. J Control Release 2012;162(2):429–37.
[139] Nelson CE, et al. Tunable Delivery of siRNA from a Biodegradable Scaffold to Promote Angiogenesis In Vivo. Adv Mater 2014;26(4):607–14.
[140] Nguyen K, Dang PN, Alsberg E. Functionalized, biodegradable hydrogels for control over sustained and localized siRNA delivery to incorporated and surrounding cells. Acta Biomater 2013;9(1):4487–95.

[141] Krebs MD, Jeon O, Alsberg E. Localized and Sustained Delivery of Silencing RNA from Macroscopic Biopolymer Hydrogels. J Am Chem Soc 2009;131(26):9204–6.
[142] Nguyen MK, et al. Sustained localized presentation of RNA interfering molecules from in situ forming hydrogels to guide stem cell osteogenic differentiation. Biomaterials 2014;35(24):6278–86.
[143] Adolph EJ, et al. Enhanced performance of plasmid DNA polyplexes stabilized by a combination of core hydrophobicity and surface PEGylation. J Mater Chem B 2014;.
[144] Fine D, et al. Silicon Micro- and Nanofabrication for Medicine. Advanced Healthcare Materials 2013;2(5):632–66.
[145] Zhang M, et al. Polycation-functionalized nanoporous silicon particles for gene silencing on breast cancer cells. Biomaterials 2014;35(1):423–31.
[146] Vinogradov SV, Bronich TK, Kabanov AV. Nanosized cationic hydrogels for drug delivery: preparation, properties and interactions with cells. Adv Drug Deliv Rev 2002;54(1):135–47.
[147] Vinogradov S, Batrakova E, Kabanov A. Poly(ethylene glycol)–polyethyleneimine NanoGel™ particles: novel drug delivery systems for antisense oligonucleotides. Colloids Surf B Biointerfaces 1999;16(1–4):291–304.
[148] Vinogradov SV, Batrakova EV, Kabanov AV. Nanogels for oligonucleotide delivery to the brain. Bioconjug Chem 2004;15(1):50–60.
[149] Vinogradov SV, et al. Polyplex Nanogel formulations for drug delivery of cytotoxic nucleoside analogs. J Control Release 2005;107(1):143–57.
[150] Oishi M, et al. pH-Responsive PEGylated nanogels as targetable and low invasive endosomolytic agents to induce the enhanced transfection efficiency of nonviral gene vectors. Colloid Polymer Sci 2007;285(9):1055–60.
[151] Tamura A, Oishi M, Nagasaki Y. Enhanced cytoplasmic delivery of siRNA using a stabilized polyion complex based on PEGylated nanogels with a cross-linked polyamine structure. Biomacromolecules 2009;10(7):1818–27.
[152] Tamura A, Oishi M, Nagasaki Y. Efficient siRNA delivery based on PEGylated and partially quaternized polyamine nanogels: enhanced gene silencing activity by the cooperative effect of tertiary and quaternary amino groups in the core. J Control Release 2010;146(3):378–87.
[153] Mimi H, et al. Polyethyleneimine-based core-shell nanogels: a promising siRNA carrier for argininosuccinate synthetase mRNA knock-down in HeLa cells. J Control Release 2012;158(1):123–30.
[154] Lee H, et al. Target-specific intracellular delivery of siRNA using degradable hyaluronic acid nanogels. J Control Release 2007;119(2):245–52.
[155] Raemdonck K, et al. Prolonged gene silencing by combining siRNA nanogels and photochemical internalization. J Control Release 2010;145(3):281–8.
[156] Naeye B, et al. PEGylation of biodegradable dextran nanogels for siRNA delivery. Eur J Pharm Sci 2010;40(4):342–51.
[157] Sunasee R, et al. Biodegradable and Nontoxic Nanogels as Nonviral Gene Delivery Systems. Bioconjug Chem 2012;23(9):1925–33.
[158] Ahmed M, Narain R. Intracellular Delivery of DNA and Enzyme in Active Form Using Degradable Carbohydrate-Based Nanogels. Mol Pharm 2012;9(11):3160–70.
[159] Blackburn WH, et al. Peptide-functionalized nanogels for targeted siRNA delivery. Bioconjug Chem 2009;20(5):960–8.
[160] Dickerson EB, et al. Chemosensitization of cancer cells by siRNA using targeted nanogel delivery. BMC Cancer 2010;10:10.

[161] Dunn SS, et al. Reductively Responsive siRNA-Conjugated Hydrogel Nanoparticles for Gene Silencing. J Am Chem Soc 2012;134(17):7423–30.
[162] Agarwal R. Mammalian cells preferentially internalize hydrogel nanodiscs over nanorods and use shape-specific uptake mechanisms. Proc Natl Acad Sci 2013;.
[163] Friedmann T. Progress toward human gene therapy. Science 1989;244(4910):1275–81.
[164] Rosenberg SA, et al. Gene Transfer into Humans — Immunotherapy of Patients with Advanced Melanoma, Using Tumor-Infiltrating Lymphocytes Modified by Retroviral Gene Transduction. New England Journal of Medicine 1990;323(9):570–8.
[165] Blaese RM, et al. T Lymphocyte-Directed Gene Therapy for ADA– SCID: Initial Trial Results After 4 Years. Science 1995;270(5235):475–80.
[166] Marcum JA. From the molecular genetics revolution to gene therapy: Translating basic research into medicine. J Lab Clin Med 2005;146(6):312–6.
[167] Weiss, R., Boy's cancer prompts FDA to halt gene therapy, in Washington Post. 2005.
[168] Anderson WF. The Best of Times, the Worst of Times. Science 2000;288(5466):627–9.
[169] Kutzler MA, Weiner DB. DNA vaccines: ready for prime time? Nat Rev Genet 2008;9(10):776–88.
[170] Ferraro B, et al. Clinical applications of DNA vaccines: current progress. Clin Infect Dis 2011;53(3):296–302.
[171] Kanasty R, et al. Delivery materials for siRNA therapeutics. Nat Mater 2013;12(11):967–77.
[172] Ginn SL, et al. Gene therapy clinical trials worldwide to 2012 - an update. J Gene Med 2013;15(2):65–77.

Gold nanomaterials for gene therapy

8

M. Ebara, K. Uto
National Institute for Materials Science (NIMS), Tsukuba, Japan

8.1 Introduction

Colloidal gold is a suspension consisting of sub-micron gold nanoparticles (AuNPs) suspended within a solvent, most often water. AuNPs have unique optical, electronic, and thermal properties and are being incorporated in a wide variety of technologies including microscopy, electronics, diagnostics, and therapeutics. Although AuNPs have fascinated scientists for over a decade because of their huge potential in biological and therapeutic contexts, the history spans at least several thousand years because gold is one of the first metals to have been discovered. One of the most documented examples of AuNPs is medieval stained-glass artisans. The trapped AuNPs in the glass matrix generate the ruby red color in the windows. The first scientific report describing the production of colloidal AuNPs came in 1857. Michael Faraday found that the "fine particles" formed from the aqueous reduction of gold chloride by phosphorus could be stabilized by the addition of carbon disulfide, resulting in a "beautiful ruby fluid." AuNPs can be manufactured into a variety of shapes including gold nanospheres, nanorods, nanobelts, nanocages, nanoprisms, and nanostars. The chemical, optical, and electromagnetic properties of AuNPs are strongly influenced by their size and shape. For example, the rod-shaped nanoparticles have two resonances: one due to plasmon oscillation along the nanorod short axis and another due to plasmon oscillation along the long axis, which depends strongly on the nanorod aspect ratio, i.e., length-to-width ratio. When the nanorod aspect ratio is increased, the long-axis wavelength position red shifts from the visible to the near-infrared (NIR) and also progressively increases in oscillator strength. Due to these unique optical properties, AuNPs are the subject of substantial research, with enormous applications including biological imaging, electronics, and materials science. In particular, anisotropic morphologies of AuNPs are attractive for biological applications such as sensing, imaging, and diagnostics. AuNPs also play an important role in curing various diseases such as cancer by their photothermal effects. In the following sections of this chapter, we review different types of AuNPs that have been developed in the last few decades. In section 9.2, we describe the synthetic routes of AuNPs in different size and shape. The effects of their size and shape on intracellular uptake are discussed in section 9.3. In section 9.4, gene delivery applications of the AuNPs are discussed. The chapter ends with an overview of some of the future trends in applications in biotechnology and biomedicine.

Polymers and Nanomaterials for Gene Therapy. http://dx.doi.org/10.1016/B978-0-08-100520-0.00008-4
© 2016 Elsevier Ltd. All rights reserved.

8.2 Synthesis of gold nanoparticles

Today, AuNPs are one of the most widely used nanomaterials for a broad range of applications. Especially, the number of studies to control their shapes and morphologies have increased over the last decade because their optical, electronic, magnetic, and catalytic features are tunable by changing the size, shape, or surface chemistry [1]. For example, anisotropic AuNPs strongly change the scattered light polarization compared with isotropic nanoparticles, and this property can provides an excellent way to improve the sensitivity of plasmon-based bioassays diagnostics. AuNPs also show the ability to absorb, transfer, and convert light energy into heat depending on their shape, allowing for the highly specific thermal ablation of diseased tissues. In this section, fundamental techniques for synthesis of shape-controlled and structured AuNPs are reviewed.

8.2.1 Control of particle sizes

Control over the shape and size of metal nanoparticles has been one of the challenging tasks. Different approaches were developed to fabricate nanoparticles in different shapes such as spherical particles, cubes, prisms, triangles, platelets, rods and wires. The most common method employed for the synthesis of metal nanoparticles involves chemical reduction method, in which chemical precipitation was achieved by using a reducing agent and stabilizer. This is one of the most classical techniques for synthesis of metal nanoparticles. Initially, a part of the metal ions in solution are chemically reduced to yield metal atoms that subsequently form metal clusters. Then, these metal clusters act as the nucleation catalyst for the reduction of remaining metal ions present on the surface of metal clusters, leading to further growth. This process is usually employed in the presence of a stabilizing material or surfactant known as a capping agent. In aqueous media, cetyltrimethylammonium bromide (CTAB) is commonly used as a capping agent. The advantages of the reduction process are its lower cost, greater convenience, and capability of producing smaller powders. However, there is a significant challenge in obtaining a targeted shape and size by the chemical reduction method because growth and formation of new metal clusters occur simultaneously. To date, many studies have focused on improving methods for size-controlled synthesis of metal nanoparticles, including concentrations of seeds, surfactants, and reducing agents, growth temperature, time, and pH. Chen et al. reported a simple method to produce a series of multipod gold nanocrystals via solution-phase chemical reduction method at room temperature [2]. In this method, $HAuCl_4$, concentrated silver plates, and L-ascorbic acid were added to CTAB solution. Then, NaOH was rapidly added to induce the AuNP formation. During this process, the forced reduction of gold by ascorbic acid through the addition of NaOH is considered as the key step to obtain multipod AuNPs, because the use of strong reducing agent such as $NaBH_4$ generally yields spherical or rod-like shape. Interestingly, the addition of silver plates increased the yield of branched particles, while direct addition of Ag^+ ions showed a detrimental effect on branched particle formation.

Seed-mediated growth method has also found most attention as a bottom-up technique and offers the possibility to prepare higher quantities of AuNPs with ubiquitous laboratory equipment. In this method, the nucleation and growth stages are controllably separated. Therefore, the new nucleation is technically avoided during the growth. The principle of the seed-mediated growth method is the addition of small particles that act as seeds to a growth solution (see Figure 8.1). This growth solution contains additional metal ions, a structure-directing agent, and a weak reducing agent. The seed-mediated method for the synthesis of gold nanorods (AuNRs) was first introduced by Murphy's group in 2001. They demonstrated a synthetic route of homogeneous AuNRs with regulated aspect ratio in the presence of an aqueous micellar template (Figure 8.1(a)). They have used citrate-capped AuNPs as the seed and successfully controlled the aspect ratio from 1 to 25 by varying the ratio of preformed seed to metal salt [3–5]. First, $HAuCl_4$ is reduced to $HAuCl_2$ by ascorbic acid in the presence of CTAB and $AgNO_3$. Then, citrate-capped spherical AuNPs are added to this $HAuCl_2$ solution. These spherical AuNPs catalyze the reduction of Au^I to Au^0 by ascorbic acid. An increase in aspect ratio was also achieved by employing a three-step seed method. The time intervals between the solution transfers from one test tube to another are crucial and influence the resulting aspect ratio of the rods. The shorter

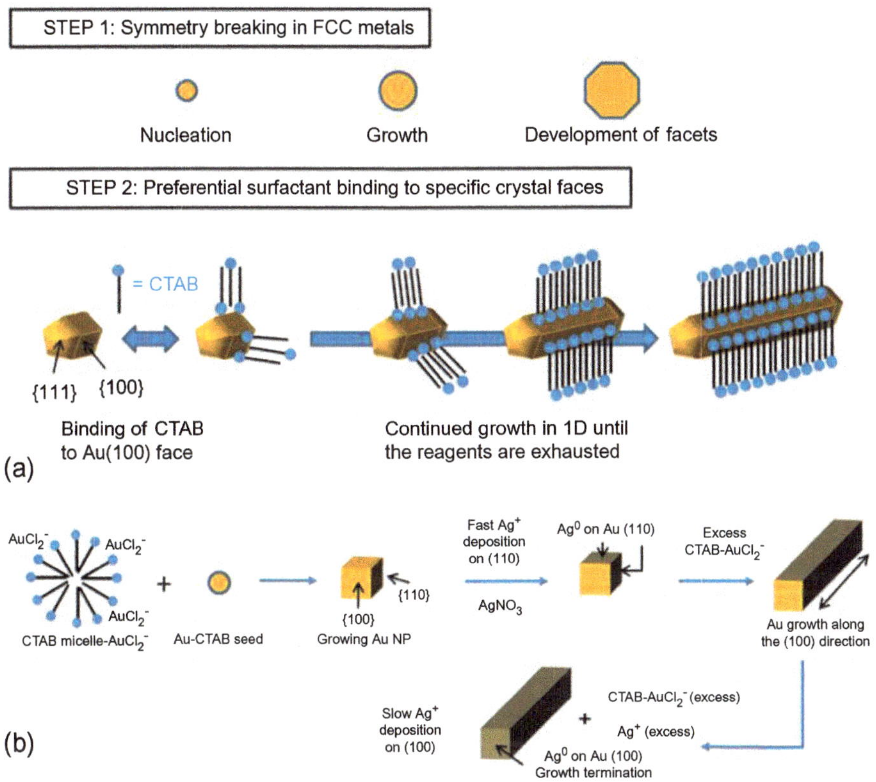

Figure 8.1 General mechanism of AuNRs growth. (a) Surface-directed metal growth by face preferential binding of surfactants [5] and (b) silver underpotential deposition on growing AuNPs for anisotropic growth [97].

the time intervals are the higher the aspect ratio of the nanorods. Nikoobakht and El-Sayed modified the protocol developed by Jana et al. and improved the yield and polydispersity of the AuNRs. They replaced the commonly used stabilizing agent, sodium citrate, by CTAB during the formation of the seeds. They successfully controlled the aspect ratio of AuNRs by the use of $AgNO_3$ because Ag ions assist binding and packing of the CTAB on these facets (Figure 8.1(b)), which result in the slow growth of the NR width [6]. Moreover, AuNRs were synthesized using binary surfactants. When benzyldimethylhexadecylammonium chloride (BDAC) was added to the solution together with CTAB, the aspect ratio of the resulting AuNRs increased compared to those prepared without BDAC.

Electrochemical methods have also been proven to have some additional advantages over chemical methods in the synthesis of size-selective or shape-controlled AuNPs. Also, this method is environmentally friendly because it avoids the use of reducing agents that usually are toxic. Svedberg first reported the electrochemical method for synthesis of Au nanomaterials in 1921 [7]. After a while, Wang's group successfully synthesized AuNRs via electrochemical oxidation/reduction within a simple two-electrode cell system that is known as the current basic electrochemical technique [8]. In this approach, two electrodes are immersed in an electrolytic solution consisting of a cationic surfactant and a structure-directing cosurfacetant (Figure 8.2a). CTAB and tetradodecylammonium bromide (TOAB) are commonly used as cationic surfactant and cosurfactant, respectively. The CTAB acts not only as the supporting

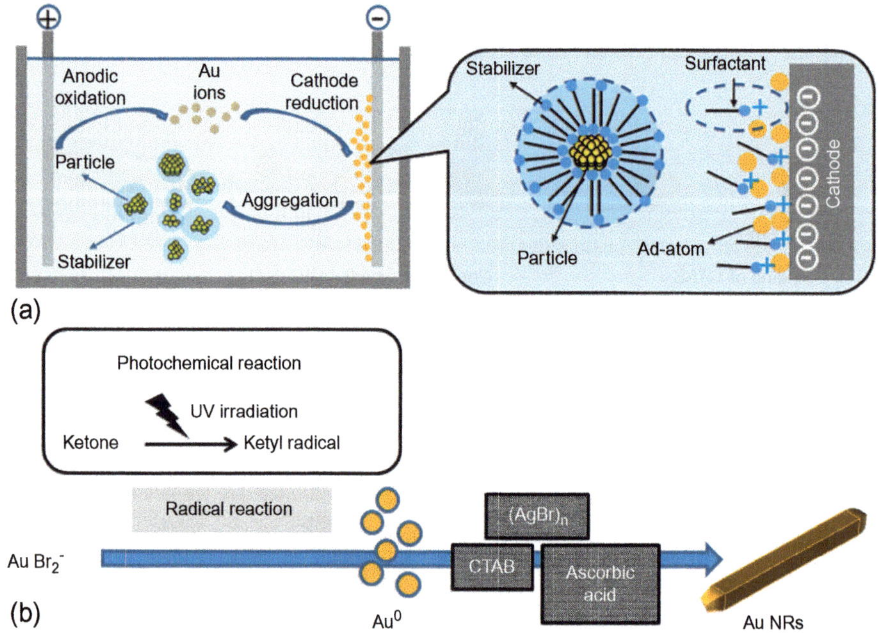

Figure 8.2 (a) Electrochemical synthesis of AuNPs using a two-electrode cell in surfactant solution [9]. (b) Schematic illustration of AuNRs formation by photochemical reaction of ketone using UV irradiation [20].

electrolyte but also as particle stabilizer to prevent the further growth. A gold metal plate is used as an anode and a platinum plate serves as a cathode. The salient features of this method include ease of operation, high yield, and the absence of undesired side products. Size-controlled AuNRs with narrow size distributions was synthesized in aqueous solution by an electrochemical method with the careful choice of an ideal stabilizer for the metallic nanoclusters and the use of a rotating platinum cathode [9]. Liu et al. recently reported that the Au nanomaterials with different shapes and sizes were effectively synthesized by adding $HAuCl_4$ as a gold precursor to an electrolyzed aqueous solution of poly(N-vinylpyrrolidone) (PVP) and KNO_3, which indicates the good reducing capacity of the PVP-containing solution after being treated by electrolysis [10]. In this approach, most AuNPs were spherical particles when a catholyte was used as the reducing agent or $HAuCl_4$, while they were plate-like NPs when an anolyte was used. They also found that the change in the pH value of electrolytes caused by the electrolysis of water would be the origin of the differences in shape and morphology of Au nanomaterials.

8.2.2 Control of particle shapes

For the preparation of size-controlled AuNPs, photochemical methods have also aroused much attention because controlled reduction of metal ions can be carried out without using excess reducing agent and the reduction reaction is uniformly performed in the solution without the needs of specific instrument. In 1995, Esumi et al. demonstrated the first physical synthetic method that is the photochemical reduction of auric acid [11]. In this system, ultraviolet (UV) light was often used as a light source that can reduce $HAuCl_4$ to for AuNRs when a cationic micelle with a rod shape derived from added surfactant was bound to $HAuCl_4$. Then, the photo-induced reduction (photoreduction) to Au^0 atoms was followed in this case by controlled aggregation. The micelle surfactant stabilizes a specific crystal face, as in the seed-mediated growth procedure, leading to anisotropic shapes. Because NaCl increases the ionic strength and decreases electrostatic repulsion between micells and particles, the addition of NaCl also affects the aspect ratio and yield of AuNRs [12]. Miranda and Ahmadi reported that Ag^+ ion also influences the aspect ratio and yield in a similar fashion with a seed-mediated growth process [13]. They also reported that UV light with a wavelength of 300 nm was optimal for obtaining these properties. The role of ascorbic acid concentration on the length and aspect ratio of AuNRs synthesized by the accelerated chemical-photochemical reduction method was also reported by the same group [14]. They found that a critical concentration of ascorbic acid exists beyond which no further elongation of nanorods occurs; instead, higher concentrations of ascorbic acid beyond the critical value leads to the formation of shorter nanorods. Thus, the photochemical method largely benefits from the improvement observed in the seed-mediated growth method, namely the use of $NaBH_4$, $AgNO_3$, ascorbic acid, and CTAB (or co-surfactants system). The advantage of these systems is that the AuNRs are obtained in high yield in a single step, unlike in the seed-mediated growth method. In this reaction system, the aspect ratio of AuNRs can be tuned by amount of $NaBH_4$ [15] and reaction temperature [16]. Adversely, the disadvantage of photochemical reduction system is requirement for long reaction time [13]. Niidome et al. addressed this problem and achieved the rapid synthesis by combination of chemical reduction of tetrachloroaurate by ascorbic acid and subsequent UV photoirradiation [17]. As a result, quick (within 30 min) and quantitative generation of AuNRs with appreciable shape changes as the reaction has been proceeded. In addition, it is well known that a ketone excited by photoirradiation undergoes an intersystem crossing to a triple state and then abstracts a hydrogen atom from an organic molecule. Because this reaction gives a ketyl radical, therefore, it can reduce the metal ions. Scaiano et al. demonstrated that aqueous, unprotected AuNPs could be synthesized from $HAuCl_4$ using a water-soluble benzoin (Irgacure-2959), which is well known as a photochemical source of strongly reducing ketyl radicals [18,19]. Niidome et al. also reported that photochemical synthesis of AuNRs in a micellar solution of CTAB was triggered by photoreactions of ketones [20]. The utilization of ketyl radicals results in an especially efficient reaction, and causes the rapid reduction of Au ions and formation of AuNPs or AuNRs (Figure 8.2(b)). Other than that, the photochemical route using organic materials such as ethylene glycol and PVP [21], or inorganic materials such as colloidal TiO_2 [22] have been investigated. In these systems, ethylene glycol serves as functional solvent and provides slowing down diffusion and ligand like effect similar with polyol compounds for metal reduction process, and PVP plays a role of a capping material. Colloidal TiO_2 serves both photocatalyst to initiate the reaction and the stabilizing agent for intermediate Au nanostructures. Interestingly, the photochemical process using colloidal TiO_2 can kinetically control the crystal evolution at various growth states by tuning the irradiation intensity, therefore, the ultimate AuNPs with different shapes and structures can be obtained.

The sonochemical approaches such as ultrasound have become a useful option to produce metal nanoparticles due to the low cost and effectiveness of the procedure. The main idea of the ultrasound method is that applied ultrasound endows chemical changes due to acoustic cavitation phenomena involving the formation, growth, and implosive collapse of bubbles in liquid. It is said that this effect generates localized hot spots having a temperature of roughly 5000 °C, pressures of about 500 atm, and a lifetime of a few microseconds [23]. These extreme conditions can drive chemical reactions such as oxidation, reduction, dissolution, and decomposition, which have been exploited to prepare a variety of metal, oxide, sulfide and carbide nanoparticles [24], and surface synthesis of NPs or NRs [25]. However, metal NPs synthesized by sonochemical reduction often have multiple shapes and wide size distribution. These extreme conditions can be improved by utilizing the surfactants and/or addition of additive such as alcohol compounds to control the particles shape and size in sonochemical approach [26,27]. From the results of previous studies, the sonochemical AuNPs formation occurs the following mechanism [28,29]: (i) sonochemical formation of reducing radicals (generation from water and organic pyrolysis), (ii) organic additives are then adsorbed on the Au cluster surface, and finally reduction of Au^{III} to Au^0 and the formation of AuNP. Zhang et al. successfully demonstrated the synthesis of single crystalline gold nanobelts by sonochemical method in the presence of α-D-glucose [30]. This simple and environmentally benign method can produce unique Au nanostructures with a width of 30–50 nm and a length of several micrometers under ambient conditions.

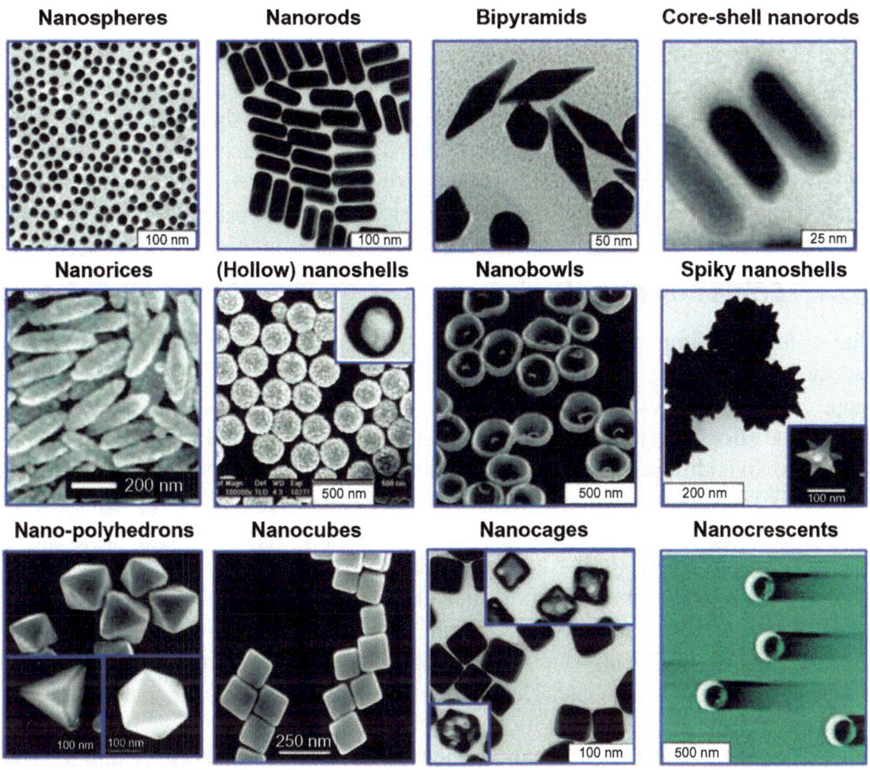

Figure 8.3 Various types of Au nanomaterials with different shapes such as spheres, rods, bipyramids, rices, and bowls, core-shell and hollow structure and so on [98].

As described in this section, synthetic methodology and technology of AuNPs are getting more advancement in past few decades. These methods are extended to preparation of many other unique gold nanostructures including nanorods [31], bipyramids[32], polyhedron (tetrahedral, octahedral, and cuboctahedra] [33], nanocrescents [34], nanocubes [33] and composite materials such as core-shell nanorods (gold nanorods surrounded by silver nanoshells) [35], nanorice (gold-coated Fe2O3 nanorods) [36], SiO_2/Au Nanoshell (or hollow nanoshell [37]), nanobowls with bottom cores [38], spiky SiO_2/Au nanoshell (or gold nanostar) [39], and gold-silver nanocages [40] (Figure 8.3). These AuNPs showed interesting electrical, optical, and plasmonic features based on their structures compared to that of quasispherical particles (nanospheres).

8.3 Intracellular uptake of gold nanoparticles

The successful delivery of therapeutic genes to the targeted cells and their availability at the intracellular site of action are crucial requirements for successful gene therapy [41]. It is also well known that the *in vitro* toxicity of a nanoparticle system is strongly related to its cell uptake behavior. Therefore, assessing the relationship between the size and shape of AuNPs and intracellular uptake behavior is especially important to design and develop more effective non-viral vectors beyond traditional ones for gene delivery. In this section, we focus on the effects of size and shape as well as surface properties such as charge or ligand type of AuNPs on the intracellular uptake behavior. Moreover, pharmacokinetics and toxicology of AuNPs (*in vivo* kinetics), which are notably important parameters for gene delivery, are also discussed.

8.3.1 Effects of particle size

One of the most important parameters of target geometry in intracellular delivery is particle size. As described above, homogeneous AuNPs with well-regulated size and shape could be synthesized by using different techniques. This advantage enables exploring the effects of structural and geometrical features of AuNPs on their intracellular uptake. In addition, AuNPs have other advantages tjat address both qualitative and quantitative studies of cellular uptake behavior because they are easy to characterize by the techniques of UV-vis spectrophotometer, inductively coupled plasma atomic emission spectroscopy (ICP-AES), and transmission electron microscopy (TEM). Chan and co-workers have demonstrated the intracellular uptake of different sized and shaped AuNPs using a HeLa cell in 2006 [42]. They used AuNPs with sizes varying from 14 to 100 nm, which were synthesized by a solution-based approach and stabilized by citric acid ligand. HeLa cells were incubated with the particles for 6 h and the concentrations of Au within the cells were quantitatively analyzed by ICP-AES. The uptake half-life for the different-sized AuNPs was 2.10, 1.90, and 2.24 h at a rate of 622, 1294, and 417 nanoparticles per hour for the 14, 50, and 74 nm AuNPs, respectively. The maximum uptake by a cell occurred at a nanoparticle size of 50 nm (Figure 8.4(a)). Osaki et al. also suggested that 50 nm semiconductor nanoparticles entered cells via receptor-mediated endocytosis more efficiently than smaller nanoparticles [43]. Interestingly, AuNPs were trapped in vesicles in the cytoplasm and did not enter the nucleus. The number of nanoparticles per vesicle diameter was also related to the size of AuNPs. When working with nanoparticles, the effect of aggregation on the cellular uptake has to be considered because aggregation frequently occurs in biological media such as blood, saliva, or cell culture media. Although surface modification of AuNPs with hydrophilic polymers such as PEG stabilizes the particles, they are stable only in the short term and eventually aggregate both *in vitro* and *in vivo* [44,45]. Chan and co-workers first addressed the effect of aggregation of AuNP on cellular uptake and toxicity [46]. They have successfully developed a simple approach to produce transferrin-coated AuNP aggregates of different sizes and characterized their uptake and toxic behavior in different cells (HeLa, MDA-MB-435, and A549). There was a 25% decrease in uptake of aggregated nanoparticles with HeLa and A549 cell in comparison to monodisperse nanoparticles. For MDA-MB 435, on the other hand, a two-fold increase in cell uptake for the largest aggregates was observed. These results suggest that multiple cellular responses are possible when nanomaterials are aggregated. Since the aggregation appears to be an inevitable occurrence *in vivo*, this study is very important for not only highlights the need for more thorough studies

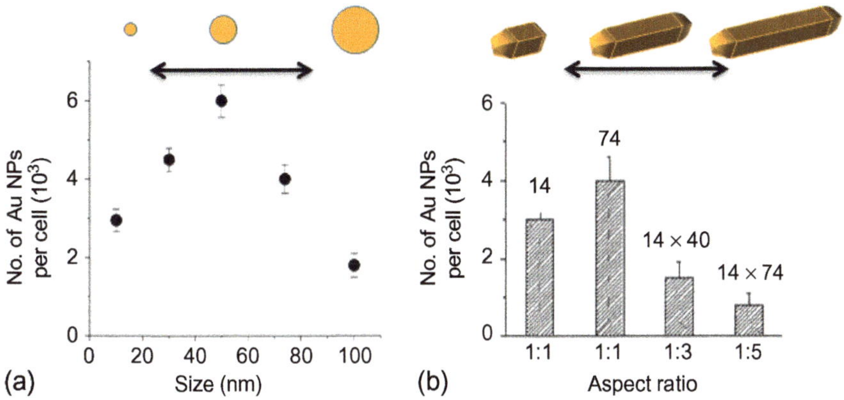

Figure 8.4 Effects of (a) particle size and (b) shape on cellular uptake by HeLa cells [42].

regarding the downstream effects and toxicity of long-circulating particles that are prone to aggregate but also provides a synthetic technique for designing particle aggregates. Similar cell type dependent uptake for AuNP system but with different sizes has been reported in other research groups [47,48].

Many theoretical models have provided insights into the dynamics of receptor-mediated endocytosis based on energetic and kinetic considerations, primarily seen in virus budding. For example, Lerner and Oster approached the question of how viruses bud from the plasma membrane of its host [49]. They investigated some possible kinetic processes including thermal fluctuations of the plasma membrane, hydrodynamic interactions, and diffusion of the glycoprotein spikes positioned on virus surface, and proposed that the discreteness of membrane wrapping via ligand-receptor binding results in a corrugated energy landscape for nanoparticle wrapping, which in turn governs the kinetics of endocytosis. Gao et al. studied how a cell membrane containing diffusive mobile receptors wraps around a ligand-coated cylindrical or spherical particle [50]. The particle with the diameter of several tens to hundreds nanometer can enter or exit cells via wrapping, and an optimal particle size exists for the smallest wrapping time that is the rate of endocytosis was limited by the diffusion of receptors toward the particles. Zhang et al. reported a thermodynamic model for receptor-mediated endocytosis of ligand-coated nanoparticles [51]. They identified an optimal particle radius at which the cellular uptake reaches a maximum of several thousand at physiologically relevant parameters, and showed that the cellular uptake of NPs was regulated by membrane tension, and can be elaborately controlled by particle size. The optimal radius for endocytosis was of the order of 25 to 30 nm, which is in good agreement with the experimental results reported by Chen's group. Shan et al. first investigated the size dependent force of endocytosis in single AuNPs by HeLa cells [52]. They used atomic force microscopy (AFM) with a fine cantilever and the corresponding force spectroscopy to detect the interaction forces. They successfully detected the interaction force of the living HeLa cell endocytosing the single AuNP with the diameters ranging from 4 to 17 nm by AFM, and revealed that both uptake and unbinding force values were dependent on the size of AuNPs. Single NP level analysis open the new way to examine the interaction between nanoparticles with small size and living cells, their interaction force as well as the mechanism. Hao et al. employed fluorescence microscopy to investigate the internalization mechanism of small-size AuNPs by living HeLa cells [53]. They used AuNPs with diameter of about 4.5 nm and revealed that caveolae-mediated endocytosis was the dominant pathway for the intracellular delivery of small-size AuNPs. These works will pave the way to exploring the internalization mechanism of nanoparticles with different targeting ligands and their overall roles in drug and gene delivery application [54].

8.3.2 Effects of particle shape

This section focuses on the effects of nanoparticle shape on cellular uptake. Chan and co-workers compared intracellular uptake of spherical and rod-shaped AuNPs [42]. The nanoparticle uptake was dependent upon shape and the uptake of rod-shaped AuNPs was lower than their spherical counterpart. For example, cells took up 500 and 375% more 74 and 14 nm spherical AuNPs than 74 x 14 nm rod-shape structures, respectively. The cellular uptake of rod-shaped structures with lower aspect ratio (1:3) is greater than higher aspect ratio (1:5) nanoparticles (Figure 8.4(b)). Bartczak et al. reported the interaction between endothelial cells and AuNPs with different morphologies but the same ligand capping [55]. They prepared four different shapes of AuNPs including spherical, rodlike, hollow gold, and core/shell (silica/gold and stabilized with monocarboxy (1-mercaptoundec-11-yl) hexaethylene glycol (OEG) to minimize the possibility of nonspecific interactions with biological molecules such as proteins in serum. The OEG-coated hollow and the core/shell particles were taken up the least by endothelial cells. On the other hand, the nanorods were taken up 100 times more than the hollow particles. Schaeublin et al. addressed the shape effect of AuNPs on biological response in human keratinocyte (HaCaT) cell [56]. They prepared both spherical nanoparticles and nanorods with diameters of 20 nm and length by width of 43.8 x 16.7 nm, respectively. The spherical particles were found to be nontoxic even at the highest dose of 100 μg/mL, while AuNRs were found to be cytotoxic with a significant decrease in cell viability occurring at 25 μg/mL and higher. The nanorods also caused significant production of reactive oxygen species (ROS) and up-regulated several genes involved in cellular stress and toxicity. These results indicate that shape appears to play a key role in mediating the cellular response to gold nanomaterials. The interactions of microglia and neurons with AuNPs of three morphologies, spheres, rods, and urchins have been also studied by Hutter et al. [57]. They showed that AuNPs were internalized by microglia with no sign of acute cytotoxicity and that the extent of internalization depended on their shape. In fact, rod- and urchin-shaped AuNPs caused transient microglia activation in animal experiment in transgenic mice. These data provide evidence that tuning of AuNP geometry can serve as a modality in regulating nanoparticle-cell interactions.

Cho et al. reported the effect of sedimentation and diffusion on cellular uptake of AuNPs with different morphologies [58]. They hypothesized that although *in vitro* experiments typically measure the uptake of nanoparticles by exposing cells at the bottom of a culture plate to a suspension of nanoparticles, the concentration of nanoparticles on the cell surface may be higher than the initial bulk concentration, and this could lead to increased uptake by cells. From this perspective, they used upright and inverted cell-culture configurations to show that cellular uptake of AuNPs depends on the sedimentation and diffusion velocities of the nanoparticles. Their results demonstrated that the cellular uptake of NPs strongly depended on the ratio of sedimentation to diffusion velocities, regardless of size, shape, density, surface ligand, and initial concentration.

8.3.3 Effects of surface property

As described above, nanomaterial shape and size contribute significantly to their interaction with cells. Those physical parameters of nanoparticles can affect their non-specific uptake in cells, with potential to induce cellular responses. However, these studies provide limited information regarding the cellular processes involved in nanoparticle trafficking and their ensuing functional impact. From this perspective, Chithrani et al. studied the uptake mechanism of transferrin-coated AuNPs into different mammalian cells such as HeLa (ovarian cancer), SNB19 (brain tumor), and STO (fibroblast) cells [59]. Transferrin is known as the protein that carries iron into cells through the receptor-medicated endocytosis. In fact, the transferrin-coated AuNPs were taken up into the cells via a receptor-mediated clathrin-dependent endocytosis pathway (Figure 8.5(a)). Of particular interest is that the fraction of rod-shaped NPs exocytosed was higher than spherical-shaped nanostructures. These studies show that the selection of the NPs can influence the accumulation concentration and rates of the NPs in a cell. Antibodies can also be attached to the surface of AuNPs. Jiang et al. showed that AuNPs coated with antibodies can regulate the process of membrane receptor internalization [60]. They synthesized AuNPs of sizes ranging from 2 to 100nm to selectively control specific interactions between Herceptin and its receptor ErbB2, a receptor tyrosine kinase overexpressed in various ovarian and breast cancers. The uptake of Herceptin-modified AuNPs in SK-BR-3 cells, which is human breast cancer cell over-expressed ErbB2 receptor, was significantly enhanced compared to that of unmodified AuNPs. The internalization of antibody-modified AuNPs highly depended on size, with the most efficient uptake occurring within the 25–50nm size range. Furthermore, this size-dependent uptake caused the downregulation of membrane receptor expression level (Figure 8.5(b)). These findings provide strong evidence that nanostructures can not only passively interact with cells, but also actively engage and mediate the molecular processes that are essential for regulating cell functions.

Although surface modification of AuNPs with specific biomolecules has many advantages for various applications as described above, new developments of physical and chemical modification techniques are crucial. Kobayashi et al. recently reported the enhancement of cellular uptake of amphiphilic AuNPs with ester-headed ligands [61]. The ester ligand coating was used as a stimuli-responsive surface to accelerate AuNP internalization into cells via an esterase-catalysed hydrolysis of ester moieties within cells. They found that AuNPs with ester-headed ligands were internalized into HeLa cells at a greater level than were ether-headed AuNPs. This technique could be applicable to delivering various functional nanoparticles into cells (Figure 8.6(a)).

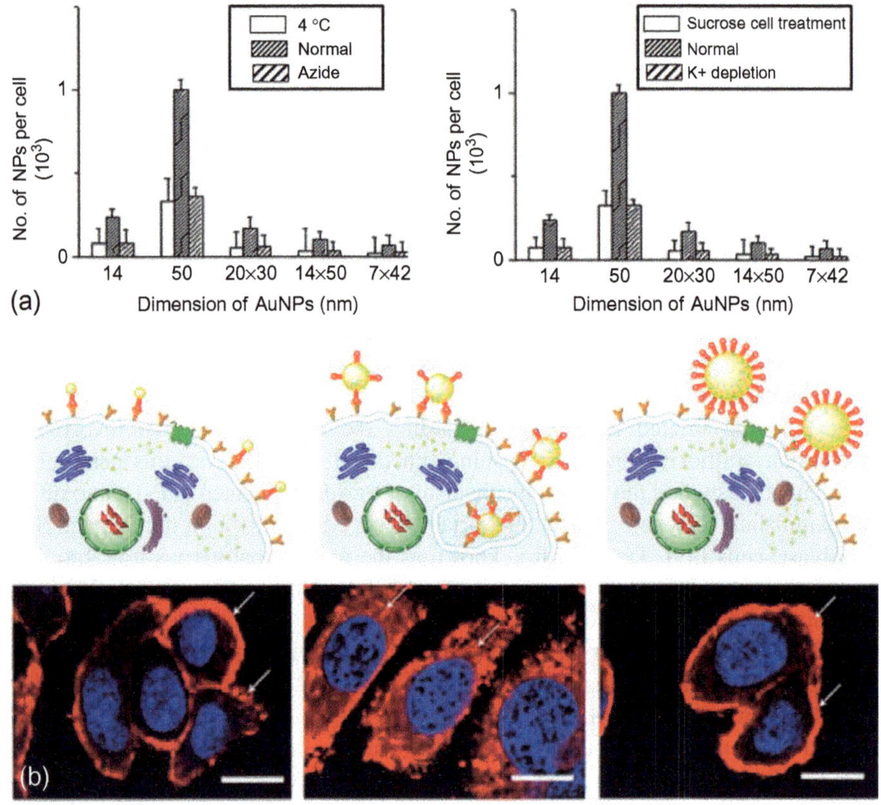

Figure 8.5 (a) Elucidating the mechanism of cellular uptake of transferrin-coated spherical AuNP and AuNR. Graphs show the evidence of cellular uptake of AuNP and AuNR through the receptor-mediated endocytosis pathway and this endocytosis pathway is dependent on clathrin. (b) Effects of particle size on downregulation of membrane ErbB2 expression. (A) (a) Fluorescence images show the ErbB2 receptor localization after treatment with different size Herception-AuNPs [60].

Pillai et al. demonstrated the adjustable cellular uptake of mixed-charge NPs [62]. They designed AuNPs functionalized with mixed self-assembled monolayer (m-SAMs) consisting of negatively and positively charged thiol compounds that were stable at both low and high pH but precipitated sharply at the pH where the charges on the particle were balanced in similar fashion to protein molecules. The degrees of cellular uptake strongly depended on the SAMs' composition and particles' net charge (Figure 8.6(b)). Remarkably, the presence of the positively charged thiols allowed for the uptake of particles having net negative charge, although numerous previous studies have shown that positively charged NPs are internalized more easily than neutral particles. The ability to adjust the precipitation point (charge equal to zero) of the m-SAM modified AuNPs could be used for pH specific biological targeting, especially of tumors, whose extracellular pH of ~6.5 is lower by about one unit than the pH of the surrounding healthy tissue under physiological conditions of pH 7.4.

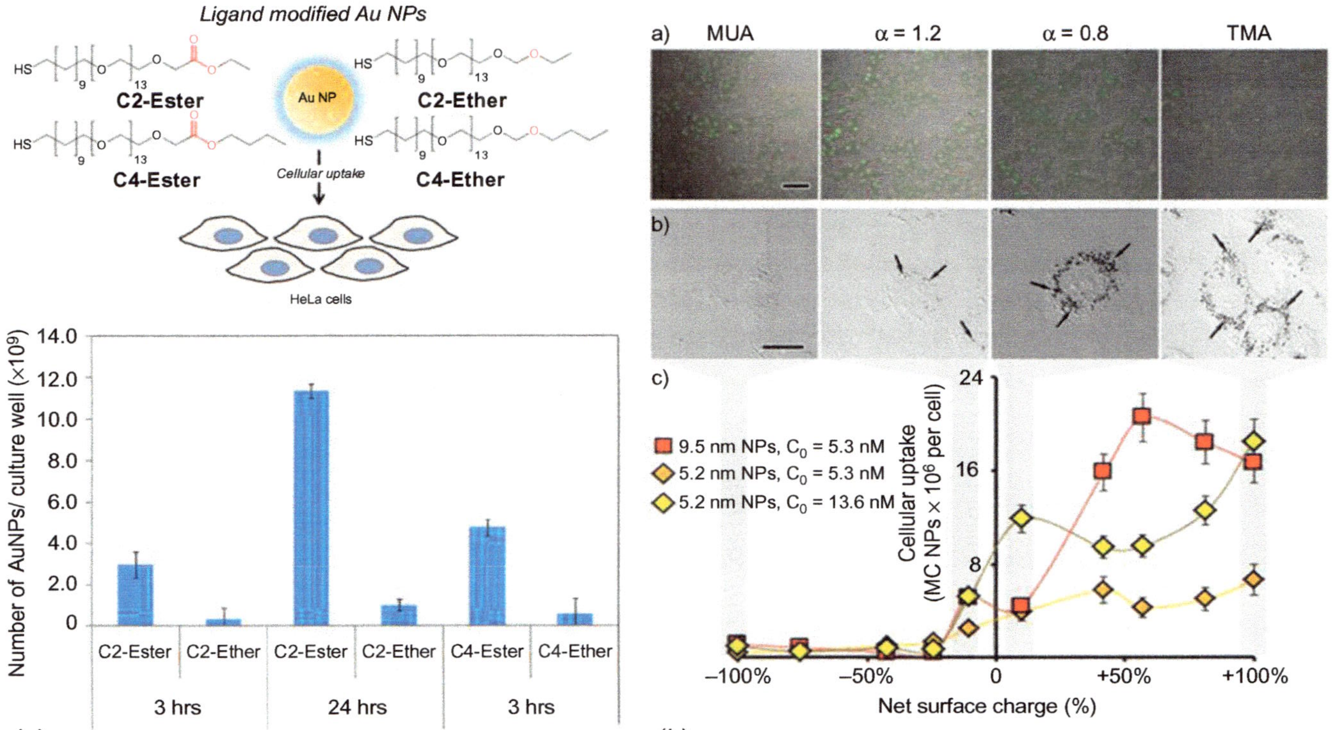

(a) (b)

Figure 8.6 (a) Chemical structure of PEG-derivative ligands for surface modification of AuNPs and number of modified AuNPs internalized into HeLa cells per well [61]. (b) Adjustable cellular uptake of mixed charged AuNPs with different size and surface compositions. (a) Confocal images of live (green)/dead (red) assay of Rat 2 cells after incubation for 24 h in the presence of 2.5 mM solution of 5.2 nm mixed charged AuNPs (α = [anionic MUA]/[cationic TMA] at surface) scale bar is 100 μm. (b) Optical images of Rat 2 cells illustrating different degrees of mixed charge NP uptake. Scale bar is 25 μm. (c) ICP data quantifying the degree of cellular uptake as a function of relative net charge on the mixed charge NPs [62].

8.4 Gene delivery applications

As we have shown, the beneficial attributes of AuNPs such as their ease of synthesis, functionalization, and shape control have made them the subject of intensive studies and biomedical applications. Over the past few years, AuNPs have been used in laboratory-based clinical diagnostic methodologies, such as detection of specific DNA and RNA sequences as well as protein biomarkers. However, the main excitement concerning AuNPs is their potential to cross over into clinical practice for use in humans. From this perspective, bioactive AuNPs have been developed for drug or gene delivery. Generally, gene therapy can be carried out by using viral and non-viral vectors to transport foreign genes into somatic cells to remedy defective genes there or provide additional biological functions. In particular, non-viral gene delivery systems can provide some potential benefits and have low toxicity. However, the current non-viral gene delivery systems still suffer from low transfection efficiency due to the difficulty of controlling the process at the nanoscale. To date, various nanoparticles such as magnetic nanoparticles, carbon nanotubes, and liposomes have been developed as non-viral carriers for gene delivery. In this section, the recent advances and current challenges of AuNPs in gene delivery applications are reviewed.

8.4.1 Non-covalent approaches

There are several strategies to immobilize genes on the surface of AuNPs. In general, these can be divided into methods based on covalent or non-covalent reactions. For the non-covalent approach, Sundaram et al. proposed a particle-mediated delivery of recombinant expression vectors to rabbit skin in 1996 [63]. They prepared Au particles with 1–3 μm in diameter coated with mammalian expression vectors encoding a cytoplasmic (β-galactosidase) or a nuclear (L1 capsid of cottontail rabbit papillomavirus) protein. Although polyclonal antibodies were successfully generated in rabbits by delivery of the particles to skin, the transfection efficiency was still low due to the lower surface-to-volume ratio compared with nano-size particles. To provide an effective scaffold for multivalent interactions with DNA, McIntosh et al. fabricated mixed monolayer protected gold clusters (MMPCs) functionalized with tetraalkylammonium ligands that can interact with the DNA backbone via charge complementarity (Figure 8.7(a)) [64]. They have shown that MMPC displayed not only the DNA-binding ability common of multiply charged systems studied, such as polymers, dendrimers, or small molecules, but also was capable of inhibiting a transcription enzyme from producing RNA products. This study demonstrated the viability of these particles for further application to biological applications, because the DNA binding occurred in a stoichiometric, non-aggregated fashion with sufficient affinity to disrupt recognition and transcription by a well-evolved enzyme. They also reported that the MMPC-bound DNA was efficiently protected from DNAse I digestion and experienced nicking/cleavage-induced morphology changes with higher concentration of DNAse I [65]. Therefore, non-covalent electrostatic immobilization of DNA onto MMPCs will serve as an efficient DNA delivery system. Moreover, DNA-MMPC assemblies demonstrated excellent glutathione-controlled DNA transcription, even

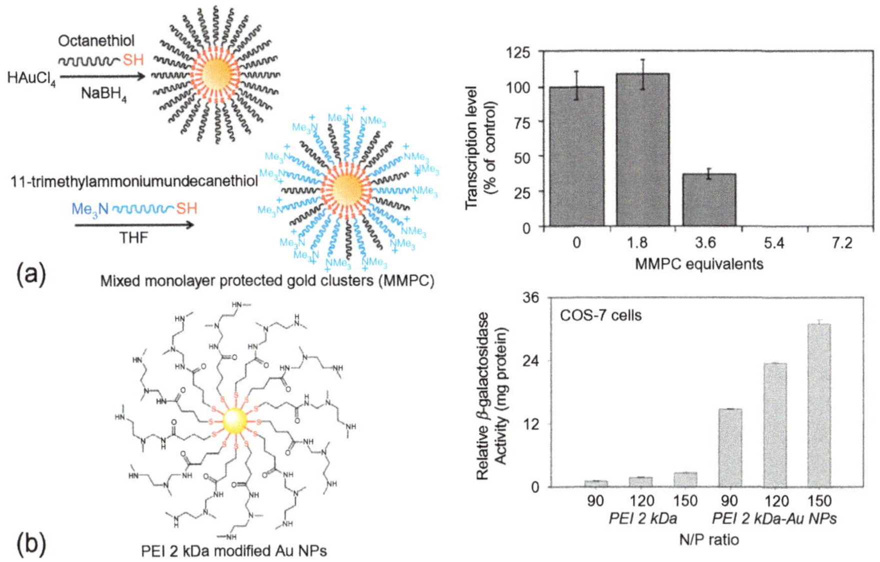

Figure 8.7 (a) Mixed monolayer protected Au clusters (MMPCs) fabricated via Murray place displacement of 2 nm octanethiol-functionalized Au NPs with 11-trimethylammoniumundecanethiol units and the inhibition of DNA transcription using MMPCs. The amount of RNA detected relative to levels produced in the absence of MMPCs (100% transcription) [64]. (b) Schematic representation of 2 kDa PEI and AuNP conjugates and transection of plasmid DNA into mammalian cells. Comparison of expression level in COS-7 cell cultures of the β-gal gene mediated in the presence of serum by PEI and the conjugates at different N/P ratio [68].

though the stability of immobilized DNA on the surface of MMPCs is quite high [66]. Immobilized DNA was released from DNA-MMPC assemblies that are stable at extracellular glutathione concentration by intracellular glutathione levels, resulting in efficient transcription. They also found that the release of DNA and recovery of DNA transcription could be tuned by not only glutathione levels but also the controlling of monolayer coverage. This release mechanism provided potential strategies for creation of MMPC-based DNA delivery vectors featuring new and unique properties, and these results strongly suggested that these non-covalent DNA-Au nanomaterial systems also provided an effective means of gene delivery in human embryonic kidney cells [67].

Thomas and Klibanov developed polymer-AuNP hybrid that covalently modified AuNP with thiolated and branched polyethyleneimine with low molecular weight (2 kDa; PEI2) and electrostatically immobilized the plasmid DNA encoding β-galactosidase onto the surface of PEI-modified AuNPs [68]. PEI2-AuNP hybrid showed a transfection efficiency against monkey kidney cells six times higher than that of free PEI with 25 kDa, which is the gold standard polymer as a polymeric and cationic vector. The high transfection efficiency of PEI2-AuNP could be further enhanced by introducing the hydrophobic dodecyl groups to PEI reported in their previous study [69]. So far, it is well known that a polycationic system using PEI is an efficient transfection vector *in vivo*, but this study clearly revealed the important finding that conjugation to AuNPs enhanced PEI's transfer into mammalian cells by synergistic actions (Figure 8.7(b)). Ow Sullivan et al. also developed a series of gene delivery vehicles based on DNA plasmid condensation with PEI-AuNP conjugates [70]. These vehicles enable systematic study of relationships between physical properties and transfection efficiency of DNA complex. As described above, AuNP is an excellent platform for further chemical modification of targeting ligands including small molecules and peptides. This characteristic, in turn, provided the potential for a combinatorial approach to assaying ligand incorporation effect. They demonstrated that complexes with a diameter of a few hundred (~200–300 nm size range) and a ratio of positive and negative charge that is slightly different with N/P ratio, of approximately 8, showed best performance for their transfection assays.

The "layer-by-layer" technique is another interesting way to modify the surface non-covalently. Elbakry et al. developed PEI/nucleic acid/PEI multilayer coated AuNPs for the delivery of nucleic acids, which are small interfering RNAs (siRNAs) [71]. They first and successfully demonstrated the layer-by-layer assembly of the oppositely charged polyelectrolytes siRNA and PEI on AuNPs and knock-down the expression of target gene. Unlike the previous system based on the electrostatic immobilization of nucleic acid, this technique offered a unique opportunity to fabricate well-defined and homogeneously distributed nanocarriers for siRNA delivery. It is well known that successful escape of carriers with nucleic acid from endosome and followed by the release of the payload into cytoplasm are very important for improving the efficiency of gene therapy. In addition, charge-reversal in polymeric carrier system has been usually used to further improve the gene delivery efficiency by enhancing endosome escape capacity [72,73]. Guo and Liang et al. also first developed charge-reversal functional Au NPs by layer-by-layer techniques for plasmid and siRNA delivery into cancerous cells [74]. Charge-reversal polyelectrolyte layer were formed by using layer-by-layer assembly of cationic PEI and PAH modified with citraconic anhydride (PAH-Cit), which was first synthesized by Lynn gropu [75], onto AuNPs modified with 11-mercaptoundecanoic acid. Since aniconic PAH-Cit can be converted readily back to cationic PAH by amide hydrolysis upon exposure to acidic environments, the obtained PEI/PAH-Cit/PEI-AuNP showed the charge-reversion ability. In fact, DNA and siRNA were then electrostatically immobilized onto the charge-reversal PEI/PAH-Cit/PEI-AuNPs. These designed charge-reversal AuNPs could both effectively enhance gene delivery efficiency and regulate gene expression in the context of RNA interference. This result strongly suggests that charge reversion in acidic environment significantly facilitates the escape of DNA or siRNA-AuNPs hybrids from endosome/lysosome and release of functional nucleic acids into cytoplasm where nucleic acid plays a function. Therefore, the layer-by-layer technique could be successfully used to form multifunctional AuNPs especially for therapeutic applications [76].

8.4.2 Covalent approaches

Although non-covalent interactions can provide interesting assemblies of AuNPs with biomolecules, AuNPs can be directly conjugated with biomolecules via covalent bonding, i.e., thiol (sulfur)-gold chemistry. In 2006, Oishi and Nagasaki et al.

developed a novel cytoplasmic delivery system of siRNA, i.e., complexation between poly(ethylene glycol)-block-poly(2-(N,N-dimethylamino)ethyl methacrylate) (PEG-PAMA) copolymer, AuNPs, and siRNA bearing a thiol group (SH-siRNA) [77]. The PEGylated AuNPs were extremely stable under physiological conditions due to the multivalent coordination and electrostatic interaction between PAMA cationic tertiary amino groups of PAMA segment in copolymer. The PEGylated AuNPs demonstrated the ability to introduce the SH-siRNA by thiol-Au chemistry. Immobilized siRNA by thiol-Au interaction can be released from the surface of AuNPs by applying thiol-containing compounds such as glutathione and dithiothreitol (DTT). Therefore, this system can facilitate the specific glutathione-mediated release of siRNA in the cytoplasm and successfully demonstrated a significant sequence-specific gene silencing by RNA interference activity in human hepatoma (HuH-7) cells. On the other hand, Rosi and Mirkin et al. demonstrated a revolutionary concept for a covalently modified oligonucleotide-AuNPs system as a vector [78]. Despite the large negative surface potential of oligonucleotide-AuNPs, these hybrids showed a higher affinity constant for complementary nucleic acids than unmodified oligonucleotide counterparts. In addition, these hybrids were extremely stable against enzymatic degradation and showed excellent cellular uptake ability and nontoxicity against cells. They used antisense DNA that inhibit or down-regulate the production of a target protein to hybridize with AuNPs. In fact, they targeted mRNA sequences coding for enhanced green fluorescent protein (EGFP) expressed in mouse endothelial (C166) cell line and clearly demonstrated a significant reduction of fluorescence intensity in cells (Figure 8.8). They also reported the advantage of this system that gene knock-down could be tuned by chemical tailoring the density DNA introduced onto the surface of AuNPs. After this report, they addressed the cellular uptake mechanism of DNA-AuNPs with large negative charge because there has been no report on the cellular internalization of negative charge DNA-AuNPs [79]. They quantified the uptake of DNA-AuNPs in mouse C-166 and human cervical cancer (HeLa) and alveolar type II epithelium-like lung carcinoma (A594) cells, and characterized the interactions of these hybrids with proteins in the extracellular environment. Cellular uptake of hybrids was strongly dependent on the density of the introduced oligonucleotide onto the surface of AuNPs and they revealed that AuNPs with higher densities of oligonucleotide lead to greater uptake [78]. They concluded this density-dependent uptake was related to the amount of adsorbed protein onto the surface of functionalized AuNPs. In fact, densely functionalized NPs adsorbed a large number of proteins on their surface. Then, they applied this basic theory and technology to develop an RNA-AuNPs hybrids system because the utilization of siRNA is a more promising way of regulating genes than that of antisense DNA as described above, and no methods have been developed for using polyvalent AuNPs and their properties to load and transport RNA through the cellular membranes [80]. They reported the synthesis and characterization of RNA-AuNPs hybrids that were densely functionalized with synthetic RNA oligonucleotides and designed to function in the RNA interference (RNAi) pathway. The hybrids were rationally designed and synthesized in the nuclease-free condition. The obtained hybrids have high surface loading of siRNA duplexes, and RNA stability can be enhanced by high local salt concentration attributed to the

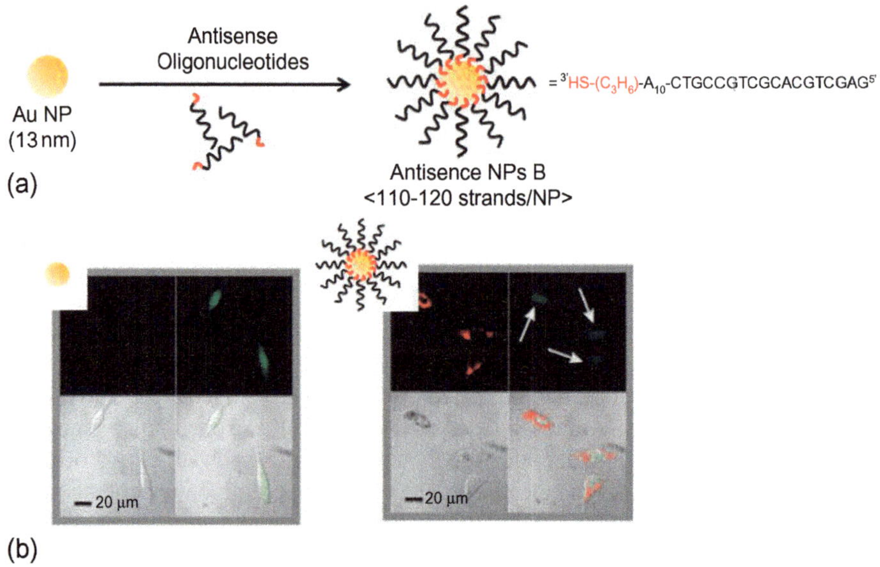

Figure 8.8 (a) Synthetic scheme of Oligonucleotide-modified AuNPs. (b) Confocal fluorescence microscopy images of EGFP knock-down cells treated with AuNPs (left) and antisense-AuNPs [78].

high-density modification. This result is very important for future applications because enhancement of RNA stability was the most challenging problem due to its chemical instability. The designed hybrids also had a half-life six times longer than that of free double-stand RNA, readily entering cells without any transfection agents. Finally, they successfully demonstrated a highly efficient gene knock-down capability in HeLa cell as a model.

Polyvalent oligonucleotide-AuNP hybrids developed by Mirkin's group possess several unique emergent properties, including enhanced cellular uptake, high oligonucleotide activity, and enzymatic nucleic acid degradation resistance, that probably originated from the dense packing and orientation of oligonucleotides on the surface of AuNPs. On the other hand, they questioned whether polyvalent nucleic acids alone readily enter cells without need for a positively charged co-carrier. To address this issue, they reported a strategy for synthesizing homogeneous, spherical high-density polyvalent DNA and DNA-RNA hybrid nanostructures with no AuNP cores. They previously discovered that propargyl-ether-modified polymer could be adsorbed and cross-linked on AuNP surfaces in water [81]. In this approach, AuNPs acted as both template and reaction catalyst. To fabricate high-density polyvalent DNA and DNA-RNA hybrid nanostructures, they synthesized new oligonucleotides with a sequence of thymine bases modified with propargyl ether groups [82]. Cross-linking took place between propargyl groups adsorbed on the particles, and the AuNP templates were then removed by oxidative dissolution. The obtained nanostructures were comprised of only cross-linked and oriented nucleic acids. They finally demonstrated that these hollow particles were capable of effecting high cellular uptake and gene regulation without need of a positively charged co-carrier in the same way in multivalent oligonucleotide-AuNPs hybrids.

Here, as for DNA uptake into specific cells, there are many internalization pathway including receptor-mediated endocytosis, pinocytosis, and phagocytosis. Apart from these processes, other compulsory techniques such as electroporation [83], naked DNA injection using a fine-needle [84] and particle bombardment [85] have been developed to facilitate gene transfection. Among them, DNA transfection through an electroporation method is based on the formation of multiple transient pores induced on the cell membrane via electric pulses, which has been extensively studied for efficient transfection of gene. This technique can be applied to covalently modified DNA-AuNP hybrids, and the system of AuNP vector joint with microelectromechanical systems for *in vitro* non-viral transfection approach have been reported by Jen et al. in 2004 [86]. They prepared AuNPs covalently modified with 21-base thiolated-oligonucleotides and evaluated their performance as delivery vehicles for the development of a non-viral transfection platform. This is the first trial of combined AuNPs as an oligonucleotide delivery vehicle to penetrate cell membranes via electroporation on the microfabricated chip, and they successfully demonstrated the highly efficient oligonucleotide transportation into mouse osteoblast (MC3T3E-1) cells. These results clearly indicate the feasibility for utilization of AuNPs as carriers for gene transfection by means of electroporation.

8.4.3 Stimuli-responsive systems

The unique chemical and physical properties of AuNPs can be also exploited in innovative ways to deliver genes to target cells. One of those approaches is triggered release of genes. Stimuli responsive carriers have been engineered to release their payload in response to local cues, including pH and enzymes, or in response to external stimuli, such as ultrasound, magnetic fields, electromagnetic radiation, and temperature. This approach is a powerful means for the development of "smart therapeutics." To date, most intracellular releases that have been reported employ glutathione or pH as a trigger [87,88]. Another example is the use of light as an external stimulus to release DNA from AuNPs in spatio-temporal manner. In 2006, a photolabile AuNP system was designed as a nucleic acid carrier by Rotello's group [89]. They prepared a positively charged AuNP with a photoactive o-nitrobenzyl ester linkage, which allowed temporal and spatial release of DNA by light (Figure 8.9). Their developed cationic photocleavable NPs successfully demonstrated effective light (Near-UV irradiation: > 350 nm)-regulated control over nucleic acid-NP interaction. The release mechanism, including multiple steps where light irradiation cleaved the nitrobenzyl linkage, followed by releasing the positively charged alkyl amine and leaving behind a negatively charged carboxylate group. This reversal in surface charge led to efficient release of nucleic acid from the NPs and resulted in a high level of recovery of DNA transcription *in vitro*. More importantly, they successfully demonstrated that effective DNA delivery and release in living mouse embryonic cells where released nucleic acid was significantly localized in cell nucleus.

The use of more complex nanoparticles such as nanorods could be valuable for *in vivo* therapy due to the increased transparency of body tissues at NIR wavelengths.

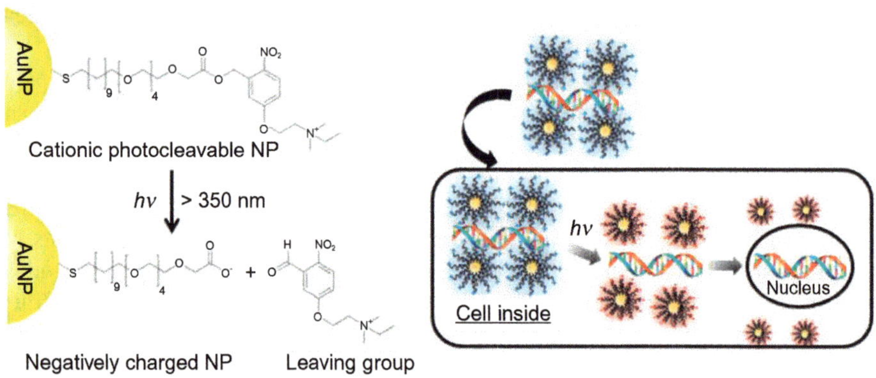

Figure 8.9 Schematic illustration of light-induced surface transformation of cationic photocleavable NPs and the release of DNA from the cationic photocleavable NPs (NP-PC)-DNA complex upon UV irradiation (>350 nm) within cell [89].

The Wang and El-Sayed group independently reported that AuNRs underwent shape transformation from rod to spherical NPs after absorption of NIR or IR light [90,91]. In 2006, Chen et al. applied this unique feature of AuNRs to remote control of localized gene expression [92]. They prepared AuNRs modified with a gene of enhanced green fluorescence protein (EGFP) for the remote control of gene expression in living HeLa cells. They successfully demonstrated that EGFP-AuNRs hybrids underwent shape transition, which resulted in the release of EGFP DNA, and the GFP expression was observed especially in cells that were locally exposed to NIR irradiation (Figure 8.10(a)). These results suggest the feasibility of co-using AuNRs and NIR irradiation to enable remote control of gene expression in specific cells. Here, the stability and gene delivery efficiency can be simultaneously enhanced by layer-by-layer coating of AuNRs. Huang and Rege et al. demonstrated the depositing polyelectrolyte multilayer composed of polystyrenesulfonate (PSS) and diglycidyl ether-polyamine based polymer onto CTAB-Au NRs enhanced the stability at least up to 1 month [93,94]. They carried out the delivery of exogenous plasmid DNA to human prostate cancer cells (PC3-PSMA) in subtoxic concentration range of coated AuNRs. Their designed coated AuNPs showed higher transfection efficiency and lower cytotoxic ability compared to that based on PEI. Therefore, their results suggested that judicious engineering of biocompatible polyelectrolytes leads to multifunctional AuNRs hybrids that possess high stability and low cytotoxicity with photothermal ablation, gene delivery, as well as optical imaging abilities on a single platform.

Barhoumi and Halsa et al. also reported light-induced release of DNA from Au nanoshell-based hybrids. This releasing mechanism is quite different than the photolabile AuNP system developed by Rotello's group. DNA dehybridization was induced by excitation of localized surface plasmons on the NP, relative to the thermal DNA dehybridization [95]. Therefore, the single-stranded DNA (ssDNA) was released from its surface when illuminated with plasmon-resonant light. On the other hand, Braun et al. reported a laser-activated gene silencing system using siRNA-Au nanoshell [96]. They demonstrated the pulsed NIR laser triggered release of siRNA from gold

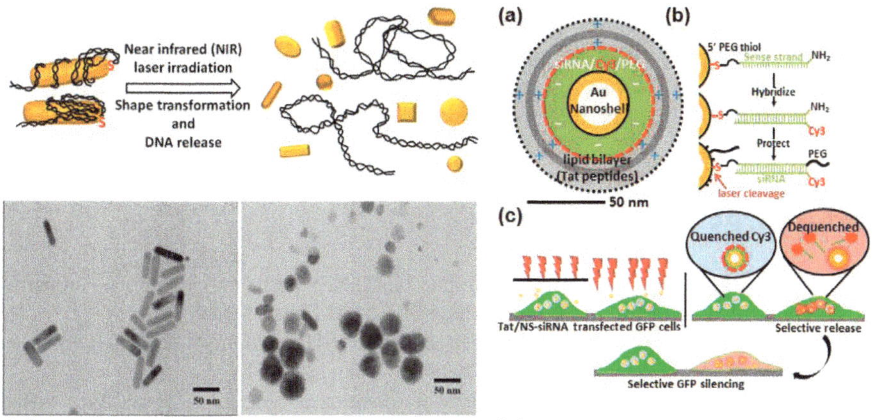

(a) (b)

Figure 8.10 (a) Schematic illustration of NIR-responsive AuNRs. After laser irradiation, the EGFP-AuNR conjugates undergo shape transformation that resulted in the release of EGFP DNA [92]. (b) (a) Diagram of Tat-lipid-coated nanoshell – siRNA used for transfection and selective release of siRNA. (b) The sense strand is composed of 5′ HS-$(CH_2)_6$-PEG_6-RNA-$(CH_2)_6$-NH_2, which is coated onto Au, hybridized, then backfilled with HS-PEG-NH_2, HS-$(CH_2)_3SO_3$, amines reacted with NHS-PEG_7-OCH_3. The 3′ Cy3 on the 27 base antisense strand is quenched due to proximity to gold. (c) Scheme of gene knock-down by laser irradiation. Transfection using the Tat-lipid allows uptake of nanoshell – siRNA, followed by patterned exposure to release siRNA into the endosome volume or into the cytoplasm, and tracked using Cy3 dequenching upon release. After 1-2 days, laser-dependent GFP silencing is observed [96].

nanoshells. They also functionalized Au nanoshells with cell-penetrating peptide (TAT) to enhance the cellular uptake. NIR laser-induced release from Au nanoshells occurred through bond cleavage of surface linker of siRNA. It was found to be power and time dependent, while the escape of siRNA from endosome occurred above a critical pulse energy was attributed to local heating and cavitation. They succeeded not only temporal but also spatial control over the siRNA delivery into mouse cells (C-166) by combining with photomasking approach (Figure 8.10(b)).

8.5 Summary

Although gene therapy is becoming one of the most powerful tools available to improve the human condition, the effectiveness and efficiency of gene delivery are not yet adequate because precise control over *in vivo* biodistribution and disease responsive release has been difficult to achieve. Therefore, it is highly desirable to have nanoparticles of definite/prechosen sizes and shapes. From this perspective, gold nanoparticles are proving to be useful delivery vehicles for gene therapy application. The attractive features of gold nanoparticles include their surface plasmon resonance, the controlled manner in which they interact with thiol groups, and their non-toxic nature. In particular, the combination of gold nanoparticles and laser irradiation to control the release of genes has the potential to provide useful therapeutic benefits. Since an ongoing challenge is to identify new transformations with the requisite qualities of targeting ability and biocompatibility, novel strategies for surface design of gold nanoparticles will facilitate the creation of new class of gene vehicles in the future.

References

[1] Li N, Zhao P, Astruc D. Anisotropic Gold Nanoparticles: Synthesis, Properties, Applications, and Toxicity. Angew Chem Int Ed 2014;53:1756–89.
[2] Chen S, Wang ZL, Ballato J, Foulger SH, Carroll DL. Monopod, Bipod, Tripod, and Tetrapod Gold Nanocrystals. J Am Chem Soc 2003;125:16186–7.
[3] Jana NR, Gearheart L, Murphy CJ. Wet Chemical Synthesis of High Aspect Ratio Cylindrical Gold Nanorods. J Phys Chem B 2001;105:4065–7.
[4] Jana NR, Gearheart L, Murphy CJ. Seed-Mediated Growth Approach for Shape-Controlled Synthesis of Spheroidal and Rod-like Gold Nanoparticles Using a Surfactant Template. Adv Mater 2001;13:1389–93.
[5] Murphy CJ, Sau TK, Gole AM, Orendorff CJ, Gao J, Gou L, et al. Anisotropic Metal Nanoparticles: Synthesis, Assembly, and Optical Applications. J Phys Chem B 2005;109:13857–70.
[6] Nikoobakht B, El-Sayed MA. Preparation and Growth Mechanism of Gold Nanorods [NRs] Using Seed-Mediated Growth Method. Chem Mater 2003;15:1957–62.
[7] Svedberg T. The Formation of Colloids. J. & A. Churchill; 1921.
[8] Yu YY, Chang SS, Lee CL, Wang CRC. Gold nanorods: Electrochemical synthesis and optical properties. J Phys Chem B 1997;101:6661–4.
[9] Huang CJ, Chiu PH, Wang YH, Chen KL, Linn JJ, Yang CF. Electrochemically controlling the size of gold nanoparticles. J Electrochem Soc 2006;153:D193–8.
[10] Liu X-y, Cui C-y, Cheng Y-w, Ma H-y, Liu D. Shape control technology during electrochemical synthesis of gold nanoparticles. Int J Miner Metall Mater 2013;20:486–92.
[11] Esumi K, Matsuhisa K, Torigoe K. Preparation of Rodlike Gold Particles by UV Irradiation Using Cationic Micelles as a Template. Langmuir 1995;11:3285–7.
[12] Leontidis E, Kleitou K, Kyprianidou-Leodidou T, Bekiari V, Lianos P. Gold Colloids from Cationic Surfactant Solutions. 1. Mechanisms That Control Particle Morphology. Langmuir 2002;18:3659–68.
[13] Miranda OR, Ahmadi TS. Effects of Intensity and Energy of CW UV Light on the Growth of Gold Nanorods. J Phys Chem B 2005;109:15724–34.
[14] Miranda OR, Dollahon NR, Ahmadi TS. Critical Concentrations and Role of Ascorbic Acid [Vitamin C] in the Crystallization of Gold Nanorods within Hexadecyltrimethyl Ammonium Bromide [CTAB]/Tetraoctyl Ammonium Bromide [TOAB] Micelles. Cryst Growth Des 2006;6:2747–53.
[15] Jana NR. Gram-Scale Synthesis of Soluble, Near-Monodisperse Gold Nanorods and Other Anisotropic Nanoparticles. Small 2005;1:875–82.
[16] Zijlstra P, Bullen C, Chon JWM, Gu M. High-Temperature Seedless Synthesis of Gold Nanorods. J Phys Chem B 2006;110:19315–8.
[17] Niidome Y, Nishioka K, Kawasaki H, Yamada S. Rapid synthesis of gold nanorods by the combination of chemical reduction and photoirradiation processes; morphological changes depending on the growing processes. Chem Commun 2003;2376–2377:.
[18] McGilvray KL, Decan MR, Wang D, Scaiano JC. Facile Photochemical Synthesis of Unprotected Aqueous Gold Nanoparticles. J Am Chem Soc 2006;128:15980–1.

[19] Marin ML, McGilvray KL, Scaiano JC. Photochemical Strategies for the Synthesis of Gold Nanoparticles from Au[III] and Au[I] Using Photoinduced Free Radical Generation. J Am Chem Soc 2008;130:16572–84.
[20] Nishioka K, Niidome Y, Yamada S. Photochemical Reactions of Ketones to Synthesize Gold Nanorods. Langmuir 2007;23:10353–6.
[21] Eustis S, Hsu H-Y, El-Sayed MA. Gold Nanoparticle Formation from Photochemical Reduction of Au3+ by Continuous Excitation in Colloidal Solutions. A Proposed Molecular Mechanism. J PhysChem B 2005;109:4811–5.
[22] Huang X, Qi X, Huang Y, Li S, Xue C, Gan CL, et al. Photochemically Controlled Synthesis of Anisotropic Au Nanostructures: Platelet-like Au Nanorods and Six-Star Au Nanoparticles. ACS Nano 2010;4:6196–202.
[23] Suslick KS. Sonochemistry. Science 1990;247:1439–45.
[24] Pol VG, Motiei M, Gedanken A, Calderon-Moreno J, Mastai Y. Sonochemical Deposition of Air-Stable Iron Nanoparticles on Monodispersed Carbon Spherules. Chem Mater 2003;15:1378–84.
[25] Gao T, Li Q, Wang T. Sonochemical Synthesis, Optical Properties, and Electrical Properties of Core/Shell-Type ZnO Nanorod/CdS Nanoparticle Composites. Chem Mater 2005;17:887–92.
[26] Barbour K, Ashokkumar M, Caruso RA, Grieser F. Sonochemistry and Sonoluminescence in Aqueous AuCl4- Solutions in the Presence of Surface-Active Solutes. J Phys Chem B 1999;103:9231–6.
[27] Qiu X-F, Zhu J-J, Chen H-Y. Controllable synthesis of nanocrystalline gold assembled whiskery structures via sonochemical route. J Cryst Growth 2003;257:378–83.
[28] Okitsu K, Mizukoshi Y, Bandow H, Yamamoto TA, Nagata Y, Maeda Y. Synthesis of Palladium Nanoparticles with Interstitial Carbon by Sonochemical Reduction of Tetrachloropalladate[II] in Aqueous Solution. J Phys Chem B 1997;101:5470–2.
[29] Belloni J, Mostafavi M, Remita H, Marignier J-L. Marie-Odile Delcourt, a Radiation-induced synthesis of mono- and multi-metallic clusters and nanocolloids. New J Chem 1998;22:1239–55.
[30] Zhang J, Du J, Han B, Liu Z, Jiang T, Zhang Z. Sonochemical Formation of Single-Crystalline Gold Nanobelts. Angew Chem Int Ed 2006;45:1116–9.
[31] Pylaev TE, Khanadeev VA, Khlebtsov BN, Dykman LA, Bogatyrev VA, Khlebtsov NG. Colorimetric and dynamic light scattering detection of DNA sequences by using positively charged gold nanospheres: a comparative study with gold nanorods. Nanotechnology 2011;22:285501.
[32] Liu, Guyot-Sionnest P. Mechanism of Silver[I]-Assisted Growth of Gold Nanorods and Bipyramids. J Phys Chem B 2005;109:22192–200.
[33] Kim F, Connor S, Song H, Kuykendall T, Yang P. Platonic Gold Nanocrystals. Angew Chem Int Ed 2004;43:3673–7.
[34] Bukasov R, Shumaker-Parry JS. Highly Tunable Infrared Extinction Properties of Gold Nanocrescents. Nano Lett 2007;7:1113–8.
[35] Khlebtsov NG, Dykman LA. Optical properties and biomedical applications of plasmonic nanoparticles. J Quant Spectrosc Radiat Transfer 2010;111:1–35.
[36] Wang H, Brandl DW, Le F, Nordlander P, Halas NJ. Nanorice: A Hybrid Plasmonic Nanostructure. Nano Lett 2006;6:827–32.
[37] Yang M, Yang X, Huai L. Synthesis and characterizations of hollow spheres and nano-spheres of Au. Appl Phys A 2008;92:367–70.
[38] Ye J, Van Dorpe P, Van Roy W, Borghs G, Maes G. Fabrication, Characterization, and Optical Properties of Gold Nanobowl Submonolayer Structures. Langmuir 1822–1827;2009:25.

[39] Sanchez-Gaytan BL, Park S-J. Spiky Gold Nanoshells. Langmuir 2010;26:19170–4.

[40] Khlebtsov BN, Khanadeev VA, Maksimova IL, Terentyuk GS, Khlebtsov NG. Silver nanocubes and gold nanocages: Fabrication and optical and photothermal properties. Nanotechnol Russia 2010;5:454–68.

[41] Khalil IA, Kogure K, Akita H, Harashima H. Uptake Pathways and Subsequent Intracellular Trafficking in Non-viral Gene Delivery. Pharmacol Rev 2006;58:32–45.

[42] Chithrani BD, Ghazani AA, Chan WCW. Determining the Size and Shape Dependence of Gold Nanoparticle Uptake into Mammalian Cells. Nano Lett 2006;6:662–8.

[43] Osaki F, Kanamori T, Sando S, Sera T, Aoyama Y. A Quantum Dot Conjugated Sugar Ball and Its Cellular Uptake. On the Size Effects of Endocytosis in the Subviral Region. J Am Chem Soc 2004;126:6520–1.

[44] Rausch K, Reuter A, Fischer K, Schmidt M. Evaluation of Nanoparticle Aggregation in Human Blood Serum. Biomacromolecules 2010;11:2836–9.

[45] van Landeghem FKH, Maier-Hauff K, Jordan A, Hoffmann K-T, Gneveckow U, Scholz R, et al. Post-mortem studies in glioblastoma patients treated with thermotherapy using magnetic nanoparticles. Biomaterials 2009;30:52–7.

[46] Albanese A, Chan WCW. Effect of Gold Nanoparticle Aggregation on Cell Uptake and Toxicity. ACS Nano 2011;5:5478–89.

[47] Sabella S, Brunetti V, Vecchio G, Galeone A, Maiorano G, Cingolani R, et al. Toxicity of citrate-capped AuNPs: an *in vitro* and *in vivo* assessment. J Nanopart Res 2011;13:6821–35.

[48] Coulter JA, Jain S, Butterworth KT, Taggart LE, Dickson GR, McMahon SJ, et al. Cell type-dependent uptake, localization, and cytotoxicity of 1.9 nm gold nanoparticles. Int J Nanomedicine 2012;7:2673–85.

[49] Lerner DM, Deutsch JM, Oster GF. How does a virus bud? Biophys J 1993;65:73–9.

[50] Gao H, Shi W, Freund LB. Mechanics of receptor-mediated endocytosis. P Natl Acad Sci USA 2005;102:9469–74.

[51] Zhang S, Li J, Lykotrafitis G, Bao G, Suresh S. Size-Dependent Endocytosis of Nanoparticles. Adv Mater 2009;21:419–24.

[52] Shan Y, Ma S, Nie L, Shang X, Hao X, Tang Z, et al. Size-dependent endocytosis of single gold nanoparticles. Chem Commun 2011;47:8091–3.

[53] Hao X, Wu J, Shan Y, Cai M, Shang X, Jiang J, et al. Caveolae-mediated endocytosis of biocompatible gold nanoparticles in living Hela cells. J Phys Condens Matter 2012;24:164–207.

[54] Dykman LA, Khlebtsov NG. Uptake of Engineered Gold Nanoparticles into Mammalian Cells. Chem Rev 2013;114:1258–88.

[55] Bartczak D, Muskens OL, Nitti S, Sanchez-Elsner T, Millar TM, Kanaras AG. Interactions of Human Endothelial Cells with Gold Nanoparticles of Different Morphologies. Small 2012;8:122–30.

[56] Schaeublin NM, Braydich-Stolle LK, Maurer EI, Park K, MacCuspie RI, Afrooz ARMN, et al. Does Shape Matter? Bioeffects of Gold Nanomaterials in a Human Skin Cell Model. Langmuir 2012;28:3248–58.

[57] Hutter E, Boridy S, Labrecque S, Lalancette-Hébert M, Kriz J, Winnik FM, et al. Microglial Response to Gold Nanoparticles. ACS Nano 2010;4:2595–606.

[58] Cho EC, Zhang Q, Xia Y. The effect of sedimentation and diffusion on cellular uptake of gold nanoparticles. Nat Nano 2011;6:385–91.

[59] Chithrani BD, Chan WCW. Elucidating the Mechanism of Cellular Uptake and Removal of Protein-Coated Gold Nanoparticles of Different Sizes and Shapes. Nano Lett 2007;7:1542–50.

[60] Jiang W, Kim BYS, Rutka JT, Chan WCW. Nanoparticle-mediated cellular response is size-dependent. Nat Nano 2008;3:145–50.

[61] Kobayashi K, Niikura K, Takeuchi C, Sekiguchi S, Ninomiya T, Hagiwara K, et al. Enhanced cellular uptake of amphiphilic gold nanoparticles with ester functionality. Chem Commun 2014;50:1265–7.
[62] Pillai PP, Huda S, Kowalczyk B, Grzybowski BA. Controlled pH Stability and Adjustable Cellular Uptake of Mixed-Charge Nanoparticles. J Am Chem Soc 2013;135:6392–5.
[63] Sundaram P, Xiao W, Brandsma JL. Particle-Mediated Delivery of Recombinant Expression Vectors To Rabbit Skin Induces High-Titered Polyclonal Antisera [and Circumvents Purification of a Protein Immunogen]. Nucleic Acids Res 1996;24:1375–7.
[64] McIntosh CM, Esposito EA, Boal AK, Simard JM, Martin CT, Rotello VM. Inhibition of DNA Transcription Using Cationic Mixed Monolayer Protected Gold Clusters. J Am Chem Soc 2001;123:7626–9.
[65] Han G, Martin CT, Rotello VM. Stability of Gold Nanoparticle-Bound DNA toward Biological, Physical, and Chemical Agents. Chem Biol Drug Des 2006;67:78–82.
[66] Han G, Chari NS, Verma A, Hong R, Martin CT, Rotello VM. Controlled Recovery of the Transcription of Nanoparticle-Bound DNA by Intracellular Concentrations of Glutathione. Bioconjugate Chem 2005;16:1356–9.
[67] Sandhu KK, McIntosh CM, Simard JM, Smith SW, Rotello VM. Gold Nanoparticle-Mediated Transfection of Mammalian Cells. Bioconjugate Chem 2001;13:3–6.
[68] Thomas M, Klibanov AM. Conjugation to gold nanoparticles enhances polyethylenimine's transfer of plasmid DNA into mammalian cells. Proc Natl Acad Sci 2003;100:9138–43.
[69] Thomas M, Klibanov AM. Enhancing polyethylenimine's delivery of plasmid DNA into mammalian cells. Proc Natl Acad Sci 2002;99:14640–5.
[70] Ow Sullivan MM, Green JJ, Przybycien TM. Development of a novel gene delivery scaffold utilizing colloidal gold-polyethylenimine conjugates for DNA condensation. Gene Ther 2003;10:1882–90.
[71] Elbakry A, Zaky A, Liebl R, Rachel R, Goepferich A, Breunig M. Layer-by-Layer Assembled Gold Nanoparticles for siRNA Delivery. Nano Lett 2009;9:2059–64.
[72] Lee Y, Miyata K, Oba M, Ishii T, Fukushima S, Han M, et al. Charge-Conversion Ternary Polyplex with Endosome Disruption Moiety: A Technique for Efficient and Safe Gene Delivery. Angew Chem Int Ed 2008;47:5163–6.
[73] Xu P, Van Kirk EA, Zhan Y, Murdoch WJ, Radosz M, Shen Y. Targeted Charge-Reversal Nanoparticles for Nuclear Drug Delivery. Angew Chem Int Ed 2007;46:4999–5002.
[74] Guo S, Huang Y, Jiang Q, Sun Y, Deng L, Liang Z, et al. Enhanced Gene Delivery and siRNA Silencing by Gold Nanoparticles Coated with Charge-Reversal Polyelectrolyte. ACS Nano 2010;4:5505–11.
[75] Liu X, Zhang J, Lynn DM. Polyelectrolyte multilayers fabricated from 'charge-shifting' anionic polymers: a new approach to controlled film disruption and the release of cationic agents from surfaces. Soft Matter 2008;4:1688–95.
[76] Wurster E-C, Elbakry A, Göpferich A, Breunig M. Layer-by-Layer Assembled Gold Nanoparticles for the Delivery of Nucleic Acids. In: Ogris M, Oupicky D, editors. Nanotechnology for Nucleic Acid Delivery: 948. Humana Press; 2013. p. 171–82.
[77] Oishi M, Nakaogami J, Ishii T, Nagasaki Y. Smart PEGylated Gold Nanoparticles for the Cytoplasmic Delivery of siRNA to Induce Enhanced Gene Silencing. Chem Lett 2006;35:1046–7.
[78] Rosi NL, Giljohann DA, Thaxton CS, Lytton-Jean AKR, Han MS, Mirkin CA. Oligonucleotide-Modified Gold Nanoparticles for Intracellular Gene Regulation. Science 2006;312:1027–30.

[79] Giljohann DA, Seferos DS, Patel PC, Millstone JE, Rosi NL, Mirkin CA. Oligonucleotide Loading Determines Cellular Uptake of DNA-Modified Gold Nanoparticles. Nano Lett 2007;7:3818–21.
[80] Giljohann DA, Seferos DS, Prigodich AE, Patel PC, Mirkin CA. Gene Regulation with Polyvalent siRNA – Nanoparticle Conjugates. J Am Chem Soc 2009;131:2072–3.
[81] Zhang K, Cutler JI, Zhang J, Zheng D, Auyeung E, Mirkin CA. Nanopod Formation through Gold Nanoparticle Templated and Catalyzed Cross-linking of Polymers Bearing Pendant Propargyl Ethers. J Am Chem Soc 2010;132:15151–3.
[82] Cutler JI, Zhang K, Zheng D, Auyeung E, Prigodich AE, Mirkin CA. Polyvalent Nucleic Acid Nanostructures. J Am Chem Soc 2011;133:9254–7.
[83] Gehl J. Electroporation: theory and methods, perspectives for drug delivery, gene therapy and research. Acta Physiol Scand 2003;177:437–47.
[84] Budker V, Budker T, Zhang G, Subbotin V, Loomis A, Wolff JA. Hypothesis: naked plasmid DNA is taken up by cells *in vivo* by a receptor-mediated process. J Gene Med 2000;2:76–88.
[85] Yang NS, Burkholder J, Roberts B, Martinell B, McCabe D. *In vivo* and *in vitro* gene transfer to mammalian somatic cells by particle bombardment. Proc Natl Acad Sci 1990;87:9568–72.
[86] Jen C-P, Chen Y-H, Fan C-S, Yeh C-S, Lin Y-C, Shieh D-B, et al. A Non-viral Transfection Approach in vitro: The Design of a Gold Nanoparticle Vector Joint with Microelectromechanical Systems. Langmuir 2004;20:1369–74.
[87] Prabaharan M, Grailer JJ, Pilla S, Steeber DA, Gong S. Gold nanoparticles with a monolayer of doxorubicin-conjugated amphiphilic block copolymer for tumor-targeted drug delivery. Biomaterials 2009;30:6065–75.
[88] Aryal S, Grailer JJ, Pilla S, Steeber DA, Gong S. Doxorubicin conjugated gold nanoparticles as water-soluble and pH-responsive anticancer drug nanocarriers. J Mater Chem 2009;19:7879–84.
[89] Han G, You C-C, Kim B-j, Turingan RS, Forbes NS, Martin CT, et al. Light-Regulated Release of DNA and Its Delivery to Nuclei by Means of Photolabile Gold Nanoparticles. Angew Chem Int Ed 2006;45:3165–9.
[90] Chang S-S, Shih C-W, Chen C-D, Lai W-C, Wang CRC. The Shape Transition of Gold Nanorods. Langmuir 1998;15:701–9.
[91] Link S, Wang ZL, El-Sayed MA. How Does a Gold Nanorod Melt?#. J Phys Chem B 2000;104:7867–70.
[92] Chen C-C, Lin Y-P, Wang C-W, Tzeng H-C, Wu C-H, Chen Y-C, et al. DNA – Gold Nanorod Conjugates for Remote Control of Localized Gene Expression by near Infrared Irradiation. J Am Chem Soc 2006;128:3709–15.
[93] Huang H-C, Barua S, Kay DB, Rege K. Simultaneous Enhancement of Photothermal Stability and Gene Delivery Efficacy of Gold Nanorods Using Polyelectrolytes. ACS Nano 2009;3:2941–52.
[94] Barua S, Joshi A, Banerjee A, Matthews D, Sharfstein ST, Cramer SM, et al. Parallel Synthesis and Screening of Polymers for Non-viral Gene Delivery. Mol Pharm 2008;6:86–97.
[95] Barhoumi A, Huschka R, Bardhan R, Knight MW, Halas NJ. Light-induced release of DNA from plasmon-resonant nanoparticles: Towards light-controlled gene therapy. Chem Phys Lett 2009;482:171–9.
[96] Braun GB, Pallaoro A, Wu G, Missirlis D, Zasadzinski JA, Tirrell M, et al. Laser-Activated Gene Silencing via Gold Nanoshell – siRNA Conjugates. ACS Nano 2009;3:2007–15.
[97] Orendorff CJ, Murphy CJ. Quantitation of Metal Content in the Silver-Assisted Growth of Gold Nanorods. J Phys Chem B 2006;110:3990–4.
[98] Dykman L, Khlebtsov N. Gold nanoparticles in biomedical applications: recent advances and perspectives. Chem Soc Rev 2012;41:2256–82.

Fabrication and development of magnetic particles for gene therapy

9

S. Uthaman[1], M. Muthiah[1], I.-K. Park[1], C.-S. Cho[2]
[1]Department of Biomedical Science and BK21 PLUS Center for Creative Biomedical Scientists, Chonnam National University Medical School, Republic of Korea; [2]Department of Agricultural Biotechnology and Research Institute for Agriculture and Life Sciences, Seoul National University, Seoul Republic of Korea

9.1 Introduction

Gene therapy involves the treatment of a genetic disease by the insertion of one or more corrective coding sequences into the genetic material of patient's cells. The expression of the new gene(s) can then alter the transcription of deoxyribonucleic acid (DNA) into ribonucleic acid (RNA) to express specific proteins for disease treatment. The success of gene delivery depends on the appropriate vehicle to carry the therapeutic cargo (DNA or RNA) from the serum proteins [1], and to exhibit excellent tissue-targeting capability and high cell specificity (to avoid toxicity to other cells) [2]. Moreover, the external genes must be efficiently released from their encapsulating particles within specific intracellular compartments. The recent advances in nanotechnology have enabled researchers to prepare novel nanovehicles as excellent gene delivery vehicles. Compared with virus-based systems, these nanovehicles possess highly tunable surface properties, size, and composition. Moreover, a variety of nanovehicles have been synthesized using lipids, polymers, and inorganic materials. Among these, polymer- or lipid-based nanovehicles are best suited for delivering therapeutic genes, due to their high reproducibility, facile manufacturing, easy functional modification, and non-oncogenesis [3].

Currently, there are three techniques for delivering genes into suitable host tissues or cells: viral vectors, non-viral vectors, and electroporation. While viral vectors are highly efficient (80–90%), they have distinct drawbacks in integrating nucleic acid sequences into the host genome, possibly leading to unexpected outcomes such as inadequate expression of the gene of interest or immunoreactivity. Another very efficient method for introducing foreign genes into host cells is electroporation. Even though a transfection rate of 50–70% can be achieved by this method, more than half of the recipient cells die from electric stimulation. On the other hand, while the use of transfection reagents might be preferable for clinical use, due to high cell viability (80–90%), these reagents do not effectively (i.e., only 20–30%) transfect host cells.

Magnetic particles of sizes 5–500 nm can be easily manipulated under external magnetic fields. Such particles usually consist of materials with magnetic properties such

Polymers and Nanomaterials for Gene Therapy. http://dx.doi.org/10.1016/B978-0-08-100520-0.00009-6
© 2016 Elsevier Ltd. All rights reserved.

as iron, cobalt, nickel, and other chemical compounds. From the early 1960s, among metal oxides, iron oxides have been especially used for magnetic separation. In 1978, Ogushi and co-workers [4] reported for the first time that iron oxide could shorten the T2 relaxation time (spin-spin relaxation time) of water and thereafter, iron oxides have been extensively pursued as magnetic resonance (MR) imaging contrast agents. Upon application of a magnetic field, the reactivity of the ferromagnetic material described by the hysteresis loop is characterized by two parameters, remanence and coercivity. It was previously found that as the particle size is reduced, coercivity reaches a maximum and then decreases toward zero. When the coercivity becomes zero, the particles are referred to as super-paramagnetic materials, which get magnetized by a magnetic field and then demagnitized upon removal of the magnetic field [5]. This special property of super-paramagnetic nanomaterial provides a unique advantage for working in biological systems, as upon the introduction into living systems, the particles are active only in the presence of an external magnetic field. Magnetic particles have been extensively used for a long time as a contrast-enhancing agent for MRI [6], and have also been applied for gene therapy. However, the surface of the particles must be modified to develop magnetic particles as effective gene carriers, through electrostatic interactions between the particle surface and therapeutic molecule, or by conjugation of a cleavable linker. Recently, the potential applications of magnetic particles for gene delivery have been widely explored, due to their unique characteristics. This chapter specifically emphasizes RNAi-based gene delivery using magnetic particles and in particular, short interfering RNAs (siRNAs) and microRNAs (miRNAs).

9.2 Synthesis of magnetic nanoparticles

There are currently several methods for synthesizing magnetic nanoparticles. Some of them involve one step, while others involve a multistep synthesis procedure. However, none of these procedures provides a universal solution for the preparation of all types of magnetic particles because each method has its own advantages and disadvantages. Most of these synthesis procedures use basic inorganic chemistry, especially iron chemistry, with the magnetic core material consisting of magnetite, maghemite, iron-based metal oxides, and iron alloys. Commonly used methods for the synthesis of magnetic particles are discussed in the following sections.

9.2.1 Precipitation

Precipitation of iron oxides is a relatively simple chemical method for the identification of iron (II or III) ions in an aqueous solution. In general, there are two types of precipitation methods involved, namely, wet precipitation and co-precipitation. In wet precipitation, one of the oldest methods for the preparation of magnetic particles [7], small-sized particles are prepared by carefully controlling the pH of an iron salt solution. Iron oxide particles (Fe_3O_4) can also be prepared by the co-precipitation method [8], in which a stoichiometric solution of two Fe^{2+} and Fe^{3+} ions is mixed together, followed by the addition of a base. However, these methods also have drawbacks, in that the obtained particle sizes depend upon pH, and broader particle sizes are generally obtained through wet precipitation. Since a large quantity of water is involved in particle synthesis, process scale-up is very difficult. Moreover, oxidation of the iron (II) precursor must also be avoided for successful synthesis of Fe_3O_4. Finally, it is virtually impossible to simultaneously precipitate along with having a protective coating over the magnetic particles, since pH control is a very delicate process.

9.2.2 Reverse micelles

One of the classic examples of surfactant chemistry is the formation of micelles [9], which are generally made at a critical micellular concentration of surfactant molecule having a hydrophilic head and a long hydrophobic tail. While normal micelles are generally formed in aqueous medium, reverse micelles are formed in an oily medium such as a hexane. Iron-oxide-based magnetic particles are prepared by slowly adding an inorganic precursor of iron (III) chloride, dissolved in aqueous medium, to the oily medium, followed by the addition of pH regulators [10–14].

Magnetic particles prepared by a reverse micelle method usually have homogeneous sizes, and inorganic coating material can also be added to the micelles during synthesis. Hence, with this method of synthesis, the size of the particles can be controlled, and an inorganic coating to the particles can also be provided [15–20]. Even though this synthesis method aids in controlling size during the development of the magnetic nanoparticles, there are also disadvantages. First, it is extremely difficult to coat the magnetic particles with an organic coating because the unused monomers remain in the organic phase of the micellar solution. Second, the size of the prepared particles is entirely dependent on the size of the micelles, since it is not possible to prepare particles outside the size range of the micelles, which is 20–500 nm [21–26]. Third, it is extremely difficult to scale up synthesis, due to the large amount of organic solvents used during micelle preparation.

9.2.3 Thermal decomposition

Recent advances in the fabrication of semiconductor and metallic nanocrystals have paved the way for the synthesis of high-quality and mono-dispersed metal oxide particles, via the decomposition of organometallic complexes at high temperature [27]. Among iron-oxide particles containing organometallic complexes, ferric and ferrous fatty acid complexes are widely used because of their low preparation cost, low toxicity, and easily scaled-up synthesis for mass production [28–31]. Although this synthesis method is very popular in industry (because no solvent is involved in the synthesis), a drawback of this technique is the difficulty in controlling size.

9.2.4 Liquid phase reduction

The liquid phase reduction method is usually applied in the reduction of magnetic or non-magnetic metal oxides to magnetic metal (or metal alloys) by powerful reducing agents, such as $NaBH_4$ and $LiAlH_4$. $NaBH_4$ is widely used as a reducing agent because it is soluble in both water and methanol [32–35]. Even though these hydrides are moisture-sensitive and difficult to handle, liquid phase reduction has some advantages over other synthesis methods due to these hydrides being very strong in action under mild conditions, and they can also penetrate into coating polymers.

9.3 Surface modification of magnetic particles by polymers

The ideal polymers for surface modification of particles should be biocompatible, have high affinity for metal oxides, and should not be immunogenic. Once the particles enter the physiological environment, they interact with hydrophobic surfaces, resulting in aggregation due to hydrophobic-hydrophobic interactions via opsonization [36] and immediate clearance by the mononuclear phagocytic system (MPS). However, if hydrophobic particles are coated with hydrophilic polymers, interactions with plasma proteins can be prevented, resulting in increased *in vivo* circulation (due to reduced uptake by the MPS) [37]; this also allows the synthesis of multifunctional particles by further conjugation of drugs, genes, or imaging agents. To that end, surface modification of magnetic particles by natural polymers (proteins and polycarbohydrate) [38–45], and synthetic organic polymers (polyethylene glycol, polyvinyl alcohol, poly-L-lactic acid, polyethylenimine) [46–68], for specific gene therapy applications, has now been achieved. The surface coating of the magnetic particles also prevents agglomeration, cytotoxicity, and also aids in the addition of functional groups.

9.3.1 Chitosan

Chitosan, in abundance from sea sources, is known to be biocompatible, hydrophilic, biodegradable, non-antigenic, and non-toxic [69]. Chitosan can form complexes with metal oxide surfaces, due to the presence of functional groups, thereby making the resulting particles hydrophilic, biocompatible, and stable. The cationic property of chitosan, due to the presence of positively charged amino groups, facilitates its interaction with negatively charged nucleic acids. Furthermore, chitosan can also facilitate particle movement across cellular barriers and between open tight junctions of epithelial cells [70,71]. Moreover, the presence of free hydroxyl and amino groups in chitosan are suitable for surface modification of superparamagnetic iron oxide nanoparticles (SPIONS) [47–50,57,72–75].

9.3.2 Dextran

Dextran is another successfully used polysaccharide-based polymer used for various *in vivo* applications [76,77]. Dextran-coated SPIONS are commercially used as clinical contrast agents for MRI, and they have also been shown to possess cancer nodal staging capabilities [78]. The stability and functionality of dextran-coated magnetic particles can be improved by the introduction of various groups such as carboxymethyl cross-linked to epichlorohydrin to make cross-linked iron oxide particles (CLIOs), which are more stable than dextran-coated SPIONS of the same size [79].

9.3.3 Poly (ethylene glycol) (PEG)

PEG is a condensation polymer of ethylene oxide and water with the general formula of $H\text{-}(O\text{-}CH_2\text{-}CH_2)_n\text{-}OH$. One of the common features of PEG appears to be its higher solubility in water. Moreover, it is non-toxic, odorless, and pH-neutral, allowing many applications in pharmaceutical and medical fields such as solvent dispersing, drug delivery, etc. PEG can be easily attached to other molecules (irrespective of their size) due to the presence of reactive OH groups at both its ends. It has also been widely used to enhance the solubility of hydrophobic imaging agents. PEG coating also helps increase circulation times of particles by minimizing their uptake by the reticuloendothelial system (RES) and immunogenicity, in addition to better tumor accumulation (due to enhanced permeability and retention) [36] and minimal non-specific interactions within the body. PEG coated on particles can also act as good spacers for the attachment of various biomolecules such as ligand-like proteins or antibodies, facilitating better targeting capability of the particles and their more specific accumulation within regions of interest, thus sparing normal (i.e., untargeted) cells [80]. For example, Mahmoudi et al. have synthesized SPIONS coated with cross-linked poly (ethylene glycol-co-fumarate) (PEGF) for *in vivo* imaging and drug delivery [81].

9.3.4 Poly (vinyl alcohol) (PVA)

PVA is another biocompatible and water-soluble material used for coating SPIONS. Fink et al. prepared various PVA-coated SPIONS having carboxyl (CM), thiol, and amino-modifications and tested their properties in human melanoma cells, in which it was observed that the iron content was always below the detection limit in cells exposed to PVA-SPIONS, CM-PVA SPIONS, or thiol-modified PVA SPIONS, as compared to the amino-SPIONS [82]. Specifically, cellular iron content increased after 24 h of continuous exposure, depending on the amount of amino-SPIONS. Hence, the presence of amino groups in PVA is necessary for improving the cellular uptake of SPIONS, and PVA can also resist protein adsorption and cell adhesion.

9.3.5 Poly (vinyl pyrrolidone) (PVP)

PVP is yet another biocompatible, hydrophilic neutrally charged polymer that has been used in various biomedical applications. PVP-coated SPIONS have been mostly achieved by covalent bonding to increase their stability in physiological medium [83]. Moreover, PVP-coated SPIONS synthesized by surface-initiated radical polymerization [84], showed increased and stable water dispensability for several months, enabling longer survival without separation from the contrast agent, even under *in vivo* conditions.

9.3.6 Polyethylenimine (PEI)

Cationic polyethylenimine (PEI) has been used as a potential siRNA carrier, due to its ability to interact with negatively charged molecules to form polyionic complexes [85–87]. Formation of polyplexes helps protect siRNA from enzymatic degradation and increases intracellular uptake via endocytosis. Branched PEI has primary, secondary, and tertiary amine groups with different pKa values. When the polymer is in an acidic endosomal environment, an influx of ions occurs, due to the protonation of amines and a lowering of osmotic potential [88]. However, the use of this polymer in clinical settings has been limited, due to higher cytotoxicity and non-biodegradability. To reduce cytotoxicity, many researchers have modified primary amino groups of PEI with other functional moieties such as acetyl and carboxylic groups [89]. Based on these properties, PEI-coated magnetic particles have been widely used as gene vectors [51–56,58–63,90].

9.4 Magnetic particles for gene therapy

9.4.1 Magnectofection-based gene delivery

For gene therapy applications, magnetic particles are generally used for increasing the transfection efficiencies of cultured cells, a technique known as magnetofection [91–104] in which magnetic particles and nucleic acids are mixed together and then added to the cell culture media. The nucleic acid-bound magnetic particles then move from the media to the cell surface upon the application of an external magnetic force, as shown in Figure 9.1. The principle advantage of this approach is the rapid sedimentation of the gene-therapeutic agent onto the target area, thereby reducing the time and dose of vector to achieve highly efficient transfection, with lower cell cytotoxicity.

In *in vivo* magentofection, the magnetic field is focused over the target site. This method has the potential not only to enhance transfection efficiency but also to target the therapeutic gene to a specific organ or site, as shown in Figure 9.2.

Generally, magnetic particles carrying therapeutic genes are injected intravenously. As the particles flow through the bloodstream, they are captured at the target site using very strong, high-gradient external magnets. Once they are captured, the magnetic particles carrying the therapeutic gene are taken up by the tissue, followed by release of the gene via enzymatic cleavage of cross-linked molecules or degradation of the polymer matrix. If DNA is embedded inside or within the coating material, the magnetic field must be applied to heat the particles and release the gene from the magnetic carrier [105].

9.4.2 Fabrication of magnetic nanoparticles for nucleic acid delivery

To fabricate magnetic particles for gene therapy applications, the negatively charged phosphate groups of the DNA may be attached by electrostatic interaction with positively charged molecules linked to the particle's surface. One such positively charged

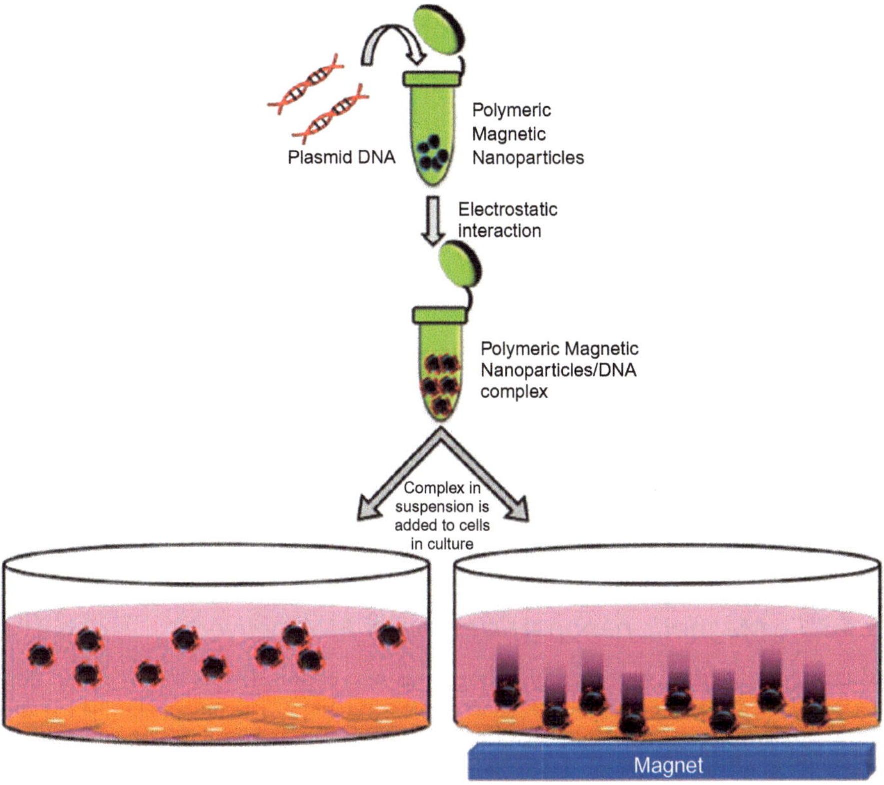

Figure 9.1 *In vitro* gene delivery using magnectofection.

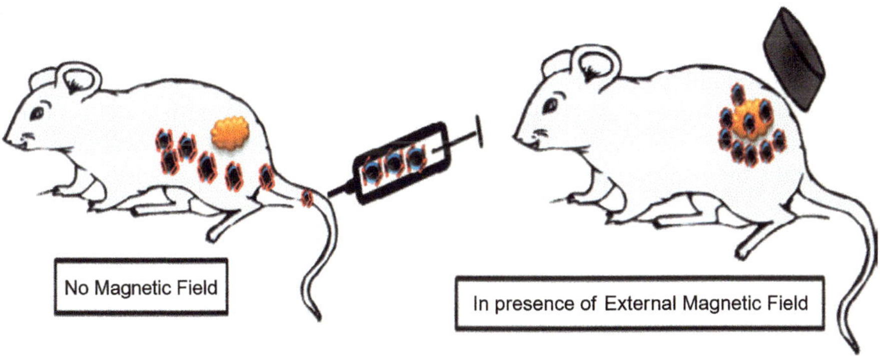

Figure 9.2 *In vivo* gene delivery using magnectofection.

molecule is PEI [106], which binds and condenses DNA due to the presence of a large number of amine groups on its chain. PEI also facilitates release of the therapeutic agent following internalization [107]. Moreover, PEI-coated magnetic particles were the first example of *in vitro* magnetic nanoparticle-mediated non-viral gene delivery [108], and they have been widely used as a coating material during the synthesis of magnetic particles for gene therapy applications [51–56,58–63,90]. The mechanism of successful transfection of PEI particles is due to the proton's "sponge effect," and its endosome-buffering activity also protects DNA from lysosomal nucleases before its transportation to the nucleus (through nuclear pores) for transgene expression [109].

Our group has also fabricated magnetic nanoparticle-embedded polymeric micelles containing PEI for tumor-specific plasmid DNA delivery [110]. Here, magnetic nanoparticle-loaded micelles, through a passive targeting mechanism, accumulated in colon cancer xenograft tumors following intravenous injection, successfully demonstrating *in vivo* delivery of genetic material.

9.4.3 Fabrication of magnetic nanoparticles for RNA delivery

RNA interference (RNAi), the silencing of gene expression by double-stranded RNA molecules (dsRNA), was first reported in 1998 by Andrew Fire and colleagues in the nematode worm *C. elegans* [111]. When a cell encounters a dsRNA transgene, an RNAi response is triggered. The two types of RNA molecules that are a central part of this mechanism are microRNAs (19–25 nucleotides long) and dsRNAs (21–22 nucleotides long) [112].

9.4.3.1 Magnetic nanoparticles for mi RNA delivery

MicroRNAs (miRNAs) are small endogenous non-coding RNAs that regulate transcriptional and post-translational gene expression in all organisms [113,114] and have gained much interest because they are conserved across all animal species as modulators of critical pathways and processes [115,116]. To develop miRNAs as a therapeutic strategy, anti-sense inhibitors can be delivered to regulate mature miRNA over-expression [117], restoring the miRNA-overexpressing cells to their normal condition [118]. However, targeted delivery of RNA-based therapeutics still remains a major challenge in cancer therapy. Anna et al. [119] developed a technique to efficiently deliver miRNAs into hMSCs using PEI bound to iron-oxide magnetic particles (MNPs), while Gomes et al. [120] synthesized a biodegradable particle (NP) containing perfluoro-1,5-crown ether (PFCE) that was able to track cells *in vivo* and release the miRNA by magnetic resonance imaging (MRI). Here, nanoparticles complexed to miRNA accumulate within the cell's endolysosomal compartment to interact with proteins that allow their biological action.

9.4.3.2 Magnetic nanoparticles for siRNA delivery

Under the influence of an external magnetic field, the magnetic properties of nanoparticles enhance their cellular uptake. For example, Sandy et al. reported the synthesis of large (11 nm) pore PEI-modified (denoted PEI-Fe-LPMSN) silica particles (200 nm in size), loaded with iron-oxide carriers, for gene delivery [121]. Such large pore sizes play a major role in the magnetic and gene delivery capabilities of nanoparticles. To enhance nanocarrier affinity for siRNA, PEI grafted on a silica surface effectively delivered siRNA-PLK1-containing PEI-Fe-LPMSNs to osteosarcoma cancer cells, inhibiting cell viability by 80%. PEI-coated SPIONS, composed of a core of iron oxide and a shell of PEI, were also reported to systemically deliver therapeutic siRNA to arthritic joints. These siRNA-loaded PEI-SPIONS displayed negligible cytotoxicity, improved siRNA stability, efficient uptake by macrophages, and the ability to specifically silence genes *in vitro*. Moreover, PEI-SPIONS delivered siRNA accumulated easily in inflamed joints, being efficiently taken up by joint macrophages and T-cells. Among other non-viral vectors, bio-reducible poly (amido amine) (rPAA) has attracted much attention as an effective siRNA carrier, due to its endosomal/lysosomal escape capacity (proton "sponge effect") and ability to disassemble. More importantly, particles made of rPAA polyplexes were stable in oxidizing extracellular environments, shielding their loaded siRNAs from enzymatic degradation in serum, although polyplex dissociation, triggered by cleavage of its disulphide-bonds, resulted in siRNA release [122]. Another strategy for siRNA delivery, reported by Medarova et al., [123], used magnetic particles first coated with the amine group-containing polymer dextran. Using traditional coupling chemistry, a peptide (myristolated polyarginine) and Cy5.5 were then conjugated to serve as a membrane translocation module and optical imaging agent, respectively, and the remaining unused amino groups accessed for siRNA attachment.

Another strategy for siRNA delivery is via a cleavable linkage to a magnetic nanoparticle. For example, Lee et al. [124] attached a siRNA to magnetic nanoparticles by disulphide bonds. Here, multifunctional magnetic nanoparticle were first coated with bovine serum albumin (BSA) whose primary groups were then converted to pyridyldisulfide by treating with N-succinmidyl-3-(2-pyridyldithio)propionate SPDP. The resulting particles were then treated with thiolated PEG functionalized with RGD peptide and Cy5-labeled thiolated siRNA, resulting in its attachment to the surface of the particles via disulphide bonds that are readily cleaved in the intracellular environment.

9.5 Conclusions and future perspectives

The fabrication of polymeric magnetic particles as carriers for gene therapy has recently gained importance. Polymer-based magnetic particles aids in the better incorporation of the therapeutic agent and a high content of available functional groups that can be utilized for further bio conjugation with cell-targeting agents. Polymer coating of magnetic particles increases their bio-compatibility, stability, and *in vivo* circulation. Moreover, for specific targeting and imaging, magnetofection-based particles are demonstrated promising candidates. Finally, these agents can fulfill a dual role as imaging agents and therapeutic agents within a single delivery vehicle. Even though polymeric magnetic particles appear to hold significant potential for targeted gene delivery, there remain many obstacles to be overcome before employing this technique in the clinic. For example, while studies in small animals showed great potential for particle targeting in larger animals and humans, it is much more difficult to target sites in areas that are farther from the magnetic source. Yet another obstacle faced by currently available polymeric magnetic particles used for gene delivery is that they are not strong enough to counter-balance blood flow to allow better accumulation at their target sites. Future research should be focused on upgrading the use of magnetic particles in other areas such as cell nanoarticulation and nanomodulation. While magnetic nanoparticle targeting is not likely to be effective in all disease scenarios, with further development it is possible for it to become an effective treatment tool for a variety of diseases.

References

[1] Niidome T, Huang L. Gene therapy progress and prospects: nonviral vectors. Gene Ther 2002;9:1647–52.

[2] Wang Z, Liu G, Zheng H, Chen X. Rigid nanoparticle-based delivery of anti-cancer siRNA: challenges and opportunities. Biotechnol Adv 2013.

[3] Park TG, Jeong JH, Kim SW. Current status of polymeric gene delivery systems. Adv Drug Deliv Rev 2006;58:467–86.

[4] Ohgushi M, Nagayama K, Wada A. Dextran-magnetite: a new relaxation reagent and its application to T2 measurements in gel systems. J Magn Reson 1978;29:599–601.

[5] Akbarzadeh A, Samiei M, Davaran S. Magnetic nanoparticles: preparation, physical properties, and applications in biomedicine. Nanoscale Res Lett 2012;7:1–13.

[6] Pankhurst QA, Connolly J, Jones S, Dobson J. Applications of magnetic particlein biomedicine. J Phys D Appl Phys 2003;36:R167.

[7] Li L, Fan M, Brown RC, Van Leeuwen J, Wang J, Wang W, et al. Synthesis, properties, and environmental applications of nanoscale iron-based materials: a review. Crit Rev Env Sci Tec 2006;36:405–31.

[8] Lu AH. Salabas EeL, Schüth F. Magnetic nanoparticles: synthesis, protection, functionalization, and application. Angew Chem Int Ed 2007;46:1222–44.

[9] Kennard R, DeSisto WJ, Giririjan TP, Mason MD. Intrinsic property measurement of surfactant-templated mesoporous silica films using time-resolved single-molecule imaging. J Chem Phys 2008;128.

[10] Shukla A, Patra MK, Mathew M, Songara S, Singh VK, Gowd GS, et al. Preparation and Characterization of Biocompatible and Water-Dispersible Superparamagnetic Iron Oxide Particle(SPIONs). Adv Sci Lett 2010;3:161–7.

[11] Bellusci M, La Barbera A, Seralessandri L, Padella F, Piozzi A, Varsano F. Preparation of albumin-ferrite superparamagnetic particleusing reverse micelles. Polym Int 2009; 58:1142–7.

[12] Duarte EL, Itri R, Lima E, Baptista MS, Berquo TS, Goya GF. Large magnetic anisotropy in ferrihydrite particlesynthesized from reverse micelles. Nanotechnology 2006;17:5549–55.

[13] Seip CT, Carpenter EE, O'Connor CJ, John VT, Li SC. Magnetic properties of a series of ferrite particlesynthesized in reverse micelles. Ieee T Magn 1998;34:1111–3.

[14] O'Connor CJ, Seip CT, Carpenter EE, Li SC, John VT. Synthesis and reactivity of nanophase ferrites in reverse micellar solutions. Nanostruct Mater 1999;12:65–70.

[15] Lewandowska-Lancucka J, Staszewska M, Szuwarzynski M, Kepczynski M, Romek M, Tokarz W, et al. Synthesis and characterization of the superparamagnetic iron oxide particlemodified with cationic chitosan and coated with silica shell. J Alloy Compd 2014;586:45–51.

[16] Du YM, Li L, Leung CW, Lai PT, Pong PWT. Synthesis and Characterization of Silica-Encapsulated Iron Oxide Nanoparticles. Ieee T Magn 2014;50.

[17] Sadjadi MS, Fathi F, Farhadyar N, Zare K. Synthesize and Characterization of Multifunctional Silica Coated Magnetic Particleusing Polyvinylpyrrolidone (PVP) as a mediator. J Nano Res-Sw 2011;16:43–8.
[18] Zhang M, Cushing BL, O'Connor CJ. Synthesis and characterization of monodisperse ultra-thin silica-coated magnetic nanoparticles. Nanotechnology 2008;19.
[19] Mohapatra S, Pal D, Ghosh SK, Pramanik P. Design of superparamagnetic iron oxide nanoparticle for purification of recombinant proteins. J Nanosci Nanotechno 2007;7:3193–9.
[20] Santra S, Tapec R, Theodoropoulou N, Dobson J, Hebard A, Tan WH. Synthesis and characterization of silica-coated iron oxide particlein microemulsion: The effect of non-ionic surfactants. Langmuir 2001;17:2900–6.
[21] Namdeo M, Saxena S, Tankhiwale R, Bajpai M, Mohan YM, Bajpai SK. Magnetic Particlefor Drug Delivery Applications. J Nanosci Nanotechno 2008;8:3247–71.
[22] McBain SC, Yiu HHP, Dobson J. Magnetic particlefor gene and drug delivery. Int J Nanomed 2008;3:169–80.
[23] Arruebo M, Fernandez-Pacheco R, Ibarra MR, Santamaria J. Magnetic particlefor drug delivery. Nano Today 2007;2:22–32.
[24] Wang MD, Shin DM, Simons JW, Nie SM. Nanotechnology for targeted cancer therapy. Expert Rev Anticanc 2007;7:833–7.
[25] Nie SM, Xing Y, Kim GJ, Simons JW. Nanotechnology applications in cancer. Annu Rev Biomed Eng 2007;9:257–88.
[26] Scherer F, Anton M, Schillinger U, Henkel J, Bergemann C, Kruger A, et al. Magnetofection: enhancing and targeting gene delivery by magnetic force in vitro and in vivo. Gene Ther 2002;9:102–9.
[27] Lin MM, Kim H-H, Kim H, Muhammed M, Kim DK. Iron oxide-based nanomagnets in nanomedicine: fabrication and applications. Nano reviews 2010;1.
[28] William WY, Falkner JC, Yavuz CT, Colvin VL. Synthesis of monodisperse iron oxide nanocrystals by thermal decomposition of iron carboxylate salts. Chem Commun 2004;2306–7.
[29] Park J, An K, Hwang Y, Park J-G, Noh H-J, Kim J-Y, et al. Ultra-large-scale syntheses of monodisperse nanocrystals. Nat Mater 2004;3:891–5.
[30] Sun S, Zeng H. Size-controlled synthesis of magnetite nanoparticles. J Am Chem Soc 2002;124:8204–5.
[31] Hyeon T. Chemical synthesis of magnetic nanoparticles. Chem Commun 2003;927–34.
[32] Senthil M, Ramesh C. Biogenic Synthesis of Fe3O4 particles uisng Tridax Procumbens leaf extract and its antibacterial activity on Pseudomonas Aeruginoas. Dig J Nanomater Bios 2012;7:1655–60.
[33] Pan Y, Gao JH, Zhang B, Zhang XX, Xu B. Colloidosome-based Synthesis of a Multifunctional Nanostructure of Silver and Hollow Iron Oxide Nanoparticles. Langmuir 2010;26:4184–7.
[34] Gao SY, Shi YG, Zhang SX, Jiang K, Yang SX, Li ZD, et al. Biopolymer-assisted green synthesis of iron oxide particleand their magnetic properties. J Phys Chem C 2008;112:10398–401.
[35] Abu Bakar M, Tan WL, Abu Bakar NHH. A simple synthesis of size-reduce magnetite nano-crystals via aqueous to toluene phase-transfer method. J Magn Magn Mater 2007;314:1–6.
[36] Lee H, Lee E, Kim DK, Jang NK, Jeong YY, Jon S. Antibiofouling polymer-coated superparamagnetic iron oxide particleas potential magnetic resonance contrast agents for in vivo cancer imaging. J Am Chem Soc 2006;128:7383–9.

[37] Kim S, Lim CK, Na J, Lee YD, Kim K, Choi K, et al. Conjugated polymer particlefor biomedical in vivo imaging. Chem Commun 2010;46:1617–9.

[38] Lee SY, Lee S, Lee J, Lee HS, Chang JH. Biomimetic magnetic particlefor rapid hydrolysis of ester compounds. Mater Lett 2013;110:229–32.

[39] Erdal E, Kavaz D, Sam M, Demirbilek M, Demirbilek ME, Saglam N, et al. Preparation and Characterization of Magnetically Responsive Bacterial Polyester Based Nanospheres for Cancer Therapy. J Biomed Nanotechnol 2012;8:800–8.

[40] Sakudo A, Onodera T. Virus capture using anionic polymer-coated magnetic beads (Review). Int J Mol Med 2012;30:3–7.

[41] Zhou YX, Sharma N, Deshmukh P, Lakhman RK, Jain M, Kasi RM. Hierarchically Structured Free-Standing Hydrogels with Liquid Crystalline Domains and Magnetic Particleas Dual Physical Cross-Linkers. J Am Chem Soc 2012;134:1630–41.

[42] Pham HL, Do HM, Tran DL, Le VH, Nguyen XP, Nguyen AT, et al. Magnetic nanoparticles: study of magnetic heating and adsorption/desorption for biomedical and environmental applications. Int J Nanotechnol 2011;8:399–413.

[43] Zhang XL, Niu HY, Pan YY, Shi YL, Cai YQ. Chitosan-Coated Octadecyl-Functionalized Magnetite Nanoparticles: Preparation and Application in Extraction of Trace Pollutants from Environmental Water Samples. Anal Chem 2010;82:2363–71.

[44] Campbell RB, Ying B, Kuesters GM, Hemphill R. Fighting Cancer: From the Bench to Bedside Using Second Generation Cationic Liposomal Therapeutics. J Pharm Sci 2009;98:411–29.

[45] Yang LL, Cao ZH, Sajja HK, Mao H, Wang LY, Geng HY, et al. Development of Receptor Targeted Magnetic Iron Oxide Particlefor Efficient Drug Delivery and Tumor Imaging. J Biomed Nanotechnol 2008;4:439–49.

[46] Muthiah M, Hieu VQ, Kim YK, Rhee JH, Kang SH, Jun SY, et al. Mannose-poly(ethylene glycol)-linked SPION targeted to antigen presenting cells for magnetic resonance imaging on lymph node. Carbohyd Polym 2013;92:1586–95.

[47] Kievit FM, Stephen ZR, Veiseh O, Arami H, Wang TZ, Lai VP, et al. Targeting of Primary Breast Cancers and Metastases in a Transgenic Mouse Model Using Rationally Designed Multifunctional SPIONs. ACS Nano 2012;6:2591–601.

[48] Kievit FM, Veiseh O, Fang C, Bhattarai N, Lee D, Ellenbogen RG, et al. Chlorotoxin Labeled Magnetic Nanovectors for Targeted Gene Delivery to Glioma. ACS Nano 2010;4:4587–94.

[49] Veiseh O, Kievit FM, Fang C, Mu N, Jana S, Leung MC, et al. Chlorotoxin bound magnetic nanovector tailored for cancer cell targeting, imaging, and siRNA delivery. Biomaterials 2010;31:8032–42.

[50] Wang CY, Ravi S, Martinez GV, Chinnasamy V, Raulji P, Howell M, et al. Dual-purpose magnetic micelles for MRI and gene delivery. J Control Release 2012;163:82–92.

[51] He L, Feng LZ, Cheng L, Liu YM, Li ZW, Peng R, et al. Multilayer Dual-Polymer-Coated Upconversion Particlefor Multimodal Imaging and Serum-Enhanced Gene Delivery. Acs Appl Mater Inter 2013;5:10381–8.

[52] Liu G, Xie J, Zhang F, Wang ZY, Luo K, Zhu L, et al. N-Alkyl-PEI-Functionalized Iron Oxide Nanoclusters for Efficient siRNA Delivery. Small 2011;7:2742–9.

[53] Kami D, Takeda S, Makino H, Toyoda M, Itakura Y, Gojo S, et al. Efficient transfection method using deacylated polyethylenimine-coated magnetic nanoparticles. J Artif Organs 2011;14:215–22.

[54] Veiseh O, Kievit FM, Mok H, Ayesh J, Clark C, Fang C, et al. Cell transcytosing polyarginine coated magnetic nanovector for safe and effective siRNA delivery. Biomaterials 2011;32:5717–25.

[55] Mok H, Veiseh O, Fang C, Kievit FM, Wang FY, Park JO, et al. pH-Sensitive siRNA Nanovector for Targeted Gene Silencing and Cytotoxic Effect in Cancer Cells. Mol Pharmaceut 2010;7:1930–9.
[56] Yiu HHP, McBain SC, Lethbridge ZAD, Lees MR, Dobson J. Preparation and characterization of polyethylenimine-coated Fe3O4-MCM-48 nanocomposite particles as a novel agent for magnet-assisted transfection. J Biomed Mater Res A 2010;92A:386–92.
[57] Kievit FM, Veiseh O, Bhattarai N, Fang C, Gunn JW, Lee D, et al. PEI-PEG-Chitosan-Copolymer-Coated Iron Oxide Particlefor Safe Gene Delivery: Synthesis, Complexation, and Transfection. Adv Funct Mater 2009;19:2244–51.
[58] Chen CB, Chen JY, Lee WC. Fast Transfection of Mammalian Cells Using Superparamagnetic ParticleUnder Strong Magnetic Field. J Nanosci Nanotechno 2009;9:2651–9.
[59] Wang XL, Zhou LZ, Ma YJ, Gu HC. Charged Magnetic Particlefor Enhancing Gene Transfection. Ieee T Nanotechnol 2009;8:142–7.
[60] Pan XG, Guan JJ, Yoo JW, Epstein AJ, Lee LJ, Lee RJ. Cationic lipid-coated magnetic particleassociated with transferrin for gene delivery. Int J Pharm 2008;358:263–70.
[61] Seo SB, Yang J, Hyung W, Cho EJ, Lee TI, Song YJ, et al. Novel multifunctional PHDCA/PEI nano-drug carriers for simultaneous magnetically targeted cancer therapy and diagnosis via magnetic resonance imaging. Nanotechnology 2007;18.
[62] Chorny M, Polyak B, Alferiev IS, Walsh K, Friedman G, Levy RJ. Magnetically driven plasmid DNA delivery with biodegradable polymeric nanoparticles. Faseb J 2007;21:2510–9.
[63] Xiang L, Bin W, Huali J, Wei J, Jiesheng T, Feng G, et al. Bacterial magnetic particles (BMPs)-PEI as a novel and efficient non-viral gene delivery system. J Gene Med 2007;9:679–90.
[64] Hong RY, Zhang SZ, Han YP, Li HZ, Ding J, Zheng Y. Preparation, characterization and application of bilayer surfactant-stabilized ferrofluids. Powder Technol 2006;170:1–11.
[65] Qiu XP, Winnik F. Preparation and characterization of PVA coated magnetic nanoparticles. Chinese J Polym Sci 2000;18:535–9.
[66] Iafisco M, Sandri M, Panseri S, Delgado-Lopez JM, Gomez-Morales J, Tampieri A. Magnetic Bioactive and Biodegradable Hollow Fe-Doped Hydroxyapatite Coated Poly(L-lactic) Acid Micro-nanospheres. Chem Mater 2013;25:2610–7.
[67] Hwang DW, Song IC, Lee DS, Kim S. Smart Magnetic Fluorescent Nanoparticle Imaging Probes to Monitor MicroRNAs. Small 2010;6:81–8.
[68] Ramanujan RV, Chong WT. The synthesis and characterization of polymer coated iron oxide microspheres. J Mater Sci-Mater M 2004;15:901–8.
[69] Shen CR, Wu ST, Tsai ZT, Wang JJ, Yen TC, Tsai JS, et al. Characterization of quaternized chitosan-stabilized iron oxide particleas a novel potential magnetic resonance imaging contrast agent for cell tracking. Polym Int 2011;60:945–50.
[70] Kuroiwa T, Noguchi Y, Nakajima M, Sato S, Mukataka S, Ichikawa S. Production of chitosan oligosaccharides using chitosanase immobilized on amylose-coated magnetic nanoparticles. Process Biochem 2008;43:62–9.
[71] Zhu LZ, Ma JW, Jia NQ, Zhao Y, Shen HB. Chitosan-coated magnetic particleas carriers of 5-Fluorouracil: Preparation, characterization and cytotoxicity studies. Colloid Surface B 2009;68:1–6.
[72] Qu JB, Shao HH, Jing GL, Huang F. PEG-chitosan-coated iron oxide particlewith high saturated magnetization as carriers of 10-hydroxycamptothecin: Preparation, characterization and cytotoxicity studies. Colloid Surface B 2013;102:37–44.

[73] Arami H, Stephen Z, Veiseh O, Zhang MQ. Chitosan-Coated Iron Oxide Particlefor Molecular Imaging and Drug Delivery. Adv Polym Sci 2011;243:163–84.
[74] Kumar A, Jena PK, Behera S, Lockey RF, Mohapatra S, Mohapatra S. Multifunctional magnetic particlefor targeted delivery. Nanomed-Nanotechnol 2010;6:64–9.
[75] Bhattarai SR, Kim SY, Jang KY, Lee KC, Yi HK, Lee DY, et al. Laboratory formulated magnetic particlefor enhancement of viral gene expression in suspension cell line. J Virol Methods 2008;147:213–8.
[76] Hsieh WJ, Liang CJ, Chieh JJ, Wang SH, Lai IR, Chen JH, et al. In vivo tumor targeting and imaging with anti-vascular endothelial growth factor antibody-conjugated dextran-coated iron oxide nanoparticles. Int J Nanomedicine 2012;7:2833–42.
[77] Anzai Y, McLachlan S, Morris M, Saxton R, Lufkin RB. Dextran-coated superparamagnetic iron oxide, an MR contrast agent for assessing lymph nodes in the head and neck. AJNR Am J Neuroradiol 1994;15:87–94.
[78] Tassa C, Shaw SY, Weissleder R. Dextran-Coated Iron Oxide Nanoparticles: A Versatile Platform for Targeted Molecular Imaging, Molecular Diagnostics, and Therapy. Accounts Chem Res 2011;44:842–52.
[79] Chen TJ, Cheng TH, Chen CY, Hsu SCN, Cheng TL, Liu GC, et al. Targeted Herceptin-dextran iron oxide particlefor noninvasive imaging of HER2/neu receptors using MRI. J Biol Inorg Chem 2009;14:253–60.
[80] Liu PF, Qin LB, Wang Q, Sun Y, Zhu MJ, Shen M, et al. cRGD-functionalized mPEG-PLGA-PLL particlefor imaging and therapy of breast cancer. Biomaterials 2012;33:6739–47.
[81] Mahmoudi M, Simchi A, Imani M, Hafeli UO. Superparamagnetic Iron Oxide Particlewith Rigid Cross-linked Polyethylene Glycol Fumarate Coating for Application in Imaging and Drug Delivery. J Phys Chem C 2009;113:8124–31.
[82] Petri-Fink A, Chastellain M, Juillerat-Jeanneret L, Ferrari A, Hofmann H. Development of functionalized superparamagnetic iron oxide particlefor interaction with human cancer cells. Biomaterials 2005;26:2685–94.
[83] Ding XY, Hong CJ, Liu Y, Gu ZL, Xing KL, Zhu AJ, et al. Pharmacokinetics, tissue distribution, and metabolites of a polyvinylpyrrolidone-coated norcantharidin chitosan nanoparticle formulation in rats and mice, using LC-MS/MS. Int J Nanomed 2012;7:1723–35.
[84] Arsalani N, Fattahi H, Nazarpoor M. Synthesis and characterization of PVP-functionalized superparamagnetic Fe3O4 particleas an MRI contrast agent. Express Polym Lett 2010;4:329–38.
[85] Mimi H, Ho KM, Siu YS, Wu AH, Li P. Polyethyleneimine-Based Core-Shell Nanogels: A Promising siRNA Carrier for Argininosuccinate Synthetase mRNA Knockdown in HeLa Cells. J Control Release 2012;158:123–30.
[86] Wiseman JW, Goddard CA, McLelland D, Colledge WH. A comparison of linear and branched polyethylenimine (PEI) with DCChol/DOPE liposomes for gene delivery to epithelial cells in vitro and in vivo. Gene Ther 2003;10:1654–62.
[87] Lee HJ, Namgung R, Kim WJ, Kim JI, Park I-K. Targeted delivery of microRNA-145 to metastatic breast cancer by peptide conjugated branched PEI gene carrier. Macromol Res 2013;21:1201–9.
[88] Yang S, May S. Release of cationic polymer-DNA complexes from the endosome: A theoretical investigation of the proton sponge hypothesis. J Chem Phys 2008;129.
[89] Breunig M, Lungwitz U, Liebl R, Goepferich A. Breaking up the correlation between efficacy and toxicity for nonviral gene delivery. P Natl Acad Sci USA 2007;104:14454–9.

[90] Veiseh O, Kievit FM, Liu V, Fang C, Stephen ZR, Ellenbogen RG, et al. In Vivo Safety Evaluation of Polyarginine Coated Magnetic Nanovectors. Mol Pharmaceut 2013;10:4099–106.
[91] Pardo J, Sosa YE, Reggiani P, Arciniegas ML, Sanchez FH, Goya RG. Magnetic field-assisted gene transfer: Studies in glial cells. Acta Bioquim Clin L 2013;47:399–406.
[92] Govindarajan S, Kitaura K, Takafuji M, Ihara H, Varadarajan KS, Patel AB, et al. Gene delivery into human cancer cells by cationic lipid-mediated magnetofection. Int J Pharm 2013;446:87–99.
[93] Yu J, Huang DY, Yousaf MZ, Hou YL, Gao S. Magnetic nanoparticle-based cancer therapy. Chinese Phys B 2013;22.
[94] Fouriki A, Dobson J. Nanomagnetic Gene Transfection for Non-Viral Gene Delivery in NIH 3T3 Mouse Embryonic Fibroblasts. Materials 2013;6:255–64.
[95] Plank C, Zelphati O, Mykhaylyk O. Magnetically enhanced nucleic acid delivery. Ten years of magnetofection-Progress and prospects. Adv Drug Deliv Rev 2011;63:1300–31.
[96] Delyagina E, Li WZ, Ma N, Steinhoff G. Magnetic targeting strategies in gene delivery. Nanomedicine 2011;6:1593–604.
[97] Ma YJ, Zhang ZJ, Wang XL, Xia WL, Gu HC. Insights into the mechanism of magnetofection using MNPs-PEI/pDNA/free PEI magnetofectins. Int J Pharm 2011;419:247–54.
[98] Ang D, Tay CY, Tan LP, Preiser PR, Ramanujan RV. In vitro studies of magnetically enhanced transfection in COS-7 cells. Mat Sci Eng C-Mater 2011;31:1445–57.
[99] Sanchez-Antequera Y, Mykhaylyk O, van Til NP, Cengizeroglu A, de Jong JH, Huston MW, et al. Magselectofection: an integrated method of nanomagnetic separation and genetic modification of target cells. Blood 2011;117:E171–81.
[100] Al-Deen FN, Ho J, Selomulya C, Ma C, Coppel R. Superparamagnetic Particlefor Effective Delivery of Malaria DNA Vaccine. Langmuir 2011;27:3703–12.
[101] Arsianti M, Lim M, Marquis CP, Amal R. Assembly of Polyethylenimine-Based Magnetic Iron Oxide Vectors: Insights into Gene Delivery. Langmuir 2010; 26:7314–26.
[102] Namgung R, Singha K, Yu MK, Jon S, Kim YS, Ahn Y, et al. Hybrid superparamagnetic iron oxide nanoparticle-branched polyethylenimine magnetoplexes for gene transfection of vascular endothelial cells. Biomaterials 2010;31:4204–13.
[103] Furlani EP. Magnetic Biotransport: Analysis and Applications. Materials 2010;3:2412–46.
[104] Sapet C, Laurent N, Le Gourrierec L, Augier S, Zelphati O. In vitro and in vivo Magnetofection (TM): a move towards gene therapy. Ann Biol Clin-Paris 2010; 68:133–42.
[105] Dobson J. Gene therapy progress and prospects: magnetic nanoparticle-based gene delivery. Gene Ther 2006;13:283–7.
[106] Abdallah B, Hassan A, Benoist C, Goula D, Behr JP, Demeneix BA. A powerful nonviral vector for in vivo gene transfer into the adult mammalian brain: Polyethylenimine. Hum Gene Ther 1996;7:1947–54.
[107] Akinc A, Thomas M, Klibanov AM, Langer R. Exploring polyethylenimine-mediated DNA transfection and the proton sponge hypothesis. J Gene Med 2005;7:657–63.
[108] Scherer F, Anton M, Schillinger U, Henke J, Bergemann C, Krüger A, et al. Magnetofection: enhancing and targeting gene delivery by magnetic force in vitro and in vivo. Gene Ther 2002;9:.
[109] Lee HJ, Nguyen YT, Muthiah M, Vu-Quang H, Namgung R, Kim WJ, et al. MR traceable delivery of p53 tumor suppressor gene by PEI-functionalized superparamagnetic iron oxide nanoparticles. J Biomed Nanotechnol 2012;8:361–71.

[110] Lee SJ, Muthiah M, Lee HJ, Lee HJ, Moon MJ, Che HL, et al. Synthesis and characterization of magnetic nanoparticle-embedded multi-functional polymeric micelles for MRI-guided gene delivery. Macromol Res 2012;20:188–96.
[111] Novina CD, Sharp PA. The RNAi revolution. Nature 2004;430:161–4.
[112] He L, Hannon GJ. Micrornas: Small RNAs with a big role in gene regulation. Nat Rev Genet 2004;5:522–31.
[113] Kim VN. MicroRNA biogenesis: Coordinated cropping and dicing. Nat Rev Mol Cell Biol 2005;6:376–85.
[114] Muthiah M, Park IK, Cho CS. Nanoparticle-mediated delivery of therapeutic genes: focus on mi RNA therapeutics. Expert Opin Drug Del 2013;10:1259–73.
[115] Allen E, Xie ZX, Gustafson AM, Carrington JC. microRNA-directed phasing during trans-acting siRNA biogenesis in plants. Cell 2005;121:207–21.
[116] Du TT, Zamore PD. microPrimer: The biogenesis and function of microRNA. Development 2005;132:4645–52.
[117] Uzumcu A, Tekguc M, Sevli S, Bayrak OF, Ozen M. The role of microrna (mi RNA)s as a novel tool in cancer diagnosis and therapy. IUBMB Life 2009;61: 298-.
[118] Shukla P, Lovren F, Pan Y, Quan A, Creighton A, Gupta N, et al. MicroRNA-145 Targeted Therapy Reduces Atherosclerosis and Improves Plaque Stability. Circulation 2011;124.
[119] Schade A, Delyagina E, Scharfenberg D, Skorska A, Lux C, David R, et al. Innovative Strategy for MicroRNA Delivery in Human Mesenchymal Stem Cells via Magnetic Nanoparticles. Int J Mol Sci 2013;14:10710–26.
[120] Gomes R, das Neves R, Cochlin L, Lima A, Carvalho R, Korpisalo P, et al. Efficient pro-survival/angiogenic mi RNA delivery by an MRI-detectable nanomaterial. ACS Nano 2013;7:3362–72.
[121] Hartono SB, Yu MH, Gu WY, Yang J, Strounina E, Wang XL, et al. Synthesis of multi-functional large pore mesoporous silica particleas gene carriers. Nanotechnology 2014;25.
[122] Piest M, Lin C, Mateos-Timoneda M, Lok M, Hennink W, Feijen J, et al. Novel poly(amido amine)s with bioreducible disulfide linkages in their diamino-units: structure effects and in vitro gene transfer properties. Journal of controlled release: official journal of the Controlled Release Society 2008;130:38–45.
[123] Medarova Z, Pham W, Farrar C, Petkova V, Moore A. In vivo imaging of siRNA delivery and silencing in tumors. Nat Med 2007;13:372–7.
[124] Lee JH, Lee K, Moon SH, Lee Y, Park TG, Cheon J. All-in-One Target-Cell-Specific Magnetic Particlefor Simultaneous Molecular Imaging and siRNA Delivery. Angew Chem Int Edit 2009;48:4174–9.

Gene therapies in clinical trials

10

K.I. Papadopoulos[1], *P. Wattanaarsakit*[2], *W. Prasongchean*[2], *R. Narain*[3]
[1]THAI StemLife, Bangkok, Thailand; [2]Faculty of Pharmaceutical Sciences, Chulalongkorn University, Bangkok, Thailand; [3]University of Alberta, Edmonton, AB, Canada

10.1 Introduction

Almost a century passed between the discovery of "nuclein" by the Swiss physician Johannes Friedrich Miescher in 1869 [1] and the discovery of the first correct double-helix model of DNA structure by James Watson and Francis Crick in 1953 [2]. Fifty years later, in 2003, the Human Genome Project (HGP) was completed and the human genome fully sequenced in 13 years, at a cost of $US 3.8 billion that ultimately generated an economic output of $US796 billion [3]. From an exorbitant cost of $US100 million in 2001, the cost of sequencing a full human genome is approaching $1,000 today [4]. In 2012, a mere 9 years after HGP completion, the first gene therapy drug was approved for human use [5]. While this inverse exponential development between genomic costs and research is continuing, life and health sciences and related disciplines are being revolutionized. A new era of quantitative and personalized medicine will most certainly dominate conventional medicine and every biological aspect of life, ranging from nutrition, exercise, environmental cues, and disease prevention, diagnosis, and treatment, all of which will be custom based on each person's genetic predisposition [6].

Genes have long been regarded as medicines [7] and in that sense genetic engineering has been exploited in the mass production of human insulin, human growth hormone, human follicle stimulating hormone, human albumin, other proteins (hemophilia clotting factors), monoclonal antibodies, and vaccines [7]. By making use of genetic engineering in humans gene therapy is a promising strategy for the cure or treatment of several inherited and acquired diseases. The specifics of the techniques involved in genetic bioengineering such as gene delivery have been dealt with extensively [8]. In the present chapter, we delve into the current and some of the future aspects of clinical applications of gene therapy in humans.

10.2 Gene therapy

Of the approximately 25,000 genes that comprise the human genome, more than 100,000 mutations have been reported up to 2011 [9]. Human disease can have diverse etiologies, ranging from pure hereditary single gene disorders (more than 2/3 of all known

Polymers and Nanomaterials for Gene Therapy. http://dx.doi.org/10.1016/B978-0-08-100520-0.00010-2
© 2016 Elsevier Ltd. All rights reserved.

diseases are monogenic, reported in more than 6,000 diseases affecting 12,000 genes), to complex etiology combining a genetic (usually polygenic) background and environmental interactions [9]. Almost 10 years after the HGP, more than 2,850 Mendelian disease genes have been identified [10]. Genome wide association studies (GWAS) have spurred a huge interest in the hypothesis-free exploration of the "genetic architecture of disease" and heritability of common diseases ranges now from 20–25% to exceeding 60% in some cases [10]. Genetics are now implicated in the origins of any disease, while environmental, malignant, and infectious predispositions show clear genetic variability. Cancers, heart disease, psychiatric disorders, and response to pharmacological treatment are dramatically diverse across genomes [10]. To date, GWAS have identified approximately 2,000 genetic loci with robust associations for more than 300 common traits and polygenic diseases, while more than 230 cancer genes identified in solid tumors are revealing new biological mechanisms and important therapeutic leads [10].

Considering the above background, gene-based therapies have the potential to alleviate, treat, or cure human disease. Diseases that have the potential to be targeted by gene therapy include monogenic disorders, cancers, infectious diseases, cardiovascular disease, neurological conditions, ocular diseases, and autoimmune conditions.

Gene therapy is made up of three components: the therapeutic gene, the vector that delivers it, and the mode of administration (Table 10.1 and Figure 10.1).

Gene-based therapy focuses on finding ways to introduce corrective genes into somatic (and germline) cells in order to: (i) correct a cell malfunction due to monogenic etiology loss of function by replacing a defective mutant gene or augmenting its product with a functional version **(gene replacement/augmentation)**; (ii) add a new corrective function to a cell by adding useful genes **(gene addition)** for complex disorders, cancer, or infectious diseases; (iii) target the host genome to alter a disease causing sequence with a preventing/corrective sequence **(gene editing)**; and/or (iv) interfere with gain of function conditions to block a harmful gene product from causing a disease **(gene expression modulation)**.

10.3 Gene replacement/augmentation

In the prototype of gene therapy, monogenic diseases served as a simple pattern where the therapeutic gene was designed to induce the expression of a functioning gene to replace a defective mutant one [11]. Most of the successes in clinical gene therapy are in this category and the defective gene is replaced/augmented *via* viral vector gene transfer either directly *in vivo* or *ex vivo* using umbilical cord or other types of hematopoietic stem cells (HSC) [11]. The first gene therapy-based drug alipogene tiparvovec (Glybera®), using adeno-associated virus (AAV) vectors engineered to express lipoprotein lipase in the muscle for the treatment of familial lipoprotein lipase (LPL) deficiency, is the first and to date only approved human gene therapy for clinical use in Europe [5]. It is regulated under European advanced therapy medicinal products (ATMPs) regulation. Remarkable success of viral vector gene augmentation therapy has been reported in human phase I/II clinical trials in Leber's Congenital Amaurosis (LCA) type II (due to mutations in the RPE65 gene) by three independent

Table 10.1 Essential components in gene therapy

Gene	Delivery/Vector		Administration	
TREATMENT AIM	Viral Oncogenicity due to insertional mutagenesis; Immune issues: neutralization, inactivation, inflammation; tropism; length of transfection	Non-viral Low cost, non-pathogenic, and non-immunogenic; Low transduction efficiency/questionable tropism	*In vivo* Directly administered in to the human body	*Ex vivo* Prepared outside human body
REPLACE/ AUGMENT	Genome of a virus is altered to contain the Therapeutic gene: Size of gene, immunogenicity, and oncogenicity are limiting factors	Natural or artificial Plasmid DNA/mRNA/siRNA/ miRNA that encode for a Gene: immunogenicity/ inactivation; low transgenic transcription/expression	Gene + Vector administered in tissue directly	Gene + Vector in cultured cells
ADD	Adenoviruses: dsDNA; 36 kb; no viral replication; no gene integration; quiescent and dividing cells	Nanoparticles (NP)/ Extracellullar Vesicles (EV)	Tissue specific delivery	Hematopoietic Stem Cells (HSC): bone marrow (BMSC), umbilical cord blood (UCB), or peripheral blood (PBSC)
EDIT	Lentivituses: ssRNA; persistent gene integration; <5 kb; quiescent and dividing cells	Lipoplexes/Polyplexes; Dendrimers/Carbon nanotubes/Magnetic NP	Intravenous (IV)	Mesenchymal Stromal Cells (MSC)
MODULATE	Retroviruses: ssRNA; persistent gene integration; 7–10 kb; replicating cells; HIV/SIV/FIV/EIAV		Intramuscular (IM)	Induced Pluripotent Stem Cells (iPSC)
COMBINED METHODS	Adeno Associated Viruses (AAV): ssDNA; 4.7 kb; stable, specific gene integration; can infect many quiescent and dividing cell types; low immunogenicity; no known human disease		Subcutaneous (SC)	Embryonic Stem Cells (ESC)

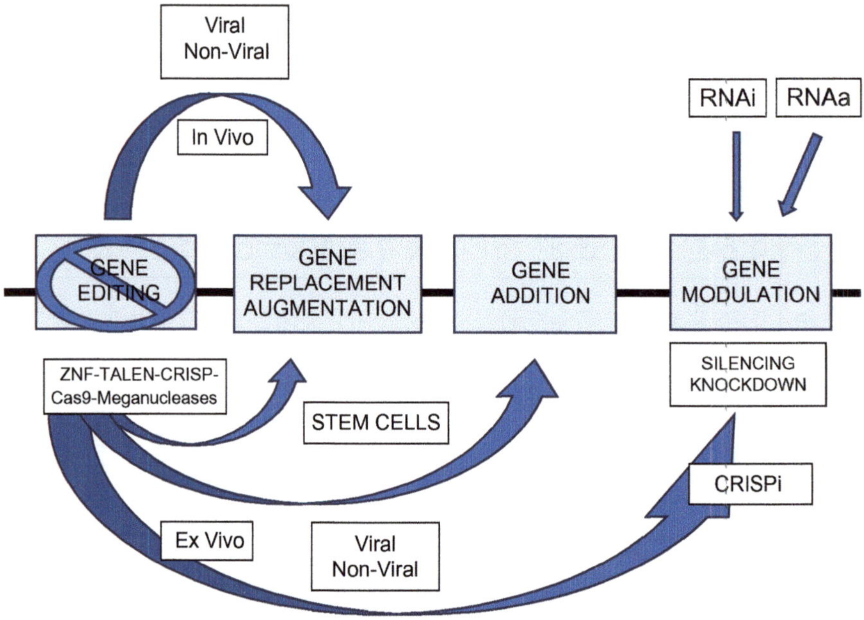

Figure 10.1 General gene therapy strategies. RNAi = RNA interference; RNAa = RNA activation; ZNF = zinc finger nucleases; TALEN = transcription activator-like effector nucleases; CRISP = clustered regularly interspaced short palindromic repeat; Cas9 = CRISP associated system; CRISPi = CRISP interference.

teams of investigators. This confirms that RPE65 can be delivered to retinal pigment epithelial cells of LCA patients by subretinal injections using recombinant AAV2/2 vectors (rAAV) resulting in clinical benefits without side effects [12]. The retina as a target organ with its defined anatomical confinement and its relative immune privileges make it suitable for application of viral vectors. Moreover, the initial results in Choroideremia, an X-linked recessive disorder of the retina and choroid that leads to blindness due to mutations in the *CHM* gene resulting in the loss of function of the Rab escort protein-1 (REP1), were recently reported and were consistent with improved rod and cone function after receiving retinal gene therapy with an AAV vector encoding REP1 (AAV REP1) [13]. These findings lent support to further assess gene therapy for the treatment of other retinal/macular degeneration diseases with vision loss, such as age-related macular degeneration (AMD), the leading cause of irreversible blindness in older adults of developed countries [14], Stardgardt macular degeneration, Usher syndrome type 1B, and retinal dystrophies. Two studies delivering a vascular endothelial growth factor (VEGF) antagonist in gene therapy, using both AAV vectors and the Flt-1 gene in different forms and administration sites, are ongoing [15]. AMD is being studied by delivering two angiostatic proteins (endostatin and angiostatin) by using an equine infectious anemia virus (EIAV)-based lentiviral gene therapy vector [16]. Stardgardt macular degeneration and Usher syndrome type 1B using EIAV-based lentiviral vector and appropriate genes encoding for ABCA4 and myosin VIIA (MYO7A), respectively, is also being studied [17]. The use of the EIAV virus is necessitated by the fact that the genes to be delivered exceed the 4.7 kilobase cargo capacity of the rAAV vectors. Another AAV2/2-based gene therapy, delivering the ND4 gene to reverse vision loss due to atrophy of optic nerve is ongoing [15]. Retinal dystrophies are now the subject of translational research after identification of more than 158 mutated genes involved in its pathology [18]. Targeted NGS (next-generation sequencing) based methods for molecular diagnosis along with vertebrate animal models offer the opportunity to study a number of rational therapies in cone dystrophy, cone rod dystrophy, LCA, macular dystrophy, and retinitis pigmentosa [18]. Recently discovered novel (over 90 different disease-causing) mutations in PRPH2 gene that encodes peripherin (a membrane glycoprotein important for the stabilization and compaction of photoreceptor) in outer segment discs may explain disease combinations [19] and pathogenesis. PRPH2 offers better treatment models in mice, as already observed with AAV and compacted DNA nanoparticles [20–22].

Newer non-viral delivery systems such as nanoparticles (cationic lipids and polymers forming lipoplexes and polyplexes) are also widely studied and could potentially overcome issues of oncogenicty, immunogenicity, gene size, and transduction and are less invasive [23–32]. The world's first induced pluripotent stem cell (iPSC) in Japanese patients to cure age-related macular degeneration is currently in progress [33].

Hemophilia A and B are X-linked recessive monogenic disorders resulting from deficiencies of factor VIII and FIX, respectively, where administration of purified clotting factor concentrates is currently the only non-curative treatment option. The conditions' monogenic nature makes them potentially amenable to gene therapy [34]. Most preclinical studies of hemophilia A gene therapy have focused on the use of viral vectors, such as adenovirus, adeno-associated virus (AAV), gamma retrovirus, and lentivirus but the large size of human Factor VIII gene and immunological reactions to viral vectors remain significant obstacles [35]. Recently, non-viral gene transfer technology using transposon vectors has shown promising results [35]. Only a minute increase in circulating FIX levels is needed for a marked clinical response, since the size of the FIX gene is small and is easy to insert in viral vectors. AAV vectors transfecting liver cells have shown promise in humans in hemophilia B [36, 37] but a lot more remains to be done in the future. Issues with immune-mediated toxicity are of concern as AAV clearly stimulate strong innate and subsequent adaptive immune responses, which can lead to both toxicity and loss of expression [36]. Recently, in 10 patients with severe hemophilia B, the infusion of a single dose of AAV8 vector resulted in long-term therapeutic factor IX expression associated with clinical improvement, without any late toxic effects reported in a follow-up period of up to 3 years [37].

Cystic Fibrosis (CF). Efficient and permanent transfer of the CFTR gene is needed in respiratory epithelium that has evolutionally developed efficient ways of fighting foreign body invasions. Viral vectors with respiratory epithelium tropism (AdV, cytoplasmic RNA viruses, lentiviruses) and non-viral modalities (cationic lipids/polymers) have been tried with varying success [38]. For newer developments in CF see the Gene Editing section.

Inherited blood disorders resulting in **Primary Immune Deficiencies (PID)** such as in X-linked severe combined immunodeficiency (X-SCID: bubble boy disease) [39, 40], adenosine deaminase (ADA) severe combined immunodeficiency (SCID) [40, 41], Wiskott-Aldrich syndrome (WAS) [40, 42], and chronic granulomatous disease (CGD) [40, 43] were among the initial targets for gene therapy trials [40]. Allogeneic hematopoietic stem cell transplantation (HSCT) using stem cell sources from the bone marrow, peripheral blood, or umbilical cord blood has been the mainstay of treatment in PID, but immunological complications such as graft vs. host reactions, and stem cell donor scarcity pose significant obstacles. Ideally, using autologous stem cells would bypass both barriers, so gene correction through *ex vivo* gene replacement/augmentation via retroviral vectors was tried initially [40]. Many of the initial setbacks (low efficiency gene delivery systems) and safety concerns (insertional mutagenesis near the LMO2 T-cell oncogene and leukoproliferative disorders in X-SCID; insertion of a retroviral vector near the EVI1-MDS1 proto-oncogenes leading to myeloprolifetative disorders in CGD) were unraveled during these trials [40], but successes have been reported as well (*ex vivo* retroviral transduction of HSC and subsequent HSCT in ADA-SCID), propelling PID gene therapy into the clinical sphere [39, 41]. Future directions include combination approaches using gene editing and *ex vivo* transfection of patient-specific induced pluripotent stem cells (iPSC) technology [43].

The potential of lentiviral vectors have been explored successfully in the past, first in X-linked adrenoleukodystrophy (X-ALD) and later in hemoglobinopathies.

Ex vivo lentiviral vector delivery of the ABCD1 gene to overcome X-ALD (a severe lipid storage disorder with brain demyelination), followed by autologous hematopoietic stem cell transplantation (HSCT), resulted in ALD protein expression [44, 45]. The same principle was repeated in Metachromatic leukodysptrophy [46], another example of autologous HSCT, where autologous CD34+ stem cells were collected from the bone marrow and were transduced *ex vivo* with a lentiviral vector encoding the human Arylsulfatase A (ARSA) cDNA (ClinicalTrials.gov Identifier: NCT01560182).

Beta (β)-thalassemia (β-thal) and **Sickle Cell Disease (SCD)** have both been or are being evaluated in gene replacement trials with varying successes. More than 300 mutations of the β-globin gene have been characterized all over the world, resulting in either absence of synthesis of β-globin chains (β^{o}-thalassemia) or a reduction in synthesis (β^{+}thalassemia) [47]. In β-thal and SCD, gene replacement therapy has been made possible through *ex vivo* manipulation of autologous hematopoietic stem cells (HSC) isolated from umbilical cord blood, bone marrow, or mobilized peripheral blood. Using viruses from the Retroviridae family, the necessary therapeutic genes are delivered into the host genome via retrovirus (RV), lentivirus (LV), and foamy virus (FV) vectors [48]. The manipulated cells are then transplanted back to the recipient using available hematopoietic stem cell transplantation [49]. The preferred vector to date is self-inactivating design LV vector (SIN-LV) that seems devoid of the initial safety concerns from insertional oncogenesis [50]. Successes in murine SCD models have been reported and human gene trials in SCD have recently begun in the United States [48].

The future holds a lot of combined approaches such as the use of newer and safer vectors as well as induced pluripotent stem cell (iPSC) technology in thalassemia [51].

10.4 Gene Addition

The gene addition principle adds a new corrective function to a cell by adding useful genes for complex disorders, such as cancer or infectious diseases [52]. Complex disorder heritability may vary from 20–25% up to 60–90% in some cases [10], while disease variability and response to pharmacological treatments are dramatically diverse across genomes [10]. To date, GWAS have identified approximately 2,000 genetic loci with robust associations for more than 300 common traits and polygenic diseases, while more than 230 cancer genes identified in solid tumors are revealing new biological mechanisms and important therapeutic leads [10].

While Glybera was the first gene therapy drug approved in Europe, Gendicine, an adenovirus (Ad) carrying the p53 tumor suppressor gene for the treatment of head and neck cancer, was the first worldwide gene treatment, albeit only approved in China [53].

Cancer. Almost 65% of all gene therapy trials focus on cancer, reflecting the need for effective cancer treatments [11]. Traditionally, autologous and allogeneic hematopoietic stem cell transplantations have been used in the past to treat hematological malignancies and to mitigate or rescue from the side effects of chemo- and radiotherapy in solid cancers [54]. T-cells enriched for anti-viral (EBV) specificity and tumor-infiltrating lymphocyte (TIL) infusions are also being used for cancer therapy [55].

Two recent strategies seem to have great potential in cancer treatment. First, the use of the suicide principle, i.e., transfecting tumor cells with suicide genes to make the tumor more susceptible to radiotherapy or chemotherapy or induce immune oncolysis [56]. Among the most studied are replication-competent herpes simplex virus (HSV) vectors delivering the thymidine kinase gene (HSV-TK) in combination with Ganciclovir used as a prodrug. Granulocyte macrophage colony stimulating factor (GM-CSF) [57] and cytosine deaminase (CD) gene of *E. coli* in combination with 5-flurocytosine (5-FC) [58] are other examples of drugs delivered by HSV-TK. Glioblastoma multiforme, neural stem cells (NSC) have been transduced *ex vivo* with Adv to express carboxyesterase-irinotecan or CD/5-FC, while mesenchymal stromal cells (MSC) expressing cytokines (such as IL-2, IL-18, IL-12 and INF-γ) resulted in increased immunity against tumor [58, 59]. HSV vectors that replicate specifically in actively dividing tumor cells have been used in Phase I-III human trials in patients with glioblastoma multiforme, a fatal form of brain cancer, and in malignant melanoma. In melanoma, T-VEC (talimogene laherparepvec, formerly known as OncoVex GM-CSF) displayed efficacy in a recent Phase III trial, when compared to standard GM-CSF treatment alone [60]. T-EVC may soon become the second FDA-approved gene therapy product used in standard patient care [57]. In head and neck cancers, MSC along with CD/5-FC/TNF-alpha inhibited tumor cell growth in animal models [58].

The second strategy is the induction of designer antigen specificity such as chimeric antigen receptors (CAR), where T-cell signaling domains are put together with fragments from antibodies specific to known surface antigens, cloned into an expression cassette, and subsequently *ex vivo* transfected to autologous T-cells [61]. For this approach to succeed, a potential CAR T-cell target must be expressed on a majority of cells in the majority of patients and be absent from normal tissues. An example of this approach is the use of CD19-specific CAR T-cell therapy in B-cell hematologic malignancies [62]. Multiple antigen targets are currently being evaluated in multiple myeloma, Hodgkin's disease, acute myeloid leukemia, melanomas, and gliomas [61]. Issues related to specificity and adverse events such as tumor lysis syndrome, cytokine storm, and healthy tissue innocent bystander effects will need to be addressed [63].

Heart Disease. Gene therapy provides great opportunities for treatment of cardiovascular disease, heart failure, coronary artery disease, myocardial infarction, intracranial atherosclerosis/stroke, arrhythmia, and peripheral vascular disease. Today, heart disease is the third most popular target in gene therapy trials [64]. Genes that seem intuitively appropriate to target are the ones involved in the pathogenesis of heart failure and induce beta adrenoreceptor dysregulation, and calcium handling such as the sarcoplasmic reticulum (SR) calcium ATPase 2a (SERCA2a) [64]. In heart failure (HF), intracoronary infusion of rAAV1-SERCA2A in the CUPID 1 trial showed encouraging results in improving cardiac function and reducing hospitalizations without any safety concerns after 3 years [65]. Laboratory studies have shown contractility improvement and ventricular arrhythmia reduction in animal models of HF [66]. The CUPID2 trial is currently ongoing [67] (ClinicalTrials.gov identifier: NCT01643330). Angiogenesis is being treated using both viral and non-viral delivery systems such as polymeric gene carriers [68]. The genes for angiogenic growth factors such as fibroblast growth factor (FGF), vascular endothelial growth factor (VEGF), and hepatocyte and platelet growth factors (HPDGF) show considerable efficacy, but are not approved commercially [64, 69]. In coronary artery disease (CAD), therapeutic angiogenesis increases collateral vessel formation and is used as an alternative to drug-eluting stents [64]. Another option to modulate CAD and arthrosclerosis could be through miRNA modulation [70]. Several miRNAs have been linked to cardiac hypertrophy, myocardial infarction, and atherosclerosis. Targeted miRNA delivery through nanoparticles could induce anti-angiogenesis within atherosclerotic plaques and prevent plaque rupture and stroke [71]. Moreover, newer ways to combat coronary re-stenosis including gene-eluting stents (GESs) are being studied [72]. Finally, gene-based biological pacemakers could herald newer approaches in arrhythmia management [73]. *In vivo* experiments in a dog model have reported successful and functional electroporation of the HCN2 pacemaker gene in MSC [74].

Peripheral Arterial Disease (PAD). Gene therapy using addition of proangiogenic genes such as VEGF (naked plasmid DNA, adenovirus, plasmid/liposomes), fibroblast growth factor 1 (FGF-1: naked pDNA), and hepatocyte growth factor (HGF: naked pDNA) has been evaluated in the past 15 years. The effectiveness of these therapies remains controversial, with some clinical trials showing promising results but larger randomized studies are needed [75]. In the latest meta-analysis between FGF-1/HGF and placebo no differences were observed [76].

Bone and Joint Degenerative Conditions. Trauma or osteoarthritis (OA), both with a high prevalence and economic impact, create full thickness chondral defects that cannot efficiently heal, thus leading to significant long-term disability. Osteoarthritis (OA) is characterized by local inflammation, slow progressive degradation of the extracellular matrix (ECM), and loss of a chondrogenic phenotype in articular cartilage [77]. The targets for gene therapy in OA must include inflammatory pathways, ECM component production, resident cell proliferation and renewal, prevention of osteophyte formation, and modification and modulation of genetic and epigenetic effects [78]. Viral [79] and non-viral gene therapy approaches combining cell and tissue engineering (articular chondrocytes, MSC, iPSC with solid, hydrogel, or gene-activated scaffolds) with non-viral based gene delivery (IGF-1, FGF, BMP 2/7, TGF-β, angiogenesis inhibitors, IL-1Ra) [80–84] are expertly reviewed in Madry [85].

Wound healing may involve viral transfection and non-viral gene delivery systems. Adenovirus Platelet Derived Growth Factor–beta (PDGF-B), lentivirus Stromal Derived Growth Factor-1alpha (SDGF-1a), rAAV2 VEGF-A, and retrovirus PDGF-B/IGF-1 based systems have been evaluated in animal models [86]. Non-viral delivery can be established via naked DNA application, gene guns, electroporation, and lipofection with cationic liposomes. In addition, biomaterial-mediated transfection with multiple growth factors in the presence or absence of stem cell sources may pave the way to future developments [86]. RNAi therapies targeting miRNAs (several miRNAs have emerged as regulators in wound healing) for treatment of both normal and chronic wounds, including diabetic foot ulcers and pressure ulcers are being evaluated [87] (Lai et.al, 2014).

Infectious Disease. Despite tremendous efforts infectious diseases like HIV, Malaria, Dengue fever, and Influenza, cost millions of human lives. Vectored immuno prophylaxis (VIP) involves passive immunization by viral vector-mediated delivery of genes encoding broadly neutralizing antibodies (bnAbs) for *in vivo* expression. VIP holds great potential to alleviate the burden from many infectious diseases [88, 89]. Antibody gene transfer can be accomplished by Ad or AAV transfection and has been studied to treat infectious (e.g., HIV, Yersinia, anthrax, influenza), and non-infectious diseases (e.g., melanoma, ovarian cancer, addiction disorders such as cocaine/nicotine) [89–91].

Moreover, resistant viral infections could become amenable to treatment by RNAi; siRNa systemic delivery via nanoparticulate systems is being explored in phase 0-II studies in HIV, EBOV, RSV, HBV, and HCV [92].

Approaches using gene editing with site-specific nucleases will be discussed in the appropriate chapter.

Parkinson's Disease. Loss of the substantia nigra dopaminergic neurons projecting into nucleus caudatus and putamen leads to Parkinson's disease. The condition is currently responsive to oral L-Dopa treatment, but long-term fluctuations in efficacy and side effects necessitate newer treatment options. Early approaches employed the use of subthalamic viral vector (AAV-GAD) injection, which was well tolerated and safely transferred the glutamic acid decarboxylase (*GAD*) gene [93], as compared to sham injections [94]. Current gene therapy approaches employ the principle of gene addition to increase the production of dopamine in the striatum using rAAV or Lentiviruses (EIAV) [95]. Prosavin uses Oxford BioMedica's Lenti-vector delivery system to transfer three genes (aromatic amino acid dopa decarboxylase (AADC), tyrosine hydroxylase (TH), and GTP-cyclohydrolase 1 (GCH)) in one unit, to the striatum in the brain, reprogramming transduced cells to secrete dopamine [96]. Others use AAV vectors to deliver glutamic acid decarboxylase (GAD) and Neurturin. All approaches have shown no adverse side effects due to surgery and encouraging results in motor function improvement. Future approaches may include combinations: ESC, MSC, and neural stem cells (NSC), reprogrammed or edited and transfected to regenerate dopaminergic neurons [97].

10.5 Gene Editing

Targeted genetic modifications in a host genome are possible through gene editing using programmable/engineered/chimeric nucleases able to induce targeted, site-specific DNA double-strand breaks (DSB) in the genome and stimulate physiological DNA repair through endogenous mechanisms such as non-homologous end joining (NHEJ) and homology-directed repair (HDR) [98]. Physiological DNA repair occurs in situations such as double-stranded breaks, which occur normally during chromosome replication and as a result of exposure to radiation, ultraviolet light, and some chemicals. NHEJ is an error-prone repair process that can result in a wide range of end-joining events that leads to somewhat random mutations at the site of DSB. HDR is a precise repair pathway that can utilize either an endogenous (second copy of the broken gene on the sister chromosome) or exogenous piece of homologous DNA as a template to repair a patch of DNA.

In real cell life, each cell has the choice between HDR and NHEJ but HDR events are much more rare than NHEJ. Success in gene editing will require techniques that increase the prevalence of HDR as HDR allows more precise gene mutations, deletions, and insertions of exogenous sequences [99].

These techniques will employ programmable/engineered/chimeric nucleases including zinc-finger nucleases (ZFNs), transcription activator-like effector (TALENs) RNA-guided engineered nucleases (RGNs), and the clustered regularly interspaced short palindromic repeat (CRISPR)–Cas9 (CRISPR-associated) system [98].

Zinc-finger nucleases (ZFNs) were used for the first time in 1996 to demonstrate that a site-specific DNA cut was possible. A zinc finger domain was coupled with the FokI endonuclease domain to create a ZFN. The zinc finger domain recognizes one nucleotide triplet, which confers the site specificity. However, zinc fingers are difficult and expensive to make, new zinc finger domains are needed for each genome locus, and they cleave inaccurately due to single nucleotide substitutions [100].

TALENs and the RGN CRISPR/Cas9 (RNA-guided engineered nucleases; clustered regularly interspaced short palindromic repeat)-associated systems are also being studied. In 2011, the TALEN was discovered during a study on Xanthomonas genus of bacteria. It was found that these bacteria secrete effector proteins (TALE) to increase their pathogenicity in plants. As the name implies they are proteins that can bind to host DNA and mimic transcription factors that activate target genes. By joining the DNA-binding domain of TALE with the catalytic domain of FokI the chimeric TALEN were created that can recognize specific nucleotide monomers like A, T, C, and G specifically [100]. Two years later in 2013, a new system, CRISPR/Cas9, was introduced [99]. This system uses the complementary interaction of a non-coding RNA (guidance or gRNA) and the target site DNA for its site specificity and thus is an RNA-guided engineered chimeric nuclease (RGN) [99]. The CRISPR/Cas system is the unique mechanism providing microorganisms protection against foreign DNA penetration. At least 11 different CRISPR-Cas systems have been described [99]. Due to the simplicity of creating a small 20-nucleotide gRNA, any site in any gene (even multiple genes with multiplexing), genome, and organism can be targeted with the CRISPR-Cas II system [99]. Hence, cell cultures with modified genomes can be created to produce disease models and to predict their treatments.

Human pluripotent stem cell generation can contribute to the production of disease models but is difficult to reprogram using existing techniques. Somatic cells can under certain conditions be genetically reprogrammed into induced pluripotent stem cells (iPSCs) by retroviral vector transfection with the transcription factors (TFs) Oct4, Sox2, Klf4, and c-Myc, (OSKM) [101]. iPSCs can thus acquire an unlimited self-renewal capacity and ability for multi-potential differentiation. To date, iPSCs have been derived from cells obtained from an increasing number of conditions among many in bone marrow disorders (Fanconi, sickle cell, thalassemia, ADA-SCID), CNS disorders (ALS, Parkinson's, Huntington's, and Alzheimer's diseases, epilepsy, autism, SMA), metabolic disorders, cystic fibrosis, Hutchinson-Gilford progeria syndrome (HGPS), ataxia-telangiectasia syndrome [102], recessive dystrophic (RDEB) [103], and junctional epidermolysis bullosa (JEB) [104] can be differentiated toward disease-specific lineages in an effort to study and evaluate their novel drugs or gene therapies [99]. Genome editing in human pluripotent stem cells using the RGN/CRISPR-Cas system increases efficiency in performing targeted gene manipulations, overexpressing genes, and introducing and repairing single point mutations [105]. Such methods can facilitate the study of human disease and propel advances in treatments [105]. Gene editing by ZFN/TALEN and type II CRISP-Cas9 systems, potentially combined with RNAi, allows investigators to manipulate virtually every gene and generate cell lines to create disease models [99]. Many cancers are due to large chromosomal rearrangements like inversions, deletions, chimeric fusions, and other genetic events and RGNs can perform editing to facilitate studies on the meaning and causality of such genomic recombination [106, 107]. Examples include lung adenocarcinoma, Ewing's sarcoma, acute myeloid leukemia, bladder cancer, and others [106, 108, 109].

Ultimately, autologous iPSC disease models created to elucidate the pathophysiological mechanisms can be corrected genetically, cultured, and eventually differentiated in many autologous cell lines that can be transplanted not only to cure genetic, epigenetic, or environmental diseases but also to delay aging and rejuvenate ailing tissues [99].

Human Immunodeficiency Virus (HIV) infection. HIV infection is human pandemic, and anti-retroviral therapy (HAART) is the only available treatment to control infection. The C-C motif chemokine receptor 5 (CCR5) is the co-receptor for HIV-1 entry and is an ideal target for research. CCR5 antagonists are being developed, and the naturally occurring homozygous delta 32 genotype that confers resistance to HIV have spurred hopes for a gene therapy breakthrough. Serendipitous allogeneic HSCT using donor stem cells with delta 32 mutations have cured one Berlin patient [110]. RNA interference has successfully been used to silence CCR5, but it requires the continuous presence of siRNA and the expression is transient [111]. CCR5 human autologous hematopoietic and progenitor stem cells have been successfully ZFN engineered [112], and CD34+ stem cells have successfully been generated from both human embryonic stem cells (hESCs) [113] and induced pluripotent stem cells (hiPSCs) [113] using specific zinc finger nucleases (ZFNs) combined with homologous recombination [113, 114]. Apart from ZFN, TALEN, and CRISPR-Cas9 systems are being evaluated in genetic manipulation of CCR5 and in excising the integrated HIV genome in order to eradicate latent infections [115].

Duchenne Muscular Dystrophy (DMD). The ability of zinc finger nucleases (ZFNs) to permanently exclude exon 51 from the resulting dystrophin transcript and to restore its reading frame in Duchenne muscular dystrophy (DMD) has been reported [116]. TALENs and CRISPR-Cas9 knocked down exon and corrected iPSCs in skeletal muscle cells resulting in successful expression of full-length functional dystrophin protein [117].

Cystic Fibrosis (CF). Recently, adult intestinal stem cells were isolated and expanded from two CF patients. Then, the CRISPR/Cas9 genome editing system was used to correct the CFTR locus by homologous recombination and the corrected allele was expressed as fully functional protein in organoids [118]. The authors suggest that this approach may present a safe complement to iPSC-based approaches and could be applied to different single-gene hereditary defects [118].

CRISPR interference (CRISPRi). CRISPRi has been proposed as an alternative, complement, and replacement for RNAi [119]. Due to the fact that it acts precisely on DNA level it can also target RNAi on RNA level [119]. Furthermore, CRISPRi could make germline gene therapy a practical possibility by reprogramming spermatogonial stem cells, creating knockout lines, transgenic animals, human disease models, and new livestock strains with great efficacy [100, 120].

10.6 Short RNA or DNA-based Gene Expression Modulation

Gene expression RNA silencing is a natural defense mechanism in eukaryotes aimed at protecting the host genome from invasion by foreign elements, such as viruses [121]. Anti-sense oligodeoxynucleotides (ODNs), small interfering RNAs (siRNAs), microRNAs (miRNAs), and decoy ODNs are short sequences of nucleotides that are able to modulate gene expressions [122, 123].

RNA interference (RNAi) mimics the epigenetic mechanism of RNA silencing, inhibiting special gene activity at the post-transcriptional level and thus reducing the expression of relevant proteins [123]. RNAi is a highly conserved gene-silencing mechanism triggered by a long dsRNA in the cytoplasm, which is processed by the RNAse type-III enzyme, called DICER [124]. The resulting RNA fragments, either as small interfering RNAs (siRNAs: 22 nucleotide in length) or microRNAs (miRNAs: 21-25 nt long), are incorporated in the RNA-induced silencing complex (RISC). The dsRNA strands are separated, and one strand guides RISC to the complimentary region of target RNA to suppress it either by mRNA cleavage or mRNA translation block [125]. miRNA and siRNA exhibit important differences. miRNA are exclusively endogenously produced from specific genes and control genes that do not produce them with varying complementarity while siRNA may originate from viruses, endogenous dsRNA, or tranfected exogenous dsRNA and will only control the genes that have produced them with perfect complementarity. Both miRNA and siRNA will ultimately affect their gene expression regulatory effect (knock-down) via transcriptional repression and mRNA cleavage but siRNA are also involved in anti-viral defense and maintenance of genome integrity [125].

MicroRNAs (miRNAs or miRs) are small non-coding RNA molecules, comprising a class of 21–25 nucleotides, which can down-regulate the gene expression in translational repression. Up to 1048 miRNAs have been documented to date [125] that can target around 60% of all genes [126]. miRNAs are profusely present in all human cells and their production is strictly regulated. They seem to exhibit decisive regulatory functions associated with a wide variety of human diseases [122].

They are implicated widely in a broad range of physiological and pathophysiological processes such as cancer (breast, colon, lung, pancreas, and stomach, as well as in chronic lymphocytic leukemia (CLL), acute myeloid leukemia (AML), glioblastoma, and myeloma), Parkinson's disease, epilepsy, schizophrenia, Down's syndrome, liver cirrhosis, heart disease, and metabolic conditions including type 2 diabetes and obesity [128, 129]. Exercise and dietary constituents such as vitamins (A, D, E, B), selenium, fatty acids, polyphenols, isoflavones, indoles, and isothiicyanates are reported to modulate miRNA profiles and serve a role in cancer etiology and prevention [126, 129, 130]. RNAi therapeutics is an exciting field with great potential but safe and efficient delivery systems for siRNA delivery are required [124]. Naked siRNA, lipoplexes, and polyplexes are being evaluated for their efficient delivery of siRNA in eyes, lung, kidney, and brain [124]. Viral diseases, castration resistant prostate cancer, acute and chronic myeloid leukemia, preeclampsia, and advanced melanoma are being treated with anti-sense therapy [92, 131]. Bevasiranib is the first siRNA-based drug in clinical trials that targets VEGF in wet ARMD and diabetic AMD. ALN-RSV01 is used to treat virus respiratory diseases, while CALAA-01 silences ribonucleotide reductase subunit 2 (RRM2) gene, in advanced cancers [92]. Lipid-based carriers of siRNA therapeutics can target liver and are being assessed in clinical trials for treatment of hypercholesterolemia [123]. A phase Ib clinical trial of the first-in-human mutation-targeted siRNA Td101 against an inherited skin disorder is now complete [92].

Gene-silencing therapy of inherited retinal degenerations such as autosomal dominant retinitis pigmentosa and dominant maculopathies in which expression of a mutant allele has a gain-of-function effect on photoreceptor cells has been tried successfully by AAV2/8-mediated subretinal delivery of an miRNA-based shRNA that could efficiently and specifically silence Peripherin-2 (PRPH2) *in vivo* in an animal model [20].

Anti-sense ODNs are 10–30 nt short chains of nucleic acids, in the form of ssDNA or DNA, which work on the post-translational level similar to siRNA and miRNa [132]. When inside the cytoplasm, anti-sense ODNs will exert their action either by physical blockage (steric hindrance) or will undergo targeted degradation via RNase H activation [132]. Anti-sense ODNs are in clinics in the form of two approved drugs. Fomivirsen was approved in 1998 for the treatment of cytomegalovirus retinitis in AIDS patients and Mipomersen was approved in 2013, both targeting ApoB 100 for the treatment of homozygous familial Hypercholesterolaemia (HoFH).

Anti-cancer anti-sense ODNs are currently a very active field of research with many Phase I and II studies targeting solid tumors in the prostate, lung, pancreas, liver, brain, and head and neck areas [133]. Specific, uniform, efficient, and sustained intratumoral delivery is paramount for their success [133].

Duchenne muscular dystrophy and **spinal muscular atrophy (SMA)** are other disorders where anti-sense ODNs are being evaluated with encouraging results [134].

In **Duchenne**, genetic modifications aim to restore the original mRNA reading frame of Dystrophin to resume the production of a functional protein. Exon skipping has been accomplished by intravenous anti-sense ODN therapy [135]. The exon knock-down via TALENs or CRISPR-Cas9 in patients resulted in the production of dystrophin [118]. An emerging therapeutic approach for Duchenne muscular dystrophy is the transplantation of autologous myogenic progenitor cells genetically modified to express dystrophin using a piggyback transposon system [136]. Others have reported the successful use of recombinant AAV vector carrying a modified U7snRNA sequence to promote exon skipping [137].

SMA is one of the most frequently inherited causes of infant mortality, where afflicted patients lose the survival motor neuron 1 (SMN1) gene but retain one or more copies of the silenced SMN2. SMN2 is a homolog of SMN1 that is incorrectly spliced resulting in a protein product that is rapidly degraded [138]. Primary treatment strategies for SMA aim at increasing SMN protein levels either by the replacement of mutant SMN genes with functional copies by means of gene therapy vectors or by abolishing exon 7 skipping of SMN2 gene via anti-sense oligonucleotides (ASOs), thus promoting intron 7 inclusion [134]. ISIS-SMNrx, an oligonucleotide-based drug with very long half-life in the CSF that makes 6–9 month dosing intervals possible is currently being evaluated in a phase Ib/IIa study in SMA (ClinicalTrials.gov: NCT01839656).

Decoy ODNs. Decoy ODNs works at the pre-transcription level while siRNAs, miRNAs, and anti-sense ODNs work at the post-translation level. Chimeric decoy ODN strategies used against cardiovascular diseases and kidney fibrosis are currently being evaluated for their potential to treat liver disease [123].

RNAi and MiRNAs are more complex than initially predicted and newer findings indicate that gene expression modulation may occur in both negative and positive ways; the latter mechanism is called **RNA activation (RNAa)**, and may hold important therapeutic potential against cancers [139].

10.7 Combination Approaches

Gene therapy has the potential to permanently correct disease through the abovementioned approaches. However, as we still are putting together the pieces of our genetic puzzle, one approach will not be enough. The most prevalent human diseases are polygenic in their etiopathology and gene-gene as well as gene environment/lifestyle/nutrition/exercise interplay are widely unknown. In many monogenic conditions, e.g., Osteogenesis Imperfecta, where negative mutations of the COL1A1 gene are causative, simple augmentation of the responsible gene failed to correct the condition [140]. In other conditions, adjacent gene regulation of the responsible mutated gene may hamper the replacement/augmentation strategies. In other instances, as in Hemophilia B, the transgene may be too large to be carried by a viral vector into the cell or the condition is polygenic and involves larger chromosomal rearrangements. Moreover, imprecise targeting increases the risks of insertional mutagenesis [141]. Combinations of replacements, additions, knock-downs, and precise editing may be necessary. For *in vivo* approaches, CRISPR-Cas9 system, ZFN, and meganucleases can be delivered using retroviral or lentiviral vectors, however TALENs' repetitive DNA sequence increases the risk for rearrangements [142]. The size of TALEN and CRISPR-Cas9 may also be prohibitive for AAV vectors but AdV can be used. The choice of viral vectors will also be influenced by their strain, immunogenicity, and tropism.

For *ex vivo* approaches, delivery systems may involve many different types of engineered or reprogrammed stem cells that could be transfected by electroporation or lipofection and should be tested for efficiency and efficacy before delivery. Combination approaches will initially be needed in disease modeling and understanding of epigenetics, as well as how the environment impacts on specific genotypes [143]. The cellular component will be of particular importance as the reprogrammed cells will be autologous and the disease model will be personalized and specific. All subsequent therapies will be more effective with minimal immunological and other adverse effects [144]. Apart from autologous cellular harvest, autologous tissue engineering with genetically enhanced or corrected contents will be a reality.

Combination approaches are discussed below.

Alpha 1 Anti-trypsin Deficiency (AAT). AAT deficiency is a common, but under-diagnosed, genetic disease with more than 100 known mutations where the mutant Z gene is seen in more than 90%. In the classical form (ZZ or PiZZ), which occurs in 1 in 2,000–3,500 births, large quantities of the mutant Z protein accumulate intracellularly in the liver, rather than being efficiently secreted. The lung manifestations of AAT deficiency can be treated with intravenous AAT enzyme therapy but there is currently no treatment for the liver symptoms. AAT deficiency is an example of how combination approaches can be employed to target certain conditions [145, 146]. Guo et al. developed a specific ASO that could effectively decrease AAt-mRNa levels and could correct liver disease *in vitro* and *in vivo* [147]. Therapeutic shRNA in disease-specific iPSCs using third-generation lentiviral vectors were used to target severe AAT deficiency and efficiently knocked down disease-related genes [145].

Alzheimer's disease is a common neurodegenerative disease and the most common type of dementia accounting for up to 80% of all cases and affecting up to one in three people over the age of 85. The pathological changes in Alzheimer's disease include apoptosis of cholinergic neurons, and an increase in levels and activity of β-secretase (beta-site amyloid precursor protein cleaving enzyme, BACE1), which is the key enzyme in amyloid beta protein (Aβ) production and inflammation. At present, there is no cure or any effective treatments apart from acetylcholinesterase inhibitors, which only temporarily augment cholinergic function [148]. Previous approaches involving transplantation of rat neural stem cells transfected with nerve growth factor, choline acetyltransferase, or brain-derived neurotrophic factor have shown promising results in animal models of the disease [149–151]. Genetic modification of stem cells is an attractive approach in many complex diseases and the subsequent cellular transplantation may not only regenerate and cure affected tissues but also sustain normal function. Liu et al. [97] could downregulate the overexpression of BACE1 in neural stem cells (NSC) by introducing a short hairpin DNA oligonucleotide into the pSilenCircle plasmid to express a short hairpin RNA (shRNA) complementary to a sequence in the coding region of human BACE1 mRNA [97]. These transfected NSC were able to replace the lost nerve cells through proliferation and differentiation, and also could sustain a down-regulated expression of BACE1 protein. This resulted in a reduction in Aβ production, thus inhibiting the inflammatory reaction, and promoting proliferation, survival, differentiation, and migration of neural stem cells to the lesion area [97, 152].

Sensorineural Hearing Loss. Hearing loss is one of the most common morbidities affecting over 5% of the world's population of all ages. Hearing loss can have genetic origin or can be acquired from environmental, iatrogenic or age related pathologies (World Health Organization, 2013). In genetic hearing loss, autosomal recessive and autosomal dominant mutations have been identified in more than 60 genes of neurotransmitters such as glutamate and its transporter VGLUT3, SLC17A8, GJB2 connexin [153]. The mutant GJB2 liposome trasnfection followed by RNAi silencing of the mutant gene avoided hearing loss in an animal model [154]. Usher syndrome is the most common syndromic hearing loss, where anti-sense oligoneucleotides (ASO) have been used to correct the aberrant splicing in the Ush1c gene [155]. Acquired hearing loss due to hair cell loss, is another area of research, where miRNA therapy can be employed [156, 157]. AAV vector mediated Atoh1 gene delivery alone or in combination with neurotrophin-3 and brain derived neurotrophic factor showed promising results to treat acquired hearing loss [158].

Osteoarthritis [159]. The simultaneous use of RNAi and gene addition was described by Zhang [160] using synovial mesenchymal stromal cells (SMSCs) transduced with recombinant lentiviral and adenoviral vectors to deliver TGF-β3 and an shRNA targeting the mRNA for type I collagen.

Down Syndrome. Non-Down syndrome iPSC can be generated from patients with Down syndrome by AAV vector mediated insertion of a negative selection marker HSV thymidine kinase (TK), in an effort to eliminate the additional chromosome 21 [161]. ZFN guided insertion of the ncRNA Xist into trisomic cells, epigenetically inactivates the supernumerary chromosome. This human chromosome silencing provides the proof of concept, which may bring Down syndrome in to the realm of future gene therapy [162].

Gene Doping. While the potential of gene therapy to alleviate suffering and disease is desirable, its use in enhancing performance in sports is debatable. Genes to boost endurance, improve oxygen uptake and delivery to tissues, increase muscle mass, augment strength and power, redistribute fat, improve response to pain and speed up regeneration of injured tissues are interesting candidates for enhancing the performance of athletes. Growth hormone (GH) has profound metabolic effects on fats and carbohydrates, hypoxia-inducible factors (HIFs) and vascular endothelial growth factor (VEGFA) are associated with improved oxygen delivery in hypoxia, peroxisome proliferator-activated receptor (PPAR) influence carbohydrate-lipid metabolism, muscle fiber transformation and reprogramming, and phosphoenolpyruvate carboxykinase (PCK1) regulates gluconeogenesis. Adenosine monophosphate analogue (AMP)-(AICAR) are found to increase speed and strength in animal models, while proopiomelanocortin (POMC) are endorphin precursors that modulate pain. These genes are also being studied for the treatment of cardiovascular diseases, obesity, diabetes, and cancer [163].

Tests for gene doping detection are inadequate at present but intense research is ongoing and involves gene expression profiling for the genes involved in doping and genes regulated by a transgene, proteomic profiling, and protein expression profiling [163].

References

[1] Dahm R. Discovering DNA: Friedrich Miescher and the early years of nucleic acid research. Hum Genet 2008;122(6):565–81.

[2] Watson JD, Crick FH. Molecular structure of nucleic acids; a structure for deoxyribose nucleic acid. Nature 1953;171(4356):737–8.

[3] Tripp S, Grueber M. Economic impact of the Human Genome Project. www.battelle.org.

[4] Hayden EC. Is the $1,000 genome for real? Nature: News Explainer. doi:10.1038/nature.2014.14530.

[5] Ylä-Herttuala S. Endgame: Glybera finally recommended for approval as the first gene therapy drug in the European Union. Mol Ther 2012;20(10):1831–2. http://dx.doi.org/10.1038/mt.2012.194.

[6] Sheridan C. Milestone approval lifts Illumina's NGS from research into clinic. Nat Biotechnol 2014;32:111–2. http://dx.doi.org/10.1038/nbt0214-111.

[7] Wirth T, Parker N, Ylä-Herttuala S. History of gene therapy. Gene 2013;525(2):162–9. http://dx.doi.org/10.1016/j.gene.2013.03.137.

[8] Avise John C. The hope, hype & reality of genetic engineering: remarkable stories from agriculture, industry, medicine, and the environment. US: Oxford University Press; 2004, ISBN: 978-0-19-516950-8. p. 22.

[9] Wei X, Ju X, Yi X, et al. Identification of sequence variants in genetic disease-causing genes using targeted next-generation sequencing. PLoS One 2011;6(12):e29500. http://dx.doi.org/10.1371/journal.pone. 0029500.

[10] Lander ES. Initial impact of the sequencing of the human genome. Nature 2011;470:187–97. http://dx.doi.org/10.1038/nature09792.

[11] Wang D, Gao G. State-of-the-art human gene therapy: part II. Gene therapy strategies and clinical applications. Discov Med 2014;18(98):151–61.

[12] Maguire AM, High KA, Auricchio A, et al. Age-dependent effects of RPE65 gene therapy for Leber's congenital amaurosis: a phase 1 dose-escalation trial. Lancet 2009;374:1597–605.

[13] MacLaren RE, Groppe M, Barnard AR, et al. Retinal gene therapy in patients with choroideremia: Initial findings from a phase 1/2 clinical trial. Lancet 2014;383:1129–37. http://dx.doi.org/10.1016/S0140-6736(13)62117-0.

[14] Holz FG, Schmitz-Valckenberg S, Fleckenstein M. Recent developments in the treatment of age-related macular degeneration. J Clin Invest 2014;124(4):1430–8. http://dx.doi.org/10.1172/JCI71029.

[15] Carvalho LS, Vandenberghe LH. Promising and delivering gene therapies for vision loss. Vision Res 2014. http://dx.doi.org/10.1016/j.visres.2014.07.013. pii: S0042-6989(14)00172-2.

[16] Binley K, Widdowson PS, Kelleher M, et al. Safety and biodistribution of an equine infectious anemia virus-based gene therapy, RetinoStat(R) for age-related macular degeneration. Hum Gene Ther 2012;23:980–91. http://dx.doi.org/10.1089/hum.2012.008.
[17] Binley K, Widdowson P, Loader J, et al. Transduction of photoreceptors with equine infectious anemia virus lentiviral vectors: safety and biodistribution of StarGen for Stargardt disease. Invest Ophthalmol Vis Sci 2013;54:4061–71. http://dx.doi.org/10.1167/iovs.13-11871.
[18] Estrada-Cuzcano A, Roepman R, Cremers FP, den Hollander AI, Mans DA. Nonsyndromic retinal ciliopathies: translating gene discovery into therapy. Hum Mol Genet 2012;21(R1):R111–24.
[19] Fakin A, Zupan A, Glavač D, Hawlina M. Combination of retinitis pigmentosa and hearing loss caused by a novel mutation in PRPH2 and a known mutation in GJB2: importance for differential diagnosis of Usher syndrome. Vision Res 2012;75:71–6. http://dx.doi.org/10.1016/j.visres.2012.07.011.
[20] Georgiadis A, Tschernutter M, Bainbridge JW, Robbie SJ, McIntosh J, Nathwani AC, et al. AAV-mediated knockdown of peripherin-2 in vivo using miRNA-based hairpins. Gene Ther 2010;17(4):486–93. http://dx.doi.org/10.1038/gt.2009.162.
[21] Conley SM, Naash MI. Gene therapy for PRPH2-associated ocular disease: challenges and prospects. Cold Spring Harb Perspect Med 2014;4(11):http://dx.doi.org/10.1101/cshperspect.a017376. pii: a017376.
[22] Wang X, Wang H, Sun V, et al. Comprehensive molecular diagnosis of 179 Leber congenital amaurosis and juvenile retinitis pigmentosa patients by targeted next generation sequencing. J Med Genet 2013;50(10):674–88. http://dx.doi.org/10.1136/jmedgenet-2013-101558.
[23] Narain R, Gonzales M, Hoffman AS, Stayton PS, Krishnan KM. Synthesis of monodisperse biotinylated p(NIPAAm)-coated iron oxide magnetic nanoparticles and their bioconjugation to streptavidin. Langmuir 2007;23:6299–304.
[24] Ahmed M, Jiang X, Deng Z, Narain R. Cationic Glyco-Functionalized single walled carbon nanotubes as efficient gene delivery vehicles. Bioconjug Chem 2009;20:2017–22.
[25] Ahmed M, Narain R. The effect of polymer architecture, composition, and molecular weight on the properties of glycopolymer-based non-viral gene delivery systems. Biomaterials 2011;32(22):5279–90.
[26] Ahmed M, Bhuchar N, Ishihara K, Narain R. Well-controlled cationic water-soluble phospholipid polymer- DNA nano-complexes for gene delivery. Bioconjug Chem 2011;22(6):1228–38.
[27] Chen JT, Ahmed M, Liu Q, Narain R. Synthesis of cationic magnetic nanoparticles and evaluation of their gene delivery efficacy in hep G2 cells. J Biomed Mater Res A 2012;100(9):2342–7.
[28] Ahmed M, Narain R. The effect of molecular weight, compositions and lectin type on the properties of hyperbranched glycopolymers as non-viral gene delivery systems. Biomaterials 2012;33(15):3990–4001.
[29] Ahmed M, Jawanda M, Ishihara K, Narain R. Impact of the nature, size and chain topologies of carbohydrate-phosphorylcholine polymeric gene delivery systems. Biomaterials 2012;33(31):7858–70.
[30] Ahmed M, Narain R. Progress of RAFT based polymers in the field of gene delivery. Prog Polym Sci 2012.
[31] Sunasee R, Watanaarsakit P, Ahmed M, Lollmahomed F, Narain R. Biodegradable and non-toxic nanogels as non-viral gene delivery systems. Bioconjug Chem 2012;23(9):1925–33.

[32] Ahmed M, Wattanaarsakit P, Narain R. Cationic glyco-nanogels for epidermal growth factor receptor (EGFR) specific siRNA delivery in cancer cells. Polymer Chemistry 2013;4(13):3829–36.
[33] UMIN-CTR Clinical ARMDTrial https://upload.umin.ac.jp/cgi-open_bin/ctr/ctr.cgi?function=brows&action=brows&type=summary&recptno=R000013279&language=E.
[34] Chuah MK, Evens H, VandenDriessche T. Gene therapy for hemophilia. J Thromb Haemost 2013;11(Suppl 1):99–110. http://dx.doi.org/10.1111/jth.12215.
[35] Matsui H, Fujimoto N, Sasakawa N, Ohinata Y, Shima M, Yamanaka S, et al. Delivery of full-length factor VIII using a piggyBac transposon vector to correct a mouse model of hemophilia A. PLoS One 2014;9(8). http://dx.doi.org/10.1371/journal.pone.0104957. eCollection 2014. e104957.
[36] Cancio MI, Reiss UM, Nathwani AC, Davidoff AM, Gray JT. Developments in the treatment of hemophilia B: focus on emerging gene therapy. Appl Clin Genet 2013;6:91–101. http://dx.doi.org/10.2147/TACG.S31928.
[37] Nathwani AC, Reiss UM, Tuddenham EG, et al. Long-term safety and efficacy of factor IX gene therapy in hemophilia B. N Engl J Med 2014;371(21):1994–2004. http://dx.doi.org/10.1056/NEJMoa1407309.
[38] Griesenbach U, Alton EW. Moving forward: cystic fibrosis gene therapy. Hum Mol Genet 2013;22(R1):R52–8. http://dx.doi.org/10.1093/hmg/ddt372.
[39] Hacein-Bey-Abina S, Pai SY, Gaspar HB, et al. A modified γ-retrovirus vector for X-linked severe combined immunodeficiency. N Engl J Med 2014;371(15):1407–17. http://dx.doi.org/10.1056/NEJMoa1404588.
[40] Kohn DB. Update on gene therapy for immunodeficiencies. Clin Immunol 2010;135(2):247–54. http://dx.doi.org/10.1016/j.clim.2009.12.003.
[41] Candotti F, Shaw KL, Muul L, et al. Gene therapy for adenosine deaminase-deficient severe combined immune deficiency: clinical comparison of retroviral vectors and treatment plans. Blood 2012;120(18):3635–46. http://dx.doi.org/10.1182/blood-2012-02-400937.
[42] Aiuti A, Biasco L, Scaramuzza S, et al. Lentiviral hematopoietic stem cell gene therapy in patients with Wiskott Aldrich Syndrome. Science 2013;341(6148):1233151. http://dx.doi.org/10.1126/science.1233151.
[43] Merling RK, Sweeney CL, Chu J, et al. An AAVS1-targeted minigene platform for correction of iPSCs from all five types of chronic Granulomatous disease. Mol Ther 2014. http://dx.doi.org/10.1038/mt.2014.195 [Epub ahead of print].
[44] Cartier N, Hacein-Bey-Abina S, Bartholomae CC, et al. Hematopoietic stem cell gene therapy with a lentiviral vector in X-linked adrenoleukodystrophy. Science 2009;326:818–23.
[45] Boudes PF. Gene therapy as a new treatment option for inherited monogenic diseases. Eur J Intern Med 2014;25(1):31–6.
[46] Patil SA, Maegawa GH. Developing therapeutic approaches for metachromatic leukodystrophy. Drug Des Devel Ther 2013;7:729–45. http://dx.doi.org/10.2147/DDDT.S15467. eCollection 2013.
[47] Satpute SB, Bankar MP, Momin AA. The prevalence of β-Thalassemia mutations in south western Maharashtra. Indian J Clin Biochem 2012;27(4):389–93. http://dx.doi.org/10.1007/s12291-012-0230-y.
[48] Chandrakasan S, Malik P. Gene therapy for hemoglobinopathies: the state of the field and the future. Hematol Oncol Clin North Am 2014;28(2):199–216. http://dx.doi.org/10.1016/j.hoc.2013.12.003.
[49] Cavazzana-Calvo M, Payen E, Negre O, et al. Transfusion independence and HMGA2 activation after gene therapy of human beta-thalassaemia. Nature 2010;467:318–22.

[50] Cavazzana-Calvo M, Fischer A, Hacein-Bey-Abina S, Aiuti A. Gene therapy for primary immunodeficiencies: part 1. Curr Opin Immunol 2012;24(5):580–4. http://dx.doi.org/10.1016/j.coi.2012.08.008.
[51] Varela I, Karagiannidou A, Oikonomakis V, et al. Generation of human β-Thalassemia induced pluripotent cell lines by reprogramming of bone marrow-derived mesenchymal stromal cells using modified mRNA. Cell Reprogram 2014;16(6):447–55. http://dx.doi.org/10.1089/cell.2014.0050.
[52] Katz MG, Fargnoli AS, Williams RD, Bridges CR. Gene therapy delivery systems for enhancing viral and nonviral vectors for cardiac diseases: current concepts and future applications. Hum Gene Ther 2013;24(11):914–27. http://dx.doi.org/10.1089/hum.2013.2517.
[53] Sheridan C. Gene therapy finds its niche. Nat Biotechnol 2011;29:121–8.
[54] Vaxman I, Ram R, Gafter-Gvili A, Vidal L, Yeshurun M, Lahav M, et al. Secondary malignancies following high dose therapy and autologous hematopoietic cell transplantation-systematic review and meta-analysis. Bone Marrow Transplant 2015. http://dx.doi.org/10.1038/bmt.2014.325 [Epub ahead of print].
[55] Duong CP, Yong CS, Kershaw MH, Slaney CY, Darcy PK. Cancer immunotherapy utilizing gene-modified T cells: from the bench to the clinic. Mol Immunol 2015. http://dx.doi.org/10.1016/j.molimm.2014.12.009. pii: S0161-5890(14)00351-4.
[56] Zarogoulidis P, Darwiche K, Sakkas A, Yarmus L, Huang H, Li Q, et al. Suicide gene therapy for cancer - current strategies. J Genet Syndr Gene Ther 2013;4. pii: 16849.
[57] Goins WF, Huang S, Cohen JB, Glorioso JC. Engineering HSV-1 vectors for gene therapy. Methods Mol Biol 2014;1144:63–79. http://dx.doi.org/10.1007/978-1-4939-0428-0_5.
[58] Mavroudi M, Zarogoulidis P, Porpodis K, Kioumis I, Lampaki S, Yarmus L, et al. Stem cells' guided gene therapy of cancer: new frontier in personalized and targeted therapy. J Cancer Res Ther (Manch) 2014;2(1):22–33.
[59] Maguire CA, Ramirez SH, Merkel SF, Sena-Esteves M, Breakefield XO. Gene therapy for the nervous system: challenges and new strategies. Neurotherapeutics 2014;11(4):817–39. http://dx.doi.org/10.1007/s13311-014-0299-5.
[60] Andtbacka, et al. J Clin Oncol 2013;31. sLBA9008.
[61] Kenderian SS, Ruella M, Gill S, Kalos M. Chimeric antigen receptor T-cell therapy to target hematologic malignancies. Cancer Res 2014;74(22):6383–9.
[62] Maus MV, Grupp SA, Porter DL, June CH. Antibody-modified T cells: CARs take the front seat for hematologic malignancies. Blood 2014;123(17):2625–35. http://dx.doi.org/10.1182/blood-2013-11-492231.
[63] Cruz CR. A closer look at chimeric antigen receptor specificity. Cytotherapy 2014;16:1323–4.
[64] Scimia MC, Gumpert AM, Koch WJ. Cardiovascular gene therapy for myocardial infarction. Expert Opin Biol Ther 2014;14(2):183–95. http://dx.doi.org/10.1517/14712598.2014.866085.
[65] Zsebo K, Yaroshinsky A, Rudy JJ, Wagner K, Greenberg B, Jessup M, et al. Long-term effects of AAV1/SERCA2a gene transfer in patients with severe heart failure: analysis of recurrent cardiovascular events and mortality. Circ Res 2014;114(1):101–8. http://dx.doi.org/10.1161/CIRCRESAHA.113.302421.
[66] Hayward C, Patel H, Lyon A. Gene therapy in heart failure. Circ J 2014;78(11):2577–87.
[67] Greenberg B, Yaroshinsky A, Zsebo KM, Butler J, Felker GM, Voors AA, et al. Design of a phase 2b trial of intracoronary administration of AAV1/SERCA2a in patients with advanced heart failure: the CUPID 2 trial (calcium up-regulation by percutaneous administration of gene therapy in cardiac disease phase 2b). JACC Heart Fail 2014;2(1):84–92. http://dx.doi.org/10.1016/j.jchf.2013.09.008.

[68] Won YW, Bull DA, Kim SW. Functional polymers of gene delivery for treatment of myocardial infarct. J Control Release 2014;195:110–9. http://dx.doi.org/10.1016/j.jconrel.2014.07.041.

[69] Kaminsky SM, Rosengart TK, Rosenberg J, Chiuchiolo MJ, Van de Graaf B, Sondhi D, et al. Gene therapy to stimulate angiogenesis to treat diffuse coronary artery disease. Hum Gene Ther 2013;24(11):948–63. http://dx.doi.org/10.1089/hum.2013.2516.

[70] Aghabozorg Afjeh SS, Ghaderian SM. The role of microRNAs in cardiovascular disease. Int J Mol Cell Med 2013;2(2):50–7.

[71] Chen XY, Tang T, Wong KS. RNA interference therapy: a new solution for intracranial atherosclerosis? Ann Transl Med 2014;2(8):79. http://dx.doi.org/10.3978/j.issn.2305-5839.2014.07.11.

[72] Yin RX, Yang DZ, Wu JZ. Nanoparticle drug- and gene-eluting stents for the prevention and treatment of coronary restenosis. Theranostics 2014;4(2):175–200. http://dx.doi.org/10.7150/thno.7210. eCollection 2014.

[73] Boink GJ, Duan L, Nearing BD, et al. HCN2/SkM1 gene transfer into canine left bundle branch induces stable, autonomically responsive biological pacing at physiological heart rates. J Am Coll Cardiol 2013;61(11):1192–201. http://dx.doi.org/10.1016/j.jacc.2012.12.031.

[74] Potapova I, Plotnikov A, Lu Z, Danilo Jr. P, Valiunas V, Qu J, et al. Human mesenchymal stem cells as a gene delivery system to create cardiac pacemakers. Circ Res 2004;94(7):952–9. 16.

[75] Shimamura M, Nakagami H, Koriyama H, Morishita R. Gene therapy and cell-based therapies for therapeutic angiogenesis in peripheral artery disease. Biomed Res Int 2013;186215. http://dx.doi.org/10.1155/2013/186215.

[76] Miao YL, Wu W, Li BW, Fang WW, Liu Y, Li L, et al. Clinical effectiveness of gene therapy on critical limb ischemia: a meta-analysis of 5 randomized controlled clinical trials. Vasc Endovascular Surg 2014;48(5-6):372–7. http://dx.doi.org/10.1177/1538574414539397.

[77] Turajane T, Thitiset T, Honsawek S, Chaveewanakorn U, Aojanepong J, Papadopoulos KI. Assessment of chondrogenic differentiation potential of autologous activated peripheral blood stem cells on human early osteoarthritic cancellous tibial bone scaffold. Musculoskelet Surg 2014;98(1):35–43. http://dx.doi.org/10.1007/s12306-013-0303-y.

[78] Cucchiarini M, Madry H. Use of tissue engineering strategies to repair joint tissues in osteoarthritis: viral gene transfer approaches. Curr Rheumatol Rep 2014;16(10):449. http://dx.doi.org/10.1007/s11926-014-0449-0.

[79] Zhang F, Yao Y, SuK Fang Y, Citra F, Wang DA. Co-transduction of lentiviral and adenoviral vectors for co-delivery of growth factor and shRNA genes in mesenchymal stem cells-based chondrogenic system. J Tissue Eng Regen Med 2012. http://dx.doi.org/10.1002/term.1656.

[80] Lee JM, Im GI. SOX trio-co-transduced adipose stem cells in fibrin gel to enhance cartilage repair and delay the progression of osteoarthritis in the rat. Biomaterials 2012;33(7):2016–24.

[81] Brunger JM, Huynh NP, Guenther CM, Perez-Pinera P, Moutos FT, Sanchez-Adams J, et al. Scaffold-mediated lentiviral transduction for functional tissue engineering of cartilage. Proc Natl Acad Sci U S A 2014;111(9):E798–806.

[82] Glass KA, Link JM, Brunger JM, Moutos FT, Gersbach CA, Guilak F. Tissue-engineered cartilage with inducible and tunable immunomodulatory properties. Biomaterials 2014;35(22):5921–31.

[83] Frisch J, Venkatesan JK, Rey-Rico A, Schmitt G, Madry H, Cucchiarini M. Determination of the chondrogenic differentiation processes in human bone marrow-derived mesenchymal stem cells genetically modified to overexpress transforming growth factor-β via recombinant adeno-associated viral vectors. Hum Gene Ther 2014; [Epub ahead of print].

[84] Orth P, Rey-Rico A, Venkatesan JK, Madry H, Cucchiarini M. Current perspectives in stem cell research for knee cartilage repair. Stem Cells Cloning 2014;7:1–17. http://dx.doi.org/10.2147/SCCAA.S42880. eCollection 2014.

[85] Madry H, Cucchiarini M. Tissue-engineering strategies to repair joint tissue in osteoarthritis: nonviral gene-transfer approaches. Curr Rheumatol Rep 2014;16(10):450. http://dx.doi.org/10.1007/s11926-014-0450-7.

[86] Branski LK, Gauglitz GG, Herndon DN, Jeschke MG. A review of gene and stem cell therapy in cutaneous wound healing. Burns 2009;35(2):171–80. http://dx.doi.org/10.1016/j.burns.2008.03.009.

[87] Lai WF, Siu PM. MicroRNAs as regulators of cutaneous wound healing. J Biosci 2014;39(3):519–24.

[88] Rodrigues MM, Soares IS. Gene-therapy for malaria prevention. Trends Parasitol 2014;30(11):511–3. http://dx.doi.org/10.1016/j.pt.2014.09.005.

[89] Yang L, Wang P. Passive immunization against HIV/AIDS by antibody gene transfer. Viruses 2014;6(2):428–47. http://dx.doi.org/10.3390/v6020428.

[90] Hicks MJ, Rosenberg JB, De BP, et al. AAV-directed persistent expression of a gene encoding anti-nicotine antibody for smoking cessation. Sci Transl Med 2012;4:140–87. http://dx.doi.org/10.1126/scitranslmed.3003611.

[91] Rosenberg JB, Hicks MJ, De BP, et al. AAVrh.10-mediated expression of an anti-cocaine antibody mediates persistent passive immunization that suppresses cocaine-induced behavior. Hum. Gene Ther 2012;23:451–9.

[92] Draz MS, Fang BA, Zhang P, Hu Z, Gu S, Weng KC, et al. Nanoparticle-mediated systemic delivery of siRNA for treatment of cancers and viral infections. Theranostics 2014;4(9):872–92. http://dx.doi.org/10.7150/thno.9404. eCollection 2014.

[93] Kaplitt MG, Feigin A, Tang C, Fitzsimons HL, Mattis P, Lawlor PA, et al. Safety and tolerability of gene therapy with an adeno-associated virus (AAV) borne GAD gene for Parkinson's disease: an open label, phase I trial. Lancet 2007;369(9579):2097–105.

[94] LeWitt PA, Rezai AR, Leehey MA, et al. AAV2-GAD gene therapy for advanced Parkinson's disease: a double-blind, sham-surgery controlled, randomised trial. Lancet Neurol 2011;10(4):309–19. http://dx.doi.org/10.1016/S1474-4422(11)70039-4.

[95] Nakata Y, Yasuda T, Mochizuki H. Recent progress in gene therapy for Parkinson's disease. Curr Mol Med 2012;12(10):1311–8.

[96] Palfi S, Gurruchaga JM, Ralph GS, et al. Long-term safety and tolerability of ProSavin, a lentiviral vector-based gene therapy for Parkinson's disease: a dose escalation, open-label, phase 1/2 trial. Lancet 2014;383(9923):1138–46. http://dx.doi.org/10.1016/S0140-6736(13)61939-X.

[97] Liu Z, Li S, Liang Z, et al. Reviewed by B Patel, S Wysong, WH Xu, and Q Niu. Edited by: WJ Mu, Y Qiu, CH Li, LP Song, WJ Liu, and M Zhao. Targeting β-secretase with RNAi in neural stem cells for Alzheimer's disease therapy. Neural Regen Res 2013; 8(33): 3095–3106. doi: 10.3969/j.issn.1673-5374.2013.33.003.

[98] Gaj T, Gersbach CA, Barbas 3rd. CF. ZFN, TALEN, and CRISPR/Cas-based methods for genome engineering. Trends Biotechnol 2013;31(7):397–405. http://dx.doi.org/10.1016/j.tibtech.2013.04.004.

[99] Harrison MM, Jenkins BV, O'Connor-Giles KM, Wildonger J. A CRISPR view of development. Genes Dev 2014;28(17):1859–72. http://dx.doi.org/10.1101/gad.248252.114.

[100] Nemudryi AA, Valetdinova KR, Medvedev SP, Zakian SM. TALEN and CRISPR/Cas genome editing systems: tools of discovery. Acta Naturae 2014;6(3):19–40.

[101] Takahashi K, Yamanaka S. Induction of pluripotent stem cells from mouse embryonic and adult fibroblast cultures by defined factors. Cell 2006;126(4):663–76.

[102] Nayler S, Gatei M, Kozlov S, et al. Induced pluripotent stem cells from ataxia-telangiectasia recapitulate the cellular phenotype. Stem Cells Transl Med 2012;1:523–35.
[103] Tolar J, McGrath JA, Xia L, et al. Patient-specific naturally gene-reverted induced pluripotent stem cells in recessive dystrophic epidermolysis bullosa. J Invest Dermatol 2014;134(5):1246–54.
[104] Tolar J, Xia L, Lees CJ, et al. Keratinocytes from induced pluripotent stem cells in junctional epidermolysis bullosa. J Invest Dermatol 2013;133(2):562–5.
[105] Chiba K, Hockemeyer D. Genome editing in human pluripotent stem cells using site-specific nucleases. Methods Mol Biol 2015;1239:267–80. http://dx.doi.org/10.1007/978-1-4939-1862-1_15.
[106] Choi PS, Meyerson M. Targeted genomic rearrangements using CRISPR/Cas technology. Nat Commun 2014;5:3728. http://dx.doi.org/10.1038/ncomms4728.
[107] Sánchez-Rivera FJ, Papagiannakopoulos T, Romero R, Tammela T, Bauer MR, Bhutkar A, et al. Rapid modelling of cooperating genetic events in cancer through somatic genome editing. Nature 2014. http://dx.doi.org/10.1038/nature13906 [Epub ahead of print].
[108] Crompton BD, Stewart C, Taylor-Weiner A, et al. Cancer Discov 2014;4(11):1326–41. http://dx.doi.org/10.1158/2159-8290.CD-13-1037.
[109] Liu Y, Zeng Y, Liu L, Zhuang C, Fu X, Huang W, et al. Synthesizing AND gate genetic circuits based on CRISPR-Cas9 for identification of bladder cancer cells. Nat Commun 2014;5:5393. http://dx.doi.org/10.1038/ncomms6393.
[110] Hütter G, Nowak D, Mossner M, et al. Long-term control of HIV by CCR5 Delta32/Delta32 stem-cell transplantation. N Engl J Med 2009;360(7):692–8. http://dx.doi.org/10.1056/NEJMoa0802905.
[111] Kim SS, Peer D, Kumar P, Subramanya S, et al. RNAi-mediated CCR5 silencing by LFA-1-targeted nanoparticles prevents HIV infection in BLT mice. Mol Ther 2010;18(2):370–6. http://dx.doi.org/10.1038/mt.2009.271.
[112] Li L, Krymskaya L, Wang J, et al. Genomic editing of the HIV-1 coreceptor CCR5 in adult hematopoietic stem and progenitor cells using zinc finger nucleases. Mol Ther 2013;21(6):1259–69. http://dx.doi.org/10.1038/mt.2013.65.
[113] Yao Y, Nashun B, Zhou T, Qin L, Qin L, Zhao S, et al. Generation of CD34+ cells from CCR5-disrupted human embryonic and induced pluripotent stem cells. Hum Gene Ther 2012;23(2):238–42. http://dx.doi.org/10.1089/hum.2011.126.
[114] Tebas P, Stein D, Tang WW, et al. Gene editing of CCR5 in autologous CD4 T cells of persons infected with HIV. N Engl J Med 2014;370(10):901–10. http://dx.doi.org/10.1056/NEJMoa1300662.
[115] Manjunath N, Yi G, Dang Y, Shankar P. Newer gene editing technologies toward HIV gene therapy. Viruses 2013;5(11):2748–66. http://dx.doi.org/10.3390/v5112748.
[116] Ousterout DG, Kabadi AM, Thakore PI, et al. Correction of dystrophin expression in cells from duchenne muscular dystrophy patients through genomic excision of exon 51 by zinc finger nucleases. Mol Ther 2014;http://dx.doi.org/10.1038/mt.2014.234.
[117] Li HL, Fujimoto N, Sasakawa N, et al. Precise correction of the dystrophin gene in duchenne muscular dystrophy patient induced pluripotent stem cells by TALEN and CRISPR-Cas9. Stem Cell Reports 2015;4(1):143–54. http://dx.doi.org/10.1016/j.stemcr.2014.10.013.
[118] Schwank G, Koo BK, Sasselli V, et al. Functional repair of CFTR by CRISPR/Cas9 in intestinal stem cell organoids of cystic fibrosis patients. Cell Stem Cell 2013;13(6):653–8. http://dx.doi.org/10.1016/j.stem.2013.11.002.
[119] Gilbert LA, Larson MH, Morsut L, et al. CRISPR-mediated modular RNA-guided regulation of transcription in eukaryotes. Cell 2013;154(2):442–51. http://dx.doi.org/10.1016/j.cell.2013.06.044.

[120] Wu Y, Zhou H, Fan X, et al. Correction of a genetic disease by CRISPR-Cas9-mediated gene editing in mouse spermatogonial stem cells. Cell Res 2014. http://dx.doi.org/10.1038/cr.2014.160.

[121] Sidahmed A, Abdalla S, Mahmud S, Wilkie B. Antiviral innate immune response of RNA interference. J Infect Dev Ctries 2014;8(7):804–10. http://dx.doi.org/10.3855/jidc.4187.

[122] Srinivasan S, Selvan ST, Archunan G, Gulyas B, Padmanabhan P. MicroRNAs -the next generation therapeutic targets in human diseases. Theranostics 2013;3(12):930–42. http://dx.doi.org/10.7150/thno.7026.

[123] Kim KH, Park KK. Small RNA- and DNA-based gene therapy for the treatment of liver cirrhosis, where we are? World J Gastroenterol 2014;20(40):14696–705.

[124] Hong CA, Nam YS. Functional nanostructures for effective delivery of small interfering RNA therapeutics. Theranostics 2014;4(12):1211–32. http://dx.doi.org/10.7150/thno.8491.

[125] Lages E, Ipas H, Guttin A, Nesr H, Berger F, Issartel JP. MicroRNAs: molecular features and role in cancer. Front Biosci (Landmark Ed) 2012;17:2508–40.

[126] Parasramka MA, Ho E, Williams DE, Dashwood RH. MicroRNAs, diet, and cancer: new mechanistic insights on the epigenetic actions of phytochemicals. Mol Carcinog 2012;51(3):213–30. http://dx.doi.org/10.1002/mc.20822.

[127] Khoury S, Tran N. Circulating microRNAs: potential biomarkers for common malignancies. Biomark Med 2015;9(2):131–51. http://dx.doi.org/10.2217/bmm.14.102.

[128] Mohr AM, Mott JL. Overview of MicroRNA Biology. Semin Liver Dis 2015;35(1):3–11. http://dx.doi.org/10.1055/s-0034-1397344.

[129] Xu T, Liu Q, Yao J, Dai Y, Wang H, Xiao J. Circulating microRNAs in response to exercise. Scand J Med Sci Sports 2015 Feb 3;http://dx.doi.org/10.1111/sms.1242.

[130] Flowers E, Won GY, Fukuoka Y. MicroRNAs associated with exercise and diet: a systematic review. Physiol Genomics 2015;47(1):1–11. http://dx.doi.org/10.1152/physiolgenomics.00095.2014.

[131] Farooqi AA, Rehman ZU, Muntane J. Antisense therapeutics in oncology: current status. Onco Targets Ther 2014;7:2035–42. http://dx.doi.org/10.2147/OTT.S49652. eCollection 2014.

[132] Rayburn ER, Zhang R. Antisense, RNAi, and gene silencing strategies for therapy: mission possible or impossible? Drug Discov Today 2008;13(11-12):513–21. http://dx.doi.org/10.1016/j.drudis.2008.03.014.

[133] Moreno PM, Pêgo AP. Therapeutic antisense oligonucleotides against cancer: hurdling to the clinic. Front Chem 2014;2:87. http://dx.doi.org/10.3389/fchem.2014.00087. eCollection 2014.

[134] Mulcahy PJ, Iremonger K, Karyka E, et al. Gene therapy: a promising approach to treating SMA. Hum Gene Ther 2014;25:575–86. http://dx.doi.org/10.1089/hum.2013.186.

[135] Wu B, Cloer C, Lu P, Milazi S, Shaban M, Shah SN, et al. Exon skipping restores dystrophin expression, but fails to prevent disease progression in later stage dystrophic dko mice. Gene Ther 2014;21(9):785–93. http://dx.doi.org/10.1038/gt.2014.53.

[136] Ley D, Van Zwieten R, Puttini S, Iyer P, Cochard A, Mermod N. A PiggyBac-mediated approach for muscle gene transfer or cell therapy. Stem Cell Res 2014;13(3 Pt A):390–403. http://dx.doi.org/10.1016/j.scr.2014.08.007.

[137] Le Guiner C, Montus M, Servais L, et al. Forelimb treatment in a large cohort of dystrophic dogs supports delivery of a recombinant AAV for exon skipping in Duchenne patients. Mol Ther 2014;22(11):1923–35. http://dx.doi.org/10.1038/mt.2014.151.

[138] Monani UR, DeVivo DC. Neurodegeneration in spinal muscular atrophy: from disease phenotype and animal models to therapeutic strategies and beyond. Future Neurol 2014;9(49):65. http://dx.doi.org/10.2217/fnl.13.58.
[139] Zheng L, Wang L, Gan J, Zhang H. RNA activation: promise as a new weapon against cancer. Cancer Lett 2014;355(1):18–24. http://dx.doi.org/10.1016/j.canlet.2014.09.004.
[140] Chamberlain JR, Schwarze U, Wang PR, et al. Gene targeting in stem cells from individuals with osteogenesis imperfecta. Science 2004;303(5661):1198–201.
[141] Rothe M, Schambach A, Biasco L. Safety of gene therapy: new insights to a puzzling case. Curr Gene Ther 2014;14(6):429–36.
[142] Holkers M, Maggio I, Liu J, Janssen JM, Miselli F, Mussolino C, et al. Differential integrity of TALE nuclease genes following adenoviral and lentiviral vector gene transfer into human cells. Nucleic Acids Res 2013;41(5):http://dx.doi.org/10.1093/nar/gks1446. e63.
[143] Kanherkar RR, Bhatia-Dey N, Csoka AB. Epigenetics across the human lifespan. Front Cell Dev Biol 2014;2:49. http://dx.doi.org/10.3389/fcell.2014.00049 eCollection 2014.
[144] Kanherkar RR, Bhatia-Dey N, Makarev E, Csoka AB. Cellular reprogramming for understanding and treating human disease. Front Cell Dev Biol 2014;2:67. http://dx.doi.org/10.3389/fcell.2014.00067. eCollection 2014.
[145] Eggenschwiler R, Loya K, Wu G, Sharma AD, Sgodda M, Zychlinski D, et al. Sustained knockdown of a disease-causing gene in patient-specific induced pluripotent stem cells using lentiviral vector-based gene therapy. Stem Cells Transl Med 2013;2(9):641–54. http://dx.doi.org/10.5966/sctm.2013-0017.
[146] Guo S, Booten SL, Aghajan M, et al. Antisense oligonucleotide treatment ameliorates alpha-1 antitrypsin-related liver disease in mice. J Clin Invest 2014;124:251–61. http://dx.doi.org/10.1172/ JCI67968.
[147] Guo S, Booten SL, Watt A, Alvarado L, Freier SM, Teckman JH, et al. Using antisense technology to develop a novel therapy for α-1 antitrypsin deficient (AATD) liver disease and to model AATD lung disease. Rare Dis 2014;2:e28511. http://dx.doi.org/10.4161/rdis.28511. eCollection 2014.
[148] Hogan DB. Long-term efficacy and toxicity of cholinesterase inhibitors in the treatment of Alzheimer disease. Can J Psychiatry 2014;59(12):618–23.
[149] Lee HJ, Lim IJ, Park SW, et al. Human neural stem cells genetically modified to express human nerve growth factor (NGF) gene restore cognition in the mouse with ibotenic acid-induced cognitive dysfunction. Cell Transplant 2012;21(11):2487–96.
[150] Park D, Joo SS, Kim TK, et al. Human neural stem cells overexpressing choline acetyltransferase restore cognitive function of kainic acid-induced learning and memory deficit animals. Cell Transplant 2012;21(1):365–71.
[151] Park D, Lee HJ, Joo SS, et al. Human neural stem cells over-expressing choline acetyltransferase restore cognition in rat model of cognitive dysfunction. Exp Neurol 2012;234(2):521–6.
[152] Jabłkowski M, Szemraj M, Oszajca K, Janiszewska G, Bartkowiak J, Szemraj J. New type of BACE1 siRNA delivery to cells. Med Sci Monit 2014;20:2598–606. http://dx.doi.org/10.12659/MSM.891219.
[153] Chien WW, Monzack EL, McDougald DS, Cunningham LL. Gene therapy for sensorineural hearing loss. Ear Hear 2015;36(1):1–7. http://dx.doi.org/10.1097/AUD.0000000000000088.
[154] Maeda Y, Fukushima K, Nishizaki K, Smith RJ. In vitro and in vivo suppression of GJB2 expression by RNA interference. Hum Mol Genet 2005;14(12):1641–50.

[155] Lentz JJ, Jodelka FM, Hinrich AJ, et al. Rescue of hearing and vestibular function by antisense oligonucleotides in a mouse model of human deafness. Nat Med 2013;19(3):345–50. http://dx.doi.org/10.1038/nm.3106.

[156] Guo W, Yuan Y, Hou Z, Wang X, Yang S. Profiles of the auditory epithelia related microRNA expression in neonatal and adult rats. Eur J Med Res 2014;19(1):48.

[157] Ushakov K, Rudnicki A, Avraham KB. MicroRNAs in sensorineural diseases of the ear. Front Mol Neurosci 2013;6:52. http://dx.doi.org/10.3389/fnmol.2013.00052.

[158] Atkinson PJ, Wise AK, Flynn BO, Nayagam BA, Richardson RT. Hair cell regeneration after ATOH1 gene therapy in the cochlea of profoundly deaf adult guinea pigs. PLoS One 2014;9(7):e102077. http://dx.doi.org/10.1371/journal.pone.0102077. eCollection 2014.

[159] Tsezou A. Osteoarthritis year in review 2014: genetics and genomics. Osteoarthritis Cartilage 2014;22(12):2017–24. http://dx.doi.org/10.1016/j.joca.2014.07.024.

[160] Zhang F, YaoY SuK, Fang Y, Citra F, Wang DA. Co-transduction of lentiviral and adenoviral vectors for co-delivery of growth factor and shRNA genes in mesenchymal stem cells-based chondrogenic system. J Tissue Eng Regen Med 2012. http://dx.doi.org/10.1002/term.1656.

[161] Li LB, Chang KH, Wang PR, Hirata RK, Papayannopoulou T, Russell DW. Trisomy correction in down syndrome induced pluripotent stem cells. Cell Stem Cell 2012;11(5):615–9. http://dx.doi.org/10.1016/j.stem.2012.08.004.

[162] Jiang J, Jing Y, Cost GJ, et al. Translating dosage compensation to trisomy 21. Nature 2013;500(7462):296–300. http://dx.doi.org/10.1038/nature12394.

[163] Brzeziańska E1, Domańska D1, Jegier A2. Gene doping in sport - perspectives and risks. Biol Sport 2014;31(4):251–9. http://dx.doi.org/10.5604/20831862.1120931.

Biochemical assays used for *in vitro* and *in vivo* gene expression

11

M. Ahmed[1], *R. Narain*[2]
[1]California Institute of Technology, Pasadena, CA, USA; [2]University of Alberta, Edmonton, AB, Canada

11.1 Introduction

Gene expression profiling has opened new venues in the field of medicine. The highly sensitive and quantitative gene expression analysis assays serve as important tool in the field of medicine. The biochemical assays employed for gene expression analysis are not only focused on efficient delivery of exogenous DNA or RNA (such as small interfering ribonucleic acids, siRNA) into living organisms, but are also involved in the identification of various diseases and their cures [1–5]. Some of the distinct advantages of efficient biochemical assays for gene expression analysis include:

- Delivery of exogenous DNA and its expression [2]
- Detection of aberrant genes in diseased tissue [3]
- Detection of single nucleotide polymorphism [4]

The delivery of exogenous DNA into living cells and its role as potential therapeutics has been a topic of interest during the last three decades [5]. A crucial requirement for the successful implementation of this therapy is the development of highly sensitive and quantitative gene expression assays. The significant advancement in the field of medicine and molecular biology has further highlighted the importance of potent biochemical assays for gene expression analysis. The exploration of genetic profiles of cancerous or diseased tissue helps in the identification of aberrant genes as well as their relevance to therapeutic decision [6,7]. The gene expression analysis also permits the comparison of significant genetic variations, as compared to the insignificant ones [4]. This chapter outlines the commonly used assays employed in research and clinics for gene expression analysis and quantification. The choice of assays for a specific application is based upon the level of detection of genetic expression, as well as the quality of data needed. The variety of assays developed for specific detection of cDNA, RNA, and protein are outlined in Figure 11.1. The commonly used biochemical assays are generally developed on two basic techniques, which are electrophoresis and hybridization. The reporter gene expression systems are an exception and measure the gene expression at translation level by either optical or enzymatic approach. Reporter gene assays are also used to analyze the activity of promoter, enhancers, and transcription elements in the plasmid [8]. Electrophoretic assays allow the separation

Polymers and Nanomaterials for Gene Therapy. http://dx.doi.org/10.1016/B978-0-08-100520-0.00011-4
© 2016 Elsevier Ltd. All rights reserved.

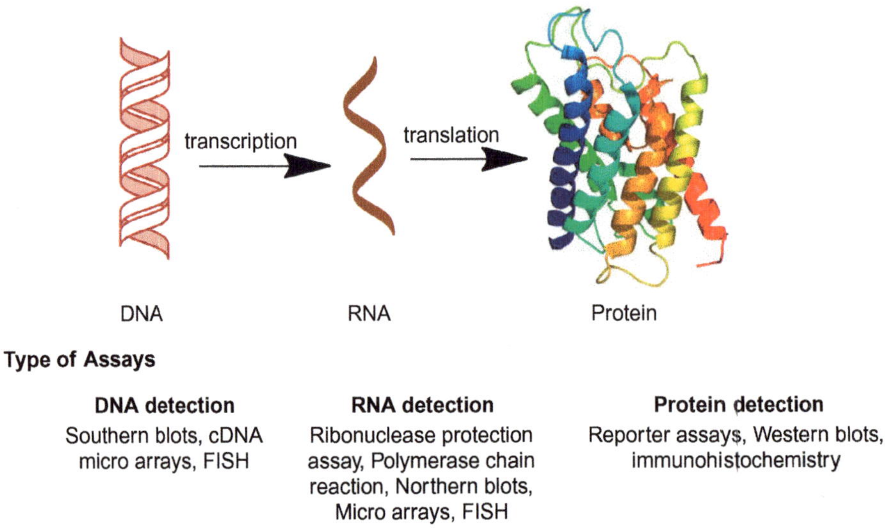

Figure 11.1 Central dogma of gene expression and list of biochemical assays used for their detection at transcription and translation level.

and identification of DNA, RNA, or proteins based upon their size and charge [9]. The hybridization-based assays are highly sensitive, and quantitative methods for gene expression analysis and are now being used in clinics to measure the extent of aberrant gene expression [10]. Polymerase chain reaction (PCR), micro-arrays, and fluorescent *in situ* hybridization (FISH) techniques are based on the hybridization of specific probes (DNA or RNA) to the gene of interest. Western blotting and immunohistochemistry are examples of gene analysis methods at translational level. These techniques are based on the hybridization of proteins (antibodies) with the protein of interest. These assays are discussed in detail.

11.2 Reporter gene assay system

The reporter gene system is one of the most commonly used assays for gene expression analysis in research. The fundamental concept of this method is that cDNA or reporter gene is introduced into the biological system, which provides a measurable signal upon modulation of gene expression. The characteristics of ideal reporter gene system are lack of endogenous expression of reporter proteins in target cells, high sensitivity, and reproducibility. The design of reporter vectors requires careful consideration and is dependent on their physiological implications. The typical reporter plasmids are replicated in *E. coli* and contain a DNA replication site *(ori)*, a gene for selection, along with ribosomal binding site, and polyadenylation signal. The coding and flanking regions, introns, and multiple cloning sites are included in the plasmids depending on the type of reporter assay and its biological significance. In addition to evaluating transfection efficacies, reporter gene expression assays are also used to monitor protein-protein interactions, protein subcellular localization, and screen libraries of genetic regulatory events. Chloramphenicol acetyltransferase (CAT) and lac Z1 [11] are among the first examples of genetic reporter assay systems used to measure the transcription efficacies in mammalian cells [12]. The characteristics of commonly used reporter systems are summarized in Table 11.1. Since then, a number of reporter assay systems have been developed and studied for the evaluation of gene expression efficacies. These assay systems include, green and red fluorescent protein (GFP, RFP), luciferase assay, alkaline phosphatase (AP), β-galactosidae (β-gal), β-glucoronidase (GUS), β-lactamase, and growth hormone (GH) [12,13].

Reporter gene systems are chosen based on the type of assay as well as the advantages and limitations associated with the system. These advantages and disadvantages are outlined in Table 11.2.

11.2.1 In vivo *gene expression*

The repetitive imaging of reporter gene expression is critical for many studies in living animals. Several imaging techniques and reporter gene assays are studied for non-invasive imaging of live animals, which include optical imaging approaches (such as fluorescence, luminescence), magnetic resonance imaging (MRI), positron emission spectroscopy (PET), and single photon emission tomography (SPECT) [14–22].

11.2.2 Optical imaging

Optical imaging is a low-cost method to study *in vivo* gene expression in small animals. The method requires the fusion of coding region of gene of interest with green or red fluorescent protein plasmid. This allows the simultaneous detection and imaging of reporter plasmid in the same animal, hence decreasing the number of animals and increasing the power of statistical analysis. As compared to fluorescence, which requires absorption and excitation (two wavelength) spectra, luminescence is a one-wavelength spectroscopy approach. Optical imaging of tissue is relatively easy near the skin surface but considerable limitations occur as the depth of tissue increases [14]. Some limitations with *in vivo* optical imaging approaches include:

- Strong absorption of incident light (near UV and visible part of spectra) by tissue
- Weak penetration of light at large wavelengths
- Light absorbance by hemoglobin and endogenous fluorophores
- Light scattering in tissue due to refractive index mismatch

Some technical approaches can be used to fix these issues. The band-pass filters near excitation wavelength can block the light outside their transmission range, while emission filters (band-pass or long-pass cut-on filters) can block the light scattered from tissue. The selection of excitation and emission filters requires careful consideration and narrow wavelengths, and no overlap should be present between excitation and emission wavelengths. However, due to larger emission spectra of filters auto-fluorescence does occur and needs to be corrected by imaging software. Using a powerful light source such as optical fibers (LED ring lights) allows

Table 11.1 **Types of reporter assay systems and their biochemical properties**

Reporter Assays	CAT	β-gal	SEAP	GFP	Luciferase	β-GUS	β-lactamse
Reporter gene origin	Transposon 9, *E.Coli*	lacZ gene, E.Coli	placental alkaline phosphatase lacking 24 amino acids	GFP, jellyfish *Aequorea victoria*	Luc gene, firefly *Photinus Pyralis, Renilla Luciferase*	Gus gene from *E.Coli*	TEM-1 β-lactamase of *E. coli*
Protein size (kDa)	25	116	variable	27	60.7	68	29
Substrate	Chloramphenicol, acetyl CoA	β-D-galactopyranoside	Phosphatase at alkaline pH	None	D-luciferin, ATP, oxygen, Mg^{2+}	β-glucuronides	PADAC and nitrocefin, CCF2
Reaction Mechanism	Enzymatic, Transfer of Acetyl CoA at position 3 of Cm by CAT	Enzymatic, Catalysis of β-galactoside by β-galactosidase	Dephosphorylation of broad range of proteins at alkaline pH	None	Enzymatic, light emission by catalysis of D-luciferin in the presence of ATP, Mg^{2+} and oxygen	Enzymatic, GUS removes terminal β-glucorunic acid residues from GAGs and glyco-conjugates	Enzymatic, an esterase cleaves ampicillin and other antibiotics, commonly known as ampicillin resistance gene
Assay format	Isotopic (^{14}C-, or ^{3}H-Cm) fluorescent (Bodipy-Cm), ELISA	Colorimetric (ONPG, CPRG), fluorescent (MUG, FDG), chemiluminiscent (1,2-dioxetane-β-gal), ELISA (X-Gal)	Colorimetric (PNPP), fluorescent (FADP), bioluminescent	Fluorescent, ELISA	Bio-luminescent	Colorimetric (X-gluc), fluorescent (4-MUG), chemiluminiscent (1,2-dioxetane aryl glucuronide, ELISA (X-gluc)	Colorimetric (PADAC, nitrofectin), fluorescent (CCF2, CCF2/AM), ELISA

Cm, chloramphenicaol; ELISA, enzyme linked immunosorbent assay; ONPG, Ortho-nitrophényl-β-galactoside; CPRG, Chlorophenol red-ß-D-galactopyranoside; MUG or FDG, fluorescent β-galactoside; PNPP, *p*-Nitrophenyl Phosphate; FADP, flavin-adenine dinucleotide phosphate; PADAC, pyridinium-2-azo-p-dimethylaniline chromophore; SEAP, secreted alkaline phosphatase; GUS, β-glucuronidase

Table 11.2 Comparison of advantages and drawbacks of different gene reporter systems

Reporter gene	Advantages	Limitations
CAT	Widely used in research	Relatively low sensitivity, high cost for TLC and isotopes
Luciferase	Fast and easy, high sensitivity, linear detection range	Luminometer required
GFP	No substrate, broad host capability	Low sensitivity, expensive microscopes, toxic at high level of expression
β-galactosidase	Easy to assay	Low sensitivity in non-chemiluminescent format
β-lactamase	Easy to assay	Relatively low sensitivity
β-GUS	Used for in situ staining of fusion proteins	Endogenous activity in mammalian cells
SEAP	Low background activity	

illumination of large samples. Similarly using highly sensitive detection cameras such as photomultiplier tubes (PMT), video cameras, and cooled CCD cameras allow improvement in signal-to-noise ratio. Cooled CCD cameras allow significant advantage over other systems when working with red fluorophores, hence allowing quantum efficiencies up to 80% along with high sensitivity. Furthermore, CCDs illuminated with back-sided chips allow broader spectral response detection in NIR part of the spectrum. In contrast, luminescence detection requires an ultra-high gain intensified CCD camera with a double micro-channel plate image intensifier to allow photon-counting efficiency and high sensitivity. The spatial resolution and imaging time is analyzed by "binning," which is simultaneous reading of neighboring pixels to improve incident flux. Binning is significantly advantageous for fluorescence spectroscopy, however improvement in CCD cameras is required to allow better imaging by luminescent assays [17]. Others have argued the successful use of GFP, as a fluorescent fluorphore for tumor imaging in animals is controversial. It is reported that fluorescent signal reduces under hypoxic conditions, a hallmark of cancer tissue [18].

11.2.3 Magnetic resonance imaging (MRI)-based reporter gene expression

MRI signal is a function of mobile protons and their relaxation time in the corresponding area. Gadolinum (Gd) and iron (Fe) are the most commonly used MR contrast enhancement agents. Gd shortens T1 (spin lattice relaxation time), while Fe shortens T2 (spin-spin relaxation time) and produce bright images in the corresponding areas. Reporter gene imaging method improves the resolution of MRI technique, by enhancing the relaxation time (T1/T2) signal. For example, Lac-Z transfected tumors were imaged in the presence of 3,4-cyclohexenoesculetin β-$_D$-galactopyranoside and ferric ion, which enhanced T2 relaxation time. A major problem with this type of imaging is restricted access of substrate in transfected cells in living animals. Ferritin is a ubiquitous protein, which stores and releases free iron to maintain homeostasis. In a C-6 glioma xenografts, ferritin was over-expressed using Tetracycline (Tet)-responsive inducible plasmid. The absence of tetracycline was accompanied by increased spin-spin relaxation rate [18,19]. Mag A gene, a gene that synthesizes magnetosome in bacteria, has gained attention for its implication in MRI therapy, providing possible new ways for non-invasive imaging. The magneotsome produced by Mag A are similar in characteristic to super-paramagnetic iron-oxide nanoparticles (SPION) [19].

The use of reporter plasmids to introduce exogenous DNA into living systems is widely used to study the gene delivery efficacies of non-viral vectors. The development of small interferring ribonucleic acid (siRNA) based therapeutics requires the establishment of sensitive approaches for the detection of gene knock-down efficacies at both transcriptional and translational levels *in vitro* and *in vivo* [23]. As discussed above, protein and nucleic acids can be detected by electrophoretic or hybridization techniques.

11.3 Electrophoretic techniques

Western, southern, and northern blots are electrophoresis-based techniques that allow the separation of DNA, RNA, and proteins, respectively. The gene expression at transcription level can be detected by northern blots and ribonuclease protection assay (RPA). Both assays require the presence of radio-labeled or fluorescent sequence specific probes for hybridization with RNA. The final product is labeled and is detected by densitometer or phosphor-imager. RPA involves the presence of ribonucleases to digest unwanted single-stranded RNAs. The quantitative accuracy depends on the type of detection method and standards used. The sensitivity of imaging also depends on the exposure time of the gel. As an alternative to radioactive probes, probes labeled with antibody or hapten can be utilized and imaging is done in the presence of substrate by the chemiluminescent method. The background subtraction of gel is required to normalize the values for both methods. The great advantage of gel-based systems are that a number of samples including housekeeping genes can be analyzed simultaneously and under the same conditions [10].

11.3.1 Western blots

Western blotting also known as immunoblotting or protein blotting, introduced by Towbin et al., is a commonly used technique for the immune detection and quantification of proteins *in vitro* [24]. Western blots allow the resolution of proteins on SDS-PAGE as a function of their molecular weights, followed by their transfer on an adsorbent membrane. The subsequent addition of antibody probes allows the detection of specific proteins on the membrane. The efficiency of protein transfer from gel to membrane depends on the gel softness and thickness. Softer and thinner

Table 11.3 Comparison of electrophoretic techniques to evaluate ribonucleic acids [10,24]

Assay	Advantages	Disadvantages
Northern blot	Most commonly used assay, large transcripts can be analyzed and retrieved, low sample to sample variance	Requires a large amount of RNA, labor intensive, limited sensitivity, large errors
Ribonuclease protection assay (RPA)	More sensitive, less labor intensive, multiple probes can be used to simultaneously track a number of genes of interest	Requires a large amount of mRNA sample

gels allow efficient transfer of proteins. Nitrocellulose and polyvinylidene difluoride (PVDF) papers are the most commonly used membranes used for the western blotting approach. The transfer of proteins from gel to membrane can be achieved by 1) simple diffusion, 2) vacuum-assisted solvent flow, and 3) electrophoretic elution. The antigen detection of blocked membranes can be achieved directly by their incubation with primary fluorescent antibodies. In another method, the antigens are probed with primary antibody followed by fluorphore or enzyme labeled secondary antibody, which further helps in the detection of primary antibody. The most commonly used enzyme-labeled secondary antibodies are horse-reddish peroxidase or phosphatase-labeled antibodies, which yield colored products in the presence of substrate. However, the quantification of proteins is the major issue with this mode of detection. Fluorescently and radioactively labeled antibodies allow the detection of antigens on western blots, with high sensitivity (Table 11.3).

Chemiluminscence is another widely used approach for the detection of antigens. Chemiluminescence and radioactive labeling both require X-ray spectroscopy for film development and are widely used approaches for protein quantification. Gold and biotin-avidin labeled antibodies are also being used for the detection of proteins [25].

11.3.2 Immunohistochemistry (IHC)

IHC is a term used to determine the protein content on the tissue of interest, by employing antibodies, followed by their detection in a manner similar to western blotting. The first step in IHC is slide preparation involving, tissue, or cell fixtures, followed by antigen retrieval and blocking. The slides are stained with antibodies of interest and are detected and analyzed due to fluorphore- or enzyme-labeled antibodies. This method is widely used to study the protein expression in tissue of treated animal. The limitations of this approach include vigorous training and experience. The acquisition, handling, and fixation of specimen are all critical factors. For example, thickness of specimen may play a critical role in data acquisition. A slice less than 3 μm in thickness may provide a very weak immune staining, while slides greater than 7 μm thick may hamper antigen site. Similarly, antigen retrieval steps such as high temperature can destroy the antigenicity of proteins. The analysis of IHC expression is usually done in a qualitative manner, hence the presence of positive and negative control slides are critical for the evaluation of results. In addition, the validity of the expression can be confirmed by using antibodies with different epitopes for the same antigen. Since there are a lot of variables in IHC, intra-laboratory reproducibility is a major issue with this technique [26].

11.4 Hybridization techniques

11.4.1 PCR and micro-arrays

PCR and micro-arrays require a relatively small amount of RNA. The accuracy of techniques and the need for a small amount of sample make them reliable tools for clinical applications.

PCR was pioneered by Kary Milus in 1983, and is based on the amplification of DNA copy number by DNA polymerase. PCR involves reverse transcription of mRNA to cDNA as the first step, which is then amplified using gene specific primers, hence providing multiple copies of the gene of interest. The amount of final product obtained by PCR is proportional to the amount of starting material in the sample, which allows the quantitation of the gene of interest [27]. The primer design is critical in ensuring the amplification of the gene of interest. The highly specific primers are designed by performing basic local alignment search tool (BLAST) on the target regions of the gene of interest. For relative and absolute quantification, the amount of mRNA added to the assay is a critical factor. The carefully calculated and pipetted pure mRNA is added to all samples to obtain high-quality data. The amount of mRNA is also normalized against a housekeeping gene such as β-actin or glyceraldehyde-3-phosphate dehydrogenase (GADPH). However, conventional PCR suffers from certain drawbacks such as accumulation of primer dimers, PCR product re-annealing, end-point saturation, and non-specific binding of DNA-polymerase to primers, which further decreases the sensitivity of the assay. Real-time PCR or RT-qPCR is another approach that allows the direct monitoring of PCR reaction as it progress, hence circumventing the problems associated with end-point saturation and providing the most accurate data on gene-expression levels. RT-qPCR is undertaken in a thermocycler, and the amount of fluorescence detected is directly proportional to the amount of product produced by PCR. The fluorescence detection can be done by using DNA intercalated dyes such as SYBR green or by using hybridization probes. SYBR green intercalates with all dsDNA and hence cannot differentiate between unwanted products such as primer-dimers and PCR product. Hybridization probes are two oligonucleotides that bind to two adjacent sides of PCR product. The oligonucleotide at 3'-end is labeled with donor probe such as fluorescin, while the oligonucleotide at 5'-end is labeled with acceptor probe, such as rhodamine. The specific binding of both probes results in fluorescence of acceptor probe by excitation of donor probe. In another approach, PCR product is labeled with a reporter dye at 5'-end and a quencher dye at 3'-end. Taq DNA polymerase with 5'-nuclease activity cleaves quencher dye and releases the amplified DNA. The mechanism of PCR is shown in Figure 11.2 [28].

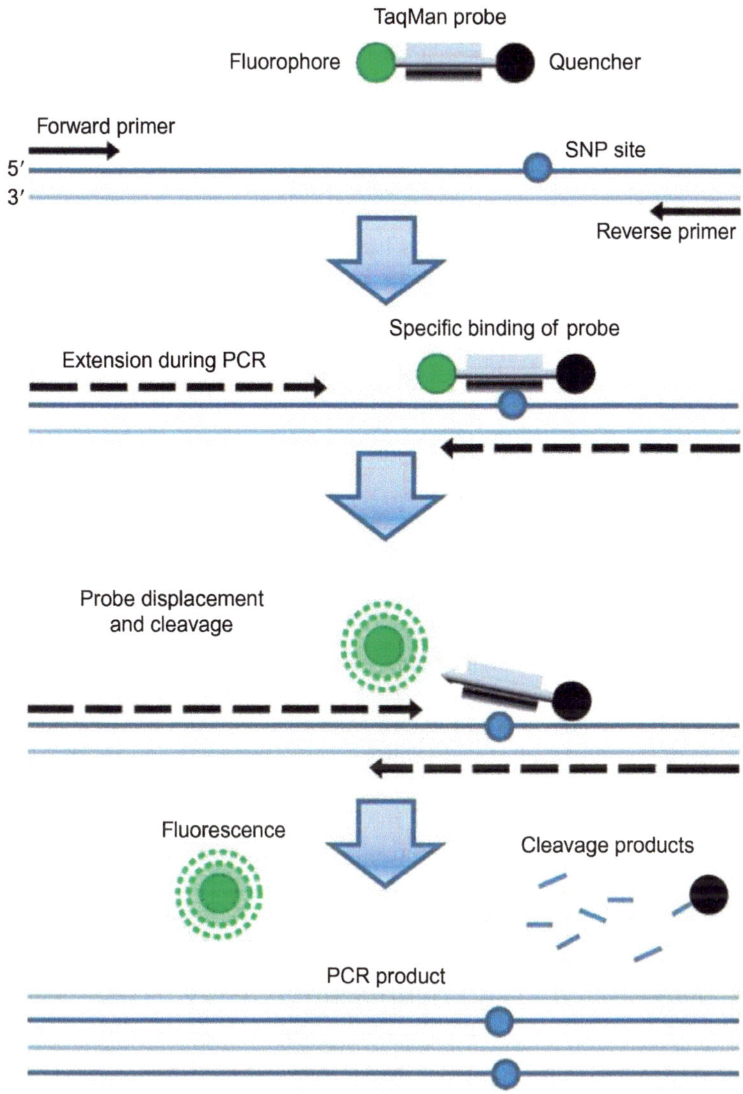

Figure 11.2 Depiction of mechanism of polymerase chain reaction (PCR) [28]

The sensitivity of PCR based assays is 4–5 folds higher than RPA assay. The individual efficiency of each sample is calculated from the slope of log of (fluorescence versus cycle number of plots) and is normalized with a housekeeping gene [10]. One of the major disadvantages of PCR is the choice of internal controls or housekeeping gene for certain experimental conditions. Glyceraldehyde-3-phosphate dehydrogenase (GADPH) and β-actin are examples of commonly used housekeeping genes. However, fluctuation in housekeeping gene expression in treated cells and careful experiment evaluation of internal controls is required for PCR assays.

Competitive PCR is an alternative type of PCR that allows the evaluation of target gene and a standard in the same reaction. A small DNA fragment (competitor) is added to the reaction. The competitor contains the same sequence for prime binding which is present on target gene. The amount of target DNA is estimated by measuring the ratio of amplified product from target gene to the product from competitor DNA. A major drawback of competitive PCR is that sizes of sample and standards need to be distinctly different for their separation by agarose gel electrophoresis. Moreover, agarose gel electrophoresis has low dynamic range and cannot resolve the issues related to heterodimers formed between sample DNA and standards. This problem was solved by using combination of competitive-PCR and matrix associated laser desorption ionization time of flight (MALDI-TOF) mass spectrometry. MALDI-TOF MS allows highly sensitive detection and quantification and distinction of heterodimers from PCR products [29].

11.4.2 Micro-arrays

Micro-array-based technologies have revolutionized the field of biological research. The micro-arrays are involved in the design of a set of sequences corresponding to the gene of interest. The sequences can be oligonucleotide probes or cDNA probes, complimentary to the transcript being analyzed. The probes are deposited on the glass, silicon, nylon, or polymeric substrates at the density of 10^3-10^5 probes/cm^2. The samples (DNA/RNA) are extracted, fluorescently labeled, and hybridized on DNA microarray substrate. The samples are washed, and fluorescence is measured by confocal laser scanning microscopy [10,28,30–32]. The general mechanism of micro-arrays is depicted in Figure 11.3. Micro-arrays are routinely used to explore changes in the genetic expressions of cancerous tissue [33]. The expression of gene of interest is quantified by measuring the ratio of fluorescence intensity of target gene and fluorescence intensity of all the genes on chip.

11.4.3 Fluorescent in situ *Hybridization Assay (FISH)*

FISH is an assay based on the principle of specific annealing of nucleic acid probes to complementary sequences in fixed tissue, followed by their detection. The earlier nucleic acid detection techniques involved radioactive, chromogenic, and gold-based probe systems. The first application of FISH was demonstrated in the 1980s, when RNA was fluorescently labeled for the detection of specific DNA. Fluorescent probes were generally prepared by enzymatic incorporation of fluorophore, or modified bases along the length of nucleic acid. To further improve the sensitivity and specificity of the system, alkyl and allyl modified bases were incorporated in the probes, followed by their chemical conjugation with fluorophores or haptens [33,34]. In comparison to northern and southern blots, FISH allows the detection of nucleic acids *in situ*. FISH is a highly specific and sensitive technique and is accepted as a general protocol in clinics to identify abnormalities in nucleic acids (Table 11.4).

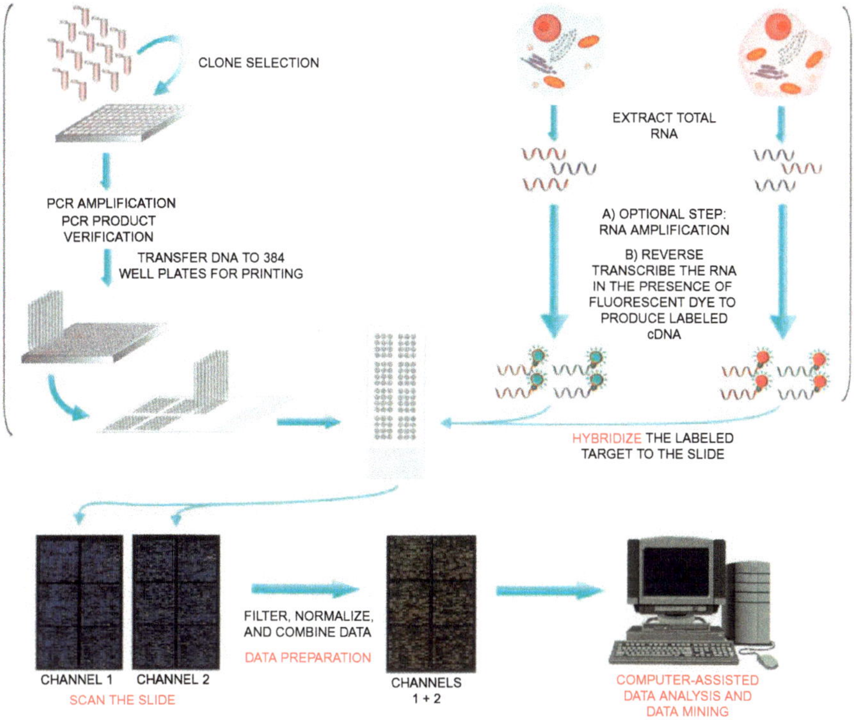

Figure 11.3 Mechanism of microarrays for DNA and RNA detection [32]

Table 11.4 **Comparison of different hybridization techniques [27–34]**

Assay	Advantages	Disadvantages
PCR	Widely used, quantitative, robust, reliable, need small amount of starting material, gold standard for gene expression analysis	Expensive reagents and instruments, choice of housekeeping gene
Micro-array	Quantitative, need small amount of material, capacity to perform whole genome profiling	Large signal to noise ratio, quantitation is difficult, relatively new techniques and rules are still emerging, high cost, statistical analysis of results is needed, false positive and negative results
FISH	*In situ* hybridization, quantitative, highly specific, imaging of many distinct targets at once	Sample to sample variations, auto fluorescence, overlapping of photons, computational methodologies are required for data analysis

11.4.4 Hybrid assays

A dual-reporter-based gene assay system was introduced to enhance the sensitivity and to minimize experimental error of luciferase assay [35]. A dual luciferase plasmid-bearing Fluc and Rluc gene was constructed. The plasmid was used for the identification of miRNA and its possible target mRNAs in cancer cells [36]. GFP- and RFP-based dual reporter fluorescence system was constructed and co-transfected with effector plasmid to determine the silencing efficacies of dsRNAi. The cells were co-expressed with a marker and translational fusion fluorescent protein bearing plasmid. The effector plasmid contains inverted repeats of target gene. Co-transfection of both plasmids resulted in lack of fluorescence of fusion protein [37]. In another approach, two fluorescent reporter constructs were used to predict siRNA efficacies with high fidelity. First, construct was based upon GFP plasmid, carrying siRNA for the gene of interest. The second construct contained 19-nucleotide target sequence with red fluorescent protein sequence. The ability of shRNA to knock down RFP expression was measured at single cell level. The ratio of intensity of two proteins was used to measure the siRNA efficacies of sequence of interest [38].

11.5 Conclusion

The development of sensitive and quantitative gene expression assays provide the basic platform for the success of novel gene delivery vectors. With the progress in the fields of molecular biology, oncology, and medicines, these gene expression assays are serving as an important tool to diagnose a number of diseases and to help in the selection of appropriate cure. Genetic profiling of transfected or diseased tissue can be done at the level of DNA, RNA, and protein. Reporter gene systems, electrophoretic approaches, and hybridization approaches are commonly used assays for gene expression analysis. Hybridization assays such as PCR, micro-arrays, and FISH are highly sensitive tools for gene expression analysis. The development of hybrid assays has further improved the sensitivity of both reporter gene assays and hybridization assays. Dual reporter plasmids are an example of hybrid assays. Moreover, the drawbacks of PCR-based assays including primer-dimer formation and heterodimers of products are encountered by development competitive PCR assays. The products produced by competitive PCR are analyzed by MALDI-TOF MS.

References

[1] Schenborn E, Groskreutz D. Reporter vectors and assays. Molec Biotech 1999;13:29–44.
[2] Welsh S, Kay SA. Reporter gene expression for monitoring gene transfer. Curr Opin Biotech 1997;8:617–22.
[3] Saparano JA, Solin LJ. Defining the clinical utility of gene expression assays in breast cancer: The intersection of science and art in clinical decision making. Am Soc Clinic Oncol 2010;1625–6.

[4] Karimi M, Goldie LC, Cruickshank MN, Moses EK, Abraham LJ. A critical assessment of the factors affecting reporter gene assays for promoter SNP function: a reassessment of _308 TNF polymorphism function using a novel integrated reporter system. Eur J Hum Genetics 2009;17:1454–62.
[5] Ahmed M, Narain R. Progress of RAFT based polymers in gene delivery. Prog Polym Sci 2013;38:767–90.
[6] Haupl T, Krenn V, Stuhlmuller B, Radbruch A, Burmester GR. Perspectives and limitations of gene expression profiling in rheumatology: new molecular strategies. Arthritis Res Ther 2004;6:140–6.
[7] Patsialou A, Wang Y, Lin J, Whitney K, Goswami S, Kenny PA, et al. Breast Cancer Res 2012;14:R139.
[8] Alcaraz-Perez F, Mulero V, Cayuela ML. Application of the dual-luciferase reporter assay to the analysis of promoter activity in zebrafish embryos. BMC Biotech 2008. http://dx.doi.org/10.1186/1472-6750-8-81.
[9] Wu M, Stockley PG, Martin WJ. An improved western blotting technique effectively reduces background. Electrophoresis 2002;23:2373–6.
[10] Roth CM. Quantifying gene expression. Curr Issues Mol Biol 2002;4:93–100.
[11] Mohler WA, Blau HM. Gene expression and cell fusion analyzed by lacZ complementation in mammalian cells. Proc Nat Acad Sci USA 1996;93:12423–7.
[12] Yang N-S, Burkholder J, Roberts B, Martinell B, McCabe AD. In vivo and in vitro gene transfer to mammalian somatic cells by particle bombardment. Proc Nat Acad Sci USA 1990;87:9568–72.
[13] Gorman CM, Moffat LF, Howard BH. Recombinant genomes which express chloramphenicol acetyltransferase in mammalian cells. Mol Cell Biol 1982;2:1044–51.
[14] Ghim C-M, Lee SK, Takayama S, Mitchell RJ. The art of reporter proteins in science: past, present and future applications. BMB Reports 2010;43:451–60.
[15] Bhaumik S, Gambhir SS. Optical imaging of renilla luciferase reporter gene expression in living mice. Proc Nat Acad Sci 2002;99:377–82. 377.full.
[16] Contag CH, Bachmann NH. Advances in In vivo bioluminescence imaging of gene expression. Annu Rev Biomed Eng 2002;4:235–60.
[17] Schuetz E, Lan L, Yasuda K, Kim R, Kocarek TA, Schuetz J, et al. Development of a real-time in vivo transcription assay: application reveals pregnane X receptor-mediated induction of CYP3A4 by cancer chemotherapeutic agents. Molec Pharmac 2002;62:439–45.
[18] Golzio M, Rols M-P, Gabriel B, Teissie J. Optical imaging of in vivo gene expression: a critical assessment of the methodology and associated technologies. Gene Ther 2004;11:S85–91.
[19] Coralli C, Cemazar M, Kanthou C, Tozer GM, Dachs GU. Limitations of the reporter green fluorescent protein under simulated tumor conditions. Cancer Res 2001;61:4784–90.
[20] Lee S-W, Lee S-H, Biswal S. Magnetic resonance reporter gene imaging. Theranostics 2012;2:403–12.
[21] Cohen B, Dafni H, Meir G, Harmelin A, Neeman M. Ferritin as an Endogenous MRI Reporter for Noninvasive Imaging of Gene Expression in C6 Glioma Tumors. Neoplasia 2005;7:109–17.
[22] Zurkiya O, Chan AW, Hu X. MagA is Sufficient for Producing Magnetic Nanoparticles in Mammalian Cells, Making it an MRI Reporter. Magn Reson Med 2008;59:1225–31.
[23] Wang J, Lu Z, Wientjes MG, Au JL-S. Delivery of siRNA therapeutics: barriers and carriers. AAPS J 2010;12:492–503.
[24] Taylor SC, Posch A. The design of a quantitative western blot experiment. Bio Med Res Intl 2014. http://dx.doi.org/10.1155/2014/361590361590.

[25] Kurien BT, Scofield RH. Western Blotting Methods 2006;38:283–93.
[26] De Matos LL, Trufelli DC, De Matos MGL, Da Silva Pinhal MA. Immunohistochemistry as an important tool in biomarkers detection and clinical practice. Biomarker Insights 2010;5:9–20.
[27] Saiki RK, et al. Enzymatic amplification of beta-globin genomic sequences and restriction site analysis for diagnosis of sickle cell anemia. Science 1985;230:1350–4.
[28] Bass C, Nikou D, Vontas J, Donnelly MJ, Williamson MS, Field LM. The vector population monitoring tool (VPMT): high-throughput DNA-based diagnostics for the monitoring of mosquito vector populations. Mat Res Treat 2010. http://dx.doi.org/10.4061/2010/190434.
[29] Ding C, Cantor CR. A high-throughput gene expression analysis technique using competitive PCR and matrix-assisted laser desorption ionization time-of-flight MS. Proc Nat Acad Sci 2003;100:3059–64.
[30] Su LJ, Wu LC, Tseng ALH. Comparison of mRNA gene expression analysis between quantitative-real time PCR Assay and DNA microarray based *assay*. Adv Biomed Eng 2011;3–5:158–67.
[31] Pulverer W, Noehammer C, Vierlinger K, Weinhaeusel A. Principles and application of microarray technology in thyroid cancer research. Updates in the Understanding and Management of Thyroid Cancer, In: Fahey Dr. Thomas J, editor; 2012. ISBN: 978-953-51-0299-1, p. 697.
[32] Macgregor PF, Squire JA. Application of microarrays to the analysis of gene expression in cancer. Clinic Chem 2002;48:1170–7.
[33] Levsky JM, Singer RH. Fluorescence in situ hybridization: past, present and future. J Cell Sci 2003;116:2833–8.
[34] Bishop R. Applications of fluorescence in situ hybridization (FISH) in detecting genetic aberrations of medical significance. Biosci Horizons 2010;3:85–95.
[35] Sherf BA, Navarro SL, Hannah RR, Wood KV. Dual-luciferaseTM reporter assay: an advanced co-reporter technology integrating firefly and renilla luciferase assays. Promega Notes Mag 1996;57:2–9.
[36] Bi Y, Zheng X, Shao C, Pan W, Jiang L, Ouyang H. Construction and application of a built-in dual luciferase reporter for microRNA functional analysis. Electron J Biotech 2011. http://dx.doi.org/10.2225/vol14-issue2-fulltext-9.
[37] Panstruga R, Kim MC, Cho MJ, Schulze-Lefert P. Testing the efficiency of dsRNAi constructs In vivo: A transient expression assay based on two fluorescent proteins. Molec Bio Rep 2003;30:135–40.
[38] Kojima S-I, Borisy GG. An image-based, dual fluorescence reporter assay to evaluate the efficacy of shRNA for gene silencing at the single-cell level. F1000Res 2014;3. http://dx.doi.org/10.12688/f1000research.3-60.v2.

Index

Note: Page numbers followed by *f* indicate figures and *t* indicate tables.

I

J

L

M

O

P

Q

R

Edwards Brothers Malloy
Ann Arbor MI. USA
July 25, 2016